GW00992118

Craig Crossen
Gerald Rhemann

Sky Vistas

Astronomy for Binoculars and Richest-Field Telescopes

SpringerWienNewYork

Craig Crossen
Vienna, Austria

Gerald Rhemann
Vienna, Austria

Printed in Austria

Springer-Verlag Wien New York is part of
Springer Science+Business Media
springeronline.com

Cover Illustration: The Lagoon Nebula M8 (lower center) and the Trifid Nebula M20 (above) in Sagittarius
Typesetting and Printing: Theiss GmbH, A-9431 St. Stefan

Printed on acid-free and chlorine-free bleached paper
SPIN: 10920541

With 48 Color Plates and 57 Figures

CIP-data applied for

ISBN 3-211-00851-9 Springer-Verlag Wien New York

Preface

Sky Vistas is an astronomy guidebook for both active and "armchair" observers. First, it is a practical observing guide to the deep-sky objects that are visible in low-power, wide-field instruments like binoculars and richest-field telescopes (RFTs). Second, it is a reader's guide to how the familiar bright nebulae and open clusters are distributed along our Milky Way Galaxy's spiral arms in the neighborhood of the Sun, and to how the brightest galaxies are distributed with respect to our Milky Way Galaxy and to the Local Galaxy Group of which our Milky Way is a member.

Not many observing guides have been written expressly for binoculars and small richest-field telescopes, fewer still that make any claim to comprehensiveness. Nevertheless such instruments are ideal for the Milky Way and for the large open clusters and bright emission nebulae along it. Indeed, the star clouds of the Milky Way, extended open clusters like the Pleiades and Hyades, and large, low-surface-brightness nebulae like the Rosette in Monoceros and the North America in Cygnus, are much better in, say, 11 × 80 giant binoculars than in conventional telescopes because conventional telescopes simply magnify them too much. There are even a good many galaxies that are so large but of such low surface brightness that they too require the very low magnifications of wide-field instruments to be seen at all. One objective of *Sky Vistas*, then, is simply to describe what can be seen with wide-field instruments, and how to find these objects. As an observing guide this book has been written to be used in conjunction with any of the fine star atlases that are available – such as Will Tirion's splendid *Sky Atlas 2000.0* or his inexpensive *Bright Star Atlas 2000.0*.

However, *Sky Vistas* describes not only *what* can be seen in the heavens with binoculars and small RFTs and how to find these objects, it also explains *why* these objects look the way they do in the eyepiece and *where* they fit in with respect to our Galaxy's spiral arms or, in the case of external galaxies, where they are with respect to our Milky Way Galaxy in intergalactic space. In other words, this book will help the active or armchair observer "see" objects with "Galactic perspective" (that is, their location within our Galaxy) or with "intergalactic perspective." Much of the pleasure of astronomical observing comes from "looking" with the mind as well as the eye. Thus *Sky Vistas* describes how the naked-eye appearance of the Milky Way itself relates to the spiral structure of our Galaxy in the vicinity of the Sun, how the familiar bright nebulae and star clusters along the Milky Way "trace" the spiral arms in the Sun's vicinity, and how the eyepiece appearance of individual open clusters relates to their age and (as a consequence) to where they are with respect to the spiral arms. And galaxies are discussed in their actual physical groupings, and their eyepiece appearances are related to their true structures.

Because the three best kinds of astronomical objects for low-power, wide-field instruments are open clusters, the Milky Way and its bright star clouds and large emission nebulae, and galaxies, individual chapters are devoted to each of these object-types, Chapter 2 to open clusters, Chapter 3 to the Milky Way, and Chapter 4 to galaxies. Double and variable stars, globular clusters, and planetary nebulae are best suited to medium and high powers in conventional telescopes: however, a few of each are good sights in small RFTs and giant binoculars, and they are covered in Chapter 5. The first chapter of this book is intended to be a refresher or review of the basic facts about stars, star clusters, nebulae, and galaxies, but should be accessible to anyone with a fundamental knowledge about astronomy. In conjunction with the "vistas" or "wide-field" slant of this book, the introductory chapter was written in a way that emphasizes how all the different types of astronomical objects relate to each other in the "global" scheme of things.

The photographs in this book are all by the veteran astrophotographer Gerald Rhemann, and comprise the first book-length anthology of his work. It is these photographs which have inspired the title for this book. The individual photos have been selected not merely to illustrate this book, but because they vividly portray the true astronomical character of the individual objects and Milky Way fields described in the text. The author wrote this book's text with prints of these photographs on his desk: consequently much of the text of this book can be considered to be extended captions for the photographs, which display a wealth of astronomical detail and information. Many of the color plates are composites of color and black-and-white originals: some of the black-and-whites are printed separately (with the text) because they emphasize internal contrasts and fine details. The wide-field Milky Way photos can also be used as "star charts" to assist the active observer in identifying the objects described in the text. Technical information has been given in the captions for a selection of the color plates and black-and-white photographs for the interest of practising astrophotographers.

The practicalities of observing are covered in so many introductory observing guides (including the author's *Binocular Astronomy*) that it has not been considered necessary to repeat them here, in what is in effect an "intermediate-level" observing guide. However, four points about observing that are particularly important for observers using binoculars and RFTs should be kept in mind:

(1) Dark skies are *extremely* important when using wide-field instruments because they do not have sufficient magnification to "spread out" light pollution and therefore artificially darken the sky background. Extended low-surface-brightness objects – like the North America and Rosette nebulae, or the Triangulum and Andromeda galaxies – are seen at all only by contrast with the sky background.

(2) It is essential to give your eyes sufficient time to dark-adapt (about a half-hour) and thereafter to avoid exposure to bright light. Use a red pen-light to consult your star chart.

(3) Like other physical activities, astronomical observing is an *acquired* skill: the sensitivity of your eyes will increase the more you observe.

(4) Do not merely glance at an object or at a field of view, but really *look at it*. Give your eyes time to adjust to the light conditions in the field of view: this will allow you to see subtle details and contrasts.

Vienna, October 2003

Craig Crossen
Gerald Rhemann
www.astrostudio.at

Contents

Chapter 4
Galaxies and Galaxy Groups

Chapter 5
Stars, Globulars, Planetaries

ACKNOWLEDGEMENTS

The authors would like to express their thanks to several individuals without whose assistance and support *Sky Vistas* could not be what it is. We hope this book is worthy of its friends.

We would like to thank Prof. Dr. H. M. Maitzen of the University of Vienna for opening the doors of the University's Institute of Astronomy to our researches. In America Dr. John Dickey of the University of Minnesota Department of Astronomy and Astrophysics has always been ready to provide encouragement and advice when asked.

The authors wish to thank Dr. M. E. Sharina of the Russian Academy of Sciences for permission to reprint the drawings that appear as Figures 4.1, 4.2, and 4.3; and Dr. Patrick Thaddeus of the Harvard-Smithsonian Center for Astrophysics for permission to reprint the chart in Figure 3.2.

Mr. Rhemann would like to especially thank Michael Jäger and Franz Kersche for their assistance in obtaining several of the astrophotos that appear in this book, and for their continuing work with him in astrophotography.

And Mr. Crossen would like to especially thank his wife, Elisabeth, and his friend, Prof. Dr. Stephan Procházka of the Oriental Institute of the University of Vienna, for providing the background of life in which a book of this length could be researched, written, and brought to print so quickly and smoothly. And he would like to thank his long-time friends, Ken and Nancy Deusterman of Andover, Minnesota, who for many years have provided that sense of a "base of operations" which have made his travel and writing possible.

It is to Ken and Nancy Deusterman that this book is dedicated.

Key to Abbreviations Used in the Data Tables

RA (2000.0) Dec	Right ascension and declination for the year 2000.0
Size, Dimen, Dimensions, Diam	Apparent size or diameter in degrees, minutes, or seconds of arc
m_V	Apparent visual magnitude
m_V*	Apparent visual magnitude of brightest physically involved star
***Spectrum**	Spectral type of brightest physically involved star
Distance	Distance in light-years
M_V	Absolute visual magnitude
A_V	Absorption in visual magnitudes due to interstellar dust
True Size	True size of object in light-years
Type [for nebulae]	EN=emission, RN=reflection, SNR=supernova remnant, DN=dark
Type [for galaxies]	Morphological type [see Chapter 1, Section 4]
Spectrum [for globular clusters]	Cluster's integrated spectral type
Color Index, C. I., B – V	[see Chapter 1, Section 1.3]
E (B – V)	Color excess [see Chapter 1, Section 3.1]

Captions to the Color Plates

Plate I

Antares (center) and Mars (below = east of Antares. North is left in this print.) Photographed in July 2001 from Namibia, Southwest Africa, when the star and planet were only about 5° apart. This color plate shows why the ancient Greeks named the bright star at the heart of the Scorpion *Antares,* "Rival of Mars." Notice the difference in color between the nebula-glow around Antares and that around the stars to its NNW and WNW, Rho (ρ) Ophiuchi and Sigma (σ) Scorpii: Antares is a ruddy-orange star embedded in a dust cloud reflecting its ruddy-orange light, Rho Oph is a blue-white star embedded in a dust cloud reflecting its blue light, but Sigma Sco is a very hot star that has ionized the hydrogen gas around it, which is fluorescing with the reddish color typical of emission nebulae. The central bulge of our Galaxy is along the lower edge of the field: it has an orange color because its brightest stars are orange K-type giants. The faint blue emission nebula toward the upper right of the field is IC 4592, a close-up of which can be seen on Plate III. Due west (above) of Antares is the granular disc of the globular cluster M4. A close-up of the Antares–Rho Oph region is shown on Plate II. This photo is a composite of two 15-minute exposures on Kodak Supra 400 film taken with a Nikon FM2 camera and a 135 mm lens at f/5.6.

Plate II

Nebulae in the Antares (lower left) and Rho (ρ) Ophiuchi (upper center) region. North is up. Like Plate I, this photo shows the contrast between the orange reflection nebula around the orange Antares, the blue reflection nebula around the blue Rho Oph, and the ruddy emission nebula around Sigma (σ) Scorpii (right center). The globular cluster M4 is conspicuous toward the lower right. The smaller globular NGC 6144 can be seen just NW of Antares.

Plate III

The reflection nebula IC 4592 ENE of Beta (β) Scorpii, the star on the top center edge of the field. The bright star on the edge of the nebula is the wide double Nu (ν) Scorpii. North is toward the lower left in this print.

Plate IV

The Milky Way from Ara NE to the North America Nebula. This panorama, shot with a "fish-eye" lens, shows a full third of the circuit of the Milky Way. Principal features and objects along it are labelled in the margin. Notice that the Great Rift does not exactly follow the central line of the Milky Way (the galactic equator), but lies to the left of the brightest clouds of the Cygnus Star Cloud toward the top of the field, arcs to the right of the Galactic bulge in the center of the field, and then returns to the Milky Way near the Norma Star Cloud.

Plate V

The Milky Way in the Tail of Scorpius. North is up. The conspicuous star-pair in the center is Lambda (λ, left) and Upsilon (υ) Scorpii. The bright star half-way to the bottom of the plate from them is Theta (θ) Sco. The large open clusters M6 and M7 are obvious NNE and ENE, respectively, of Lambda + Upsilon. In the far upper left corner can be glimpsed the Lagoon Nebula, M8 in Sagittarius. Just WNW of Lambda + Upsilon is a tight gathering of three or four small bright nebulae: this is NGC 6334, shown "close-up" on Plate VII. The bright star toward the right center edge of the field is Epsilon (ε) Sco. Below it is the E-W binary star Mu (μ) Sco. South of the Mu Sco star-pair is the Scorpio OB1 complex with the open clusters NGC 6231 and Tr 24. Sco OB1 is shown "close-up" on Plate VI.

Plate VI

Open clusters in the Tail of Scorpius: NGC 6231 (right center), NGC 6242 (left center), and NGC 6281 (lower left. North is to the left and east down.) In the center of the field is the scattered Tr 24, less a true open cluster than simply the richest part of the Scorpius OB1 association outside NGC 6231. The powerful stellar winds and radiation pressure of the hot, luminous O-type giants and supergiants in NGC 6231 have cleared much of the gas and dust out of the area (as can be seen from the richness of the star-background from NGC 6231 to Tr 24), but on the NE edge of Tr 24 is an emission nebula, IC 4628. This photo is a composite of a 25-minute exposure on hypered black-and-white Kodak TP 6415 film with 15-minute and 25-minutes exposures on color Fuji NPH 400/120 film. The exposures were obtained with a Hypergraph prime-focus 340 mm f/3.1 system and made in Namibia, Southwest Africa.

Plate VII

Emission nebulae in the Tail of Scorpius: NGC 6334 (lower right = SW) and NGC 6357 (upper left). These nebulae are in Sagittarius Rift dust clouds about 3° NW of Lambda (λ) Scorpii. They are too faint to be wide-field instrument objects, but are the major tracers of the Sagittarius-Carina Spiral Arm between the NGC 6231/Scorpius OB1 complex on the SW and the Lagoon /Sagittarius OB1 complex to the NE.

Plate VIII

The open clusters M6 (bottom center) and NGC 6383 (center-right – the loose gathering around a bright star) in the Tail of Scorpius. North is left and west up. The conspicuously orange star on the eastern corner of M6 is the cluster's lucida, the K Ib semi-regular pulsating supergiant variable BM Scorpii. The other bright stars in M6 are all blue-white B-type subgiants and main sequence objects.

Plate IX

Small nebulae in the Corona Australis Dust Cloud. North is to the left and east down. The bright star in the lower part of the field is Gamma (γ) Coronae Australis. Toward the upper edge of the field is Epsilon (ε) CrA, and toward the upper left corner the bright and large globular cluster NGC 6723, which is just across the border in Sagittarius. The largest nebulous patch is NGC 6726/27. Just to its lower right (SE) is the tiny comet-shaped NGC 6729. The dust cloud in which NGCs 6726/29 and 6729 are embedded, Bernes 158, is so isolated that it stands out well against the background star-field. The star just to the upper right (SW) of the nebulae is the binary Brs 14, a pair of blue-white B8 stars of magnitudes $6^1/_2$ and 7 separated by 13″ in PA 281° and a beautiful sight in super-giant binoculars and small telescopes.

Plate X

The Milky Way from the Tail of Scorpius NE to the Lagoon Nebula M8 (top center edge of field), illustrating the bulge of our Galaxy. The center of the bulge, toward the upper right (NW) of the field, is obscured by the dust clouds of the Sagittarius Rift (in which M8 itself is embedded). However, to the lower left of the field the curve of the periphery of the bulge is obvious, as is the yellow-orange color from its dominant stars, Population II K-type orange giants. The conspicuous star in the left-center is Epsilon (ε) Sagittarii. On the opposite side of the field from Epsilon Sgr is the small NE-SW rectangle of the open cluster M6. The larger, looser open cluster M7 is to the SE of M6 superimposed upon a very bright Milky Way background. The magnitude 1.6 Lambda (λ) Scorpii is below M6 two-thirds the way to the lower edge of the field. The brightest star in the central part of the field is Gamma (γ) Sgr, just NW of which is "Baade's Window" through our Galaxy's bulge.

Plate XI

The Lagoon (lower center) and Trifid (upper right) nebulae in their Milky Way setting. This plate has been photographed and processed to bring out the complex of emission nebulae of the region, in particular around the Lagoon itself and in the nebular complex to its ENE (the brightest patch of which is NGC 6559, on the complex's SE corner). The open cluster M21 is just NE of the Trifid; and SE of the Lagoon is the bright yellow disc of the globular cluster NGC 6544. This plate is a composite of two 50-minute exposures on Kodak Pro Gold 400/120 film taken with a Pentax SDP 125 mm refractor at f/6.4 and was photographed at Los Roques at Teneriffa in the Canary Islands in association with Franz Kersche.

Plate XII

Portrait of the Lagoon and Trifid Nebulae. A composite of two 70-minute exposures on Kodak Pro Gold HC 100/120 film taken with an Astrophysics f/8.3 125 mm refractor in association with Franz Kersche at Portillo, Teneriffa, Canary Islands. The exposures and processing emphasize the star field around the nebulae and bring out the NGC 6530 open cluster in the eastern half of the Lagoon. The star-richness of the "window" through the Lagoon/Trifid dust clouds north of the Lagoon is especially evident on this plate.

Plate XIII

"Close-up" of the Lagoon Nebula. North is to the lower left.

Plate XIV

The Small Sagittarius Star Cloud M24 (left center) and its Milky Way setting. North is left and east is down in this print. The Great Sagittarius Star Cloud is on the right edge of the field. Contrast the yellow-orange of the latter, which is from its evolved Population II K-type giants, with the bluish-white of M24: the Small Sgr Star Cloud is a segment of an inner spiral arm of our Galaxy, and therefore its brightest stars are blue Population I giants and supergiants. Just left (north) of the Great Sgr Star Cloud is the Lagoon Nebula. The star clusters due west (above) and east of M24 are M23 and M25, respectively. Conspicuous along the NW edge of M24 is the compact dark dust cloud B92; and due north of M24's northern corner is M17, the Swan Nebula. Further north is the Star-Queen Nebula, M16 in Serpens.

Plate XV

The south end of the Small Sagittarius Star Cloud M24 (along the upper edge of the field). The bright star near the lower right (SW) corner is Mu (μ) Sagittarii. The circular emission nebula in the upper right of the field (east of the southern corner of M24) is the very faint IC 1383/84. The two small bluish reflection nebulae just to its SW are NGC 6589 (NW) and NGC 6590. These nebulae are embedded in a very sharply-edged dust cloud with an extremely sharp western corner that juts over the rich background star fields of the Mu Sgr region.

Plate XVI
M16 (lower center) and NGC 6604 (upper center): two young open clusters in emission nebulae. These two complexes are both around 6,500 light-years distant in the dust clouds of the Sagittarius-Carina Spiral Arm; but the heavy dust clouds to the upper left (NW) of the field are part of the Great Rift on the inner edge of our Orion-Cygnus Spiral Arm and consequently only a few hundred l-y away.

Plate XVII
The Milky Way around Gamma (γ) Cygni, the bright star near the center of the field. Above (north) of Gamma Cyg is the tight knot of the open cluster NGC 6910. Its stars are in reality blue giants and supergiants: they look yellow because they are reddened by the heavy interstellar dust that is in this direction. Toward the lower left corner is the small but loose open cluster M29, its stars, blue B0 giants, likewise reddened by dust. The two large emission nebulae west of Gamma are IC1318 d (NW) and IC 1318 e (SE), both 10x50 binocular objects. They are part of one large H II region, but appear separate because of a superimposed NE-SW foreground dust lane. The large but faint emission area south of Gamma is catalogued as Sh 2-302.

Plate XVIII
The Milky Way north of Gamma (γ) Cygni, the bright star near the bottom-center edge of the field. The bright blue star near the field's upper left (NE) corner is Deneb. This photo reveals the wide area of very faint emission tendrils and patches north of Gamma Cyg, nebulae invisible to the eye in any telescope. However, the large triangular patch of emission glow NW of Gamma, IC 1318 b, as well as IC 1318 d and IC 1318 e just to its east, are in fact 10x50 binocular objects. The heavy NE-SW dust lane that separates IC 1318 d and e points NE into the Great Rift at a small, loose, but reasonably conspicuous gathering of faint stars. This is the heavily-obscured and heavily-reddened cluster of highly-luminous O-type giants and supergiants Cygnus OB2, which is thought to be a newly-born globular cluster–the first one discovered in our Galaxy. Cyg OB2 is only 5,700 light-years away and would be visually spectacular if its stars were not dimmed 5, 10, 15, even 20 magnitudes by dust.

Plate XIX
The North America and Pelican Nebulae in Cygnus.

Plate XX
The North America and Pelican Nebulae and the region to their south. Deneb is toward the upper right (NW) corner.

Plate XXI
The Gamma (γ) Cygni area and the region to its north. Plates XX and XXI are a photo-mosaic: two 5-minute exposures where taken with a Schmidt camera of each of four slightly-overlapping areas in the Deneb/Gamma Cygni region and the resulting composite prints "melded" by computer image-processing. The eight original negatives were on 6×6 cm Kodak Pro Gold 400/120 film.

Plate XXII
"Close-up" of the Pelican Nebula, IC 5067 in Cygnus.

Plate XXIII
Emission nebula IC 1396 in Cepheus with the "Garnet Star," Mu (μ) Cephei, on its NNE edge.

Plate XXIV
Emission nebula + open cluster NGC 7380 (lower right) in Cepheus, with the Milky Way to its NE. North is up and east to the left. The tiny bluish-green disc in the right-center of the field is the Comet Hönig. Near the top center edge of the field is the small emission nebula Sh 2-155, involved with, and at the far SW end of, the Cepheus OB3 association. The small reddish elliptical patch toward the upper left corner is the H II region NGC 7538, like NGC 7380 a major tracer of the Perseus Spiral Arm, which is between 10,000 and 13,000 light-years in this direction and which we see through a window in the dust of our own Orion-Cygnus Spiral Arm. The star-richness of the Milky Way between NGC 7380 and NGC 7538 shows how little star-obscuring dust is between ourselves and this stretch of the Perseus Arm. This plate is a composite of two 10-minute simultaneous exposures with Schmidt cameras on Kodak 100S/120 film. They were obtained on August 5, 2002 in association with Michael Jäger.

Plate XXV
Emission nebulae NGC 7822 (above = north) and Ced 214 in eastern Cepheus. The open cluster on the extreme right edge of the field is NGC 7662, an intermediate-age group at about the same distance, but not a member of, the young NGC 7822/Ced 214 complex. The small, poor open cluster on the northern edge of Ced 214 is Berkeley 59.

Plate XXVI
Emission nebula IC 1805 in eastern Cassiopeia. (North is up, east to the left.) The very scattered open cluster of young O-type stars fluorescing the nebula, Mel 15, is in the center of the field, the nebula-glow particularly bright in its immediate vicinity. The foreground open cluster NGC 1027 is near the left-center edge of the field. The bright emission patch near the upper right corner is separately catalogued as NGC 896, and the fainter area of glow across the narrow curved dust lane east of it IC 1795. The fact that emission nebulae are expanding into the cooler, denser matter around them is dramatically demonstrated by this photo: IC 1805 is brightest around its rim where its hot ionized gas is colliding with the cool, dense molecular material; and the background star-field is distinctly richer within the nebula, where the dust has been cleared away, than outside (especially toward the west). This plate is a composite of a 40-minute exposure on Kodak 100S/120 slide film with a 25-minute exposure on hypered back-and-white Kodak TP 6415, both made with a Hypergraph 340 mm prime-focus system at F/3.1.

Plate XXVII
Emission nebula IC 1848 in eastern Cassiopeia. North is left and west up. In the upper left (NW) corner of the field is the open cluster NGC 1027.

Plate XXVIII
The California Nebula, NGC 1499 in Perseus. North is left and west up. The bright star right of the nebula is the magnitude 4.0 Xi (ξ) Persei, the hot O7.5 III giant fluorescing the nebula's hydrogen gas.

Plate XXIX
Two nebulae in central Auriga: IC 405 (upper left) and IC 410 (right center). North is approximately toward the lower left. In the lower left are the two open clusters M38 and, just to its SSW, the much smaller NGC 1907. Dust is fairly thin in this direction through the Milky Way, so these objects have a large range of distances: IC 405 is about 1,600 light-years away, M38 and NGC 1907 both around 5,700 l-y distant, and IC 410 in an outer spiral arm of our Galaxy roughly 15,600 l-y from us. IC 410 is a standard H II region fluoresced by the cluster of bright O-type stars at its center. But IC 405 is a mixed emission + reflection nebula excited by the hot O9.5 V star AE Aurigae, located on the southern edge of the nebula's eastern arm: the red of the nebula is from the emission glow of its hydrogen, and the blue patches immediately around AE is dust reflecting the star's light. The small emission patch in the lower center of the field is IC 417, too small and faint for wide-field instruments. The bright star just WNW of it is Phi (ϕ) Aurigae.

Plate XXX
The Pleiades

Plate XXXI
The Pleiades (lower center), Comet Utsunomiya (upper right), and the planet Mercury (bottom left) in the evening twilight, April 30, 2002: a 30-second exposure on Kodak 100S/120 slide film with a Schmidt camera.

Plate XXXII
Open cluster M35 (upper right = NW) in Gemini and emission nebula NGC 2174 (lower left = SW) in Orion. The bright star in the center of the field is Eta (η) Geminorum. The arc of pale nebulosity just to its ENE is the supernova remnant IC 443. The scattering of brighter field stars NE of Eta Gem and north of IC 443 is the open cluster Cr 89, the oldest subgroup of the Gemini OB1 association, the youngest subgroup of which is NGC 2174 and the stars involved with it. This is a rather transparent direction through our Galaxy and therefore these objects are scattered over a large range of distances: M35 is about 2,200 light-years away, Gem OB1 around 4,900 l-y distant, and the open cluster NGC 2158 (the almost star-like partch just SW of M35) 13,000 l-y from us.

Plate XXXIII
Open clusters NGC 2264 (lower edge) and Tr 5 (right center) and the reflection nebula IC 2169 (upper edge) in northern Monoceros. North is to the left. We see the 8,000 light-year distant Tr 5 and the rich Milky Way star field around it through a "window" between the approximately 2,000 l-y distant dust clouds of IC 2169 on the top and the 3,000 l-y distant NGC 2264 dust clouds on the bottom. The blue IC 2169 is strictly a reflection nebula (as are the much smaller IC 446 to its north and NGC 2245 with NGC 2247 to its NE); but the nebula glow around S Monoceroti and the northern part of the NGC 2264 cluster has blue reflection patches from dust as well as an overall red hydrogen-emission haze. The volcano-shaped Cone Nebula is at the south end of NGC 2264.

Plate XXXIV
The Rosette Nebula (below, right) and the NGC 2264 cluster + nebula (above, left) in Monoceros.

Plate XXXV
The Rosette Nebula. North is to the left.

Plate XXXVI
The Seagull Nebula IC 2177 and open cluster M50 (upper right = NW) in south central Monoceros. Toward the upper left edge is the open cluster NGC 2353. The small cluster in the lower left corner is NGC 2345. On the very northern tip of IC 2177 is the foreground open cluster NGC 2335. Another foreground cluster, NGC 2343, is just SE of NGC 2335.

Plate XXXVII
Southern Orion from Rigel (toward the lower right corner) to Barnard's Loop. The latter is the arc of emission glow that curves from the top of the field to its far left edge and then SW almost back to Rigel. The Sword of Orion with the Orion Nebula is near the center. The three blue Belt supergiants, Delta (δ), Epsilon (ε), and Zeta (ζ) (right-to-left = WNW-ESE), are in the upper right center. North is from the Sword through the middle star of the Belt to the upper right corner. The lack of faint background stars because of the Orion giant molecular clouds is particularly striking between Zeta and Barnard's Loop. The Milky Way NE of the Loop, by contrast, is very star-rich. The small bluish reflection nebula NNE of Zeta, almost buried in the thick dust just within Barnard's Loop, is M78.

Plate XXXVIII
The Sword of Orion. The Orion Nebula M42 is in the center of the field. The bright patch projecting from its north edge is M43. The bright star near the bottom of the field is Iota (ι) Orionis. The blue reflection nebula near the northern edge is the NGC 1973/75/77 complex. Many of the faint stars scattered in the M42 glow are young pre-main sequence nebular variables still gravitationally contracting. They are members of the Trapezium Star Cluster, most of which is still obscured by dust.

Plate XXXIX
The region of Zeta (ζ) Orionis. North is up. The Horsehead Nebula, B33, is silhouetted upon the reef of emission glow, IC 434, that extends south from near Zeta. Immediately east of Zeta is the large emission nebula NGC 2024, which has a dark rift reminiscent of that across the Lagoon Nebula, M8 in Sagittarius. The smaller bright nebula NE of the Horsehead, embedded in the dust cloud from which the Horsehead projects, is NGC 2023. The multiple star Sigma (σ) Orionis is toward the right edge of the field.

Plate XL
The Lambda (λ) Orionis emission nebula in the Head of Orion. North is to the left. The bright star in the upper right (SW) corner is Gamma (γ) Orionis. The two brighter field stars south and SE of Lambda are Phi-one (φ^1) and Phi-two Orionis.

Plate XLI
The Witch-Head Nebula IC 2118 WNW of Rigel. The blue beam of light from the left is an "accident" in that it is a reflection of Rigel from the edge of the film-holder; however, it not only displays the fact that IC 2118 is indeed illuminated by Rigel's blue light, it reveals how the curvature of the nebula is centered upon the star. This plate is a composite of two 70-minute exposures on Kodak Pro Gold 400/120 film obtained with an Astrophysics 150mm refractor at f/8.3. The photography was done in collaboration with Franz Kersche.

Plate XLII
The Andromeda Galaxy and the Comet Ikeya-Zhang. This photograph is a composite of 5-minute exposures on Kodak 100S/120 color slide film and hypered Kodak black-and-white TP 4415. They were obtained with two Schmidt cameras on March 31, 2002 in association with Michael Jäger.

Plate XLIII
The Andromeda Galaxy M31 and its two satellites, M32 (below) and NGC 205 (above). North is up.

Plate XLIV
The central bulge of the Andromeda Galaxy. North is down. The prominent dust lane NW of the big spiral's central bulge can be glimpsed in supergiant binoculars: it is on the nearer side of the galaxy's disc. The dust lanes of other spirals are observed to curve into the central bulge as they do here in M31. This is a composite of a 70-minute exposure on color Kodak Pro Gold 160/120 and a 90-minute exposure on hypered black-and-white Kodak TP 4415, both made with a Hypergraph secondary focus 340mm system at f/9.0.

Plate XLV
The Triangulum Galaxy M33. North is up.

Plate XLVI
The galaxy pair M81 (below) and M82 in Ursa Major.

Plate XLVII
The Pinwheel Galaxy, M101 in Ursa Major.

Plate XLVIII
The Whirlpool Galaxy, M51 in Canes Venatici. North is to the upper left.

CHAPTER 1
INTRODUCTION

I The Stars

A star is a self-luminous spheroid of hot gas held together by gravity. The energy radiated by the vast majority of stars is derived from the thermonuclear reactions within them. The nearest star to us is the Sun, and therefore the luminosities, radii, and masses of other stars are commonly expressed in Solar units. Some stars are supergiants tens of thousands of times more brilliant than the Sun; others are feeble dwarfs with just a fraction of the Sun's size, mass, and luminosity. Stars also have a wide range of colors and therefore temperatures, from hot blue-white though warm white and yellow to cooler orange and cool reddish-orange. The coolest stars are the poppy red carbon stars. However, a star's temperature and its luminosity are not directly related: some cool stars – *red supergiants* like Antares and Betelgeuse – are tens of thousands times more luminous than the Sun simply because they are so large; and some hot stars – *the white dwarfs* – are hundreds of times fainter than the Sun because they are so small.

I.1 Star Brightness

The ancient Greek astronomers classified the stars by magnitude, meaning "size", the "largest" = brightest stars being of the 1st magnitude and the "smallest" = faintest being of the 6th magnitude. Modern photometry has shown that the brightness ratio between the typical 1st and 6th magnitude stars is 100: thus the brightness ratio for a difference of 1 magnitude is 2.512 (because 2.512 multiplied by itself 5 times is 100). This brightness ratio, though an awkward number, is no accident: it is related to the chemical physiology of the retina of the human eye.

The ancient Greek magnitude system later was extended to objects visible only in telescopes, which were given magnitudes greater than 6, and to objects brighter than the stars, which have *negative* magnitudes. The planet Venus, for example, at its brightest is near magnitude –5. The brightness of extended objects like emission nebulae, star clusters, and galaxies are expressed as *integrated magnitudes*, which are simply the magnitudes such objects would have if all their light was concentrated into a single starlike point.

Supplementing these *apparent magnitudes* are *absolute magnitudes*, which scale the intrinsic luminosities of celestial objects. A star's absolute magnitude is the apparent magnitude it would have if it was 32.6 light-years = 10 parsecs away. The absolute magnitude of the Sun is just +4.8: if it was 32.6 light-years from us, it would be barely visible to the unaided eye. The absolute magnitude of Sirius, by contrast, is +1.4; and that of Rigel, intrinsically the most luminous of the twenty-two 1st magnitude stars, is –7.1. Similarly, the *integrated absolute magnitude* of an extended object is the absolute magnitude it would have if all its light was concentrated into a stellar point 32.6 light-years away. Our Milky Way Galaxy's integrated absolute magnitude is thought to be around –20.6. The Andromeda Galaxy's absolute magnitude is –21.1.

When you *add* magnitudes, you *multiply* brightnesses, 2.5 times the brightness per magnitude, 100 times the brightness for each 5 magnitudes. For example, the difference between the absolute magnitudes of the Sun and of Rigel is +4.8 – (–7.1) = 11.9 magnitudes, so Rigel is 11.9 = 5 + 5 + 1.9 magnitudes = 100 × 100 × 6 = 60,000 times more luminous than the Sun. For rapid conversion of magnitude differences into brightness ratios, use the following approximations:

Difference of magnitude	=	Approximate brightness ratio
5		100
4		40
3		16
2		6.25
1.4		4
1		2.5
0.7		2.0
0.4		1.4

If you want to find out how much *fainter* one object is than another, you *divide*. For example: The white dwarf companion of Sirius has an absolute magnitude of just +11.5, which means that it is 11.5 – 4.8 = 6.7 = 5 + 1+ 0.7 magnitudes = 100 × 2.5 × 2 = 500 times less luminous than the Sun.

The light we perceive with our eyes is only one

narrow band of the electromagnetic spectrum, which includes (from longest to shortest wavelength) radio waves, microwaves, infrared light, visible light, ultraviolet light, soft X-rays, hard X-rays, and gamma rays. The amount of radiation emitted by celestial objects in any one of these wave bands (or even at one single wavelength) can be put into a magnitude system. For example, if one radio galaxy emits 2.5 times as much radio-wavelength energy as another radio galaxy, it is 1 magnitude brighter in radio waves. When there is any ambiguity, the waveband being measured is specified by a subscript: m_V and M_V are apparent and absolute magnitudes in visible light, m_B and M_B apparent and absolute magnitude in blue light alone, and so forth. In practice, the apparent brightness of objects in radio waves, microwaves, X-rays, and gamma rays is not expressed in magnitudes but in the actual amount of energy received from the object per unit area of the radio telescope or of the X-ray or gamma-ray detector.

The brightness a celestial object would have to an "eye" sensitive to the *entire* electromagnetic spectrum is called its *bolometric magnitude*, m_{bol}. In essence, bolometric magnitude is a measure of a celestial object's *entire* energy output. An object's *absolute bolometric magnitude*, M_{bol}, is the apparent bolometric magnitude it would have if it was 32.6 light-years distant. Bolometric absolute magnitudes are one of the best measures of the intrinsic differences between celestial objects because some types of stars, and many other types of objects, radiate very little of their energy in visible light. The Sun shines chiefly in visible light, so its bolometric absolute magnitude, +4.6, is only 0.2 mag higher than its visual absolute magnitude, +4.8. But a superhot star like Zeta (ζ) Puppis, which has a surface temperature of 52,500°K, radiates most of its energy in ultraviolet wavelengths: its visual absolute magnitude is an impressive –7.0 (a visual luminosity of 55,000 Suns), but its *bolometric* absolute magnitude is –11.2. Thus Zeta Puppis radiates 4 magnitudes ~ 40 times more energy in ultraviolet wavelengths than it does in visible light. And because Zeta Puppis' bolometric absolute magnitude is +4.6 – (–11.2) = 15.8 magnitudes greater than the Sun's, its total energy output is 5 + 5 + 5 + 0.8 = 100 × 100 × 100 × 2.2 = 2.2 *million* times greater. At the other end of the stellar temperature range, the absolute bolometric magnitudes of red supergiants like Betelgeuse and Antares are 2–2½ magnitudes greater than their visual absolute magnitudes, which means that they radiate 6 to 7 times more energy in infrared wavelengths than in visible light.

I.2. Star Distances

To determine the absolute magnitudes of the stars, we must first find their distances. This is no easy matter, but several techniques have been developed to measure the distances of the nearest stars (those up to 500 or 600 light-years from us) with no more than a 10% error either way. The best technique of stellar distance-finding is **parallax.** It uses the fact that from opposite sides of our orbit around the Sun the nearer stars appear to shift their positions with respect to the farther stars. This is a sort of celestial depth perception, exactly like what we experience because we have two eyes several inches apart in our head.

In principle, measuring a star's parallax–which is one-half the amount, in fractions of a degree, that it appears to shift its position with respect to more distant stars–and converting the result to an actual distance, is simple trigonometry. But life is not so simple in astronomy. The problem is how *extremely* small parallax angles actually are: the *largest* stellar parallax, that of Alpha Centauri, is a mere 0.75". This is the apparent size an object one inch in diameter would have at a distance of 23½ miles! On photographic plates, the shift produced by a parallax angle of 0.75" is even less than the size of the star images, which are smeared out into little discs by atmospheric turbulence (called *seeing*): thus the locations of the *centers* of the star images on *many* photographic plates of the same field must be measured to obtain a good estimate of even the *largest* parallaxes. But if these plates were obtained while the field was at widely separated locations of the sky with respect to the horizon, the actual flex of the telescope itself under the force of gravity, as well as the different amount of atmospheric refraction at different elevations above the horizon, will distort the apparent separations of the stars on the different plates.

One of the purposes of the Hipparcos Astrometric Satellite, which carried a 29-centimeter telescope and photographed the sky several times from November 1989 to March 1993, was to obtain stellar parallaxes unhindered by gravity- or atmosphere-induced distortions. However, because of the relatively small size of the telescope, the Hipparcos catalogue is complete down only to about magnitude 8. And even in outer space the photographic or CCD image of a star is smeared out by a telescope's optics (though not anywhere near as much as by atmospheric "seeing"). Thus the Hipparcos parallaxes are accurate only to about 600 light-years–which is, however, nearly 10 times farther than parallaxes can be measured by Earth-bound telescopes.

Professional astronomers express distances in terms of parallax: *one parsec* is the distance of an object that has a parallax of 1". Because parallax *decreases* with distance (the farther a star, the less it shifts its position with respect to even more distant stars), the two are the inverses of one another. For example, the distance of an object with a parallax of $^1/_2$" is 2 parsecs. Alpha Centauri has a parallax of 0.75" = $^3/_4$" and therefore is $1/^3/_4 = 4/3$ parsecs away. Astrophysicists express distances within our Galaxy in *kiloparsecs* (1000 parsecs), and distances between galaxies in *megaparsecs* (1 million parsecs). For amateur astronomers the concept of the *light-year*–the distance light travels in one year–is probably more physically meaningful than the parsec, because it can be related to the fact that light requires $1^1/_4$ seconds to get from the Moon to the Earth and about 8 minutes to get from the Sun to the Earth. A parsec is 3.26 light-years, and a light-year is 3×10^{16} meters.

Unfortunately there are certain types of stars–particularly giants, supergiants, and exotic variables–with no specimens within the range of Hipparcos parallaxes. Consequently we have no directly-measured absolute magnitudes for such stars. For supergiants and blue giants intrinsic luminosities are estimated from theoretical models for such stars. This hypothetical absolute magnitude is frequently used to estimate the distances to clusters or associations containing supergiants and blue giants, and any distance so obtained is termed a *spectroscopic parallax*. However, these stars seem to have a rather wide range of true absolute magnitudes, so the distances obtained this way are subject to large uncertainties.

I.3. Star Colors and Spectra

The differences in colors among stars correspond to true differences in temperature, blue stars like Spica being hotter than white stars like Deneb, much hotter than yellow stars like our Sun, and much, much hotter than orange stars like Arcturus and red stars like Antares. To quantify colors astronomers use an artificial number called *color index,* the difference between a star's (or other celestial object's) apparent magnitude as it appears on a blue-sensitive photographic plate and its apparent magnitude as it appears on a yellow-sensitive photographic plate:

$$\text{c. i.} = m_B - m_V = B - V.$$

By coincidence, the color index of A0 V stars like Vega chances to be near 0.0, so the color of Vega has been arbitrarily chosen to be the zero point of the B – V scale. Thus stars bluer than Vega have a negative B – V, and stars yellower a positive B – V (because the less positive a magnitude, the brighter the star. Hence in bluish stars $m_B < m_V$ and therefore $m_B - m_V = B - V < 0$.) The color index of the Sun is +0.65. The "red" supergiants Antares and Betelgeuse have color indices of +1.83 and +1.85, respectively. The color index of the very hot O4 supergiant Zeta Puppis is –0.26, just about as "blue" as a star can get.

B – V is only one (though the most common) of the color indices used by astronomers. The magnitude difference between *any* two colors can be used as a color index. Indeed, color indices have been defined that use even ultraviolet and infrared light. After B – V, the most common color index is U – B = ultra-violet magnitude – blue magnitude = $m_U - m_B$, which is very sensitive to differences in chemical composition among stars. Astrophysicists often plot color indices against each other on *color-color diagrams.* Such diagrams can reveal properties of the stars in a cluster even if we do not know the cluster's distance. (The tricky part in plotting a color-color diagram is compensating for the effects of reddening caused by the interstellar dust between the cluster and ourselves.)

Like their magnitudes, the colors of clusters and galaxies can be integrated into overall color indices. The B – V of the Andromeda Galaxy's disc, for example, is +0.8, redder even than the Sun. This is a bit surprising, given that blue supergiants are such a conspicuous component of any spiral galaxy's arms. One of the bluer spiral galaxies is M101 in Ursa Major; but even its B – V of +0.46 is not all that much bluer than the Sun's. The redness of the discs of spiral galaxies is the consequence of a combination of the young but red Betelgeuse-type supergiants that are in the spiral arms (with the blue supergiants), and the older solar-type stars that populate spiral galaxy discs. Giant elliptical galaxies like M87 in Virgo and M105 in Leo, systems completely lacking blue giants and supergiants, are significantly redder than spirals, their B – V's of around +0.9 to +1.0 corresponding to the colors of the yellow-orange K0 giants that dominate their light output. Strange to say, globular clusters, whose brightest members are also yellow-orange giants, are bluer than elliptical galaxies: the color indices of globulars range from about +0.85 down to a solar +0.6. The reason is that the yellow-orange of their K-type giants is attenuated by the bluish-white of their highly-evolved horizontal branch stars.

When a star's light is spread out into its constituent colors by a prism, it not only displays the *continuum* of red, orange, yellow, green, blue, indigo, and violet, but also dark absorption lines (called *Fraunhofer lines* after their discoverer) and sometimes bright emission bands (where the basic background continuum's light intensity is enhanced. They are termed *bands* because they are generally broad, spreading over a wavelength interval, unlike the dark *lines,* which are very narrow.) The absorption lines and emission bands are caused by the specific ions, atoms, and/or molecules in the star's atmosphere and therefore reveal the star's chemical composition. But they also indicate the star's precise temperature and atmospheric pressure: in particular, temperatures must be sufficiently high for atoms to lose electrons and become ions, or sufficiently low for atoms to combine into molecules.

The spectral sequence O, B, A, F, G, K, and M is a temperature sequence. (It is out of alphabetical order because it is a relic of the time before stellar spectra were understood.) The properties of the individual spectral types are summarized in the accompanying table:

The Stellar Spectral Types

Type	Temp (°K)	Color	Characteristic Spectral Features	Examples
WN WC	>50,000	Blue	Wolf-Rayet stars. Very strong emission bands of highly-ionized nitrogen (WN) or carbon (WC).	γ Velorum
O	>30,000	Blue	Absorption lines of ionized helium and doubly-ionized nitrogen, carbon, and oxygen. Hydrogen lines weak.	ζ Puppis
B	30,000–11,000	Blue-white	Ionized helium absent; neutral helium strongest at B2. Hydrogen strengthening. Some singly-ionized metals.	Spica Rigel
A	11,000–7,500	White	Hydrogen absorption lines strongest at A0, then weaker. Singly-ionized calcium strengthening.	Sirius Altair
F	7,500–6,000	Yellow-white	Hydrogen lines continue to weaken and ionized calcium to strengthen. Neutral metals stronger.	Canopus Procyon
G	6,000–4,500	Yellow	Ionized calcium absorption lines very strong, hydrogen lines very weak. Neutral metals abundant.	Sun Capella

The Stellar Spectral Types (continued)

K	5,000–3,800	Orange	Absorption lines of molecules strengthen. Lines of hydrogen very weak but neutral metals very strong.	Pollux Aldebaran
M	<3,800	Orange-red	TiO absorption lines very prominent. CH and CN molecules strong. "Me" stars have H emission lines.	Antares Mira
C	<3,800	Red	Carbon stars. Strong absorption lines of carbon molecules like C_2, CO, and CN. No TiO lines.	19 Psc U Hya
S	<3,800	Orange-red	Like type M but with the TiO lines replaced by lines of the oxides of zirconium, ytterbium, strontium, etc.	[No bright examples]

Each spectral type is further subdivided from 0 to 9, 0 being hotter than 9. (The hottest O-type stars, however, are just O2. But that is an accident of the spectral scale: O2 is simply as hot as a star can get.) Another artifact of 19th century astronomy is the use of the terms "earlier" and "later" with reference to the spectral sequence: O-type stars are said to be the "earliest" spectral type; K-type stars are "earlier" than M-type stars; an A7 star like Altair is "later" than an A3 star like Fomalhaut; and so forth. Though in a very general way O and B stars are in fact younger than A and "later" stars, the terms "early" and "late" were applied to the spectral sequence before the details of stellar evolution were understood.

In the early 20th century two astronomers, Ejnar Hertzsprung and Henry Norris Russell, independently plotted stellar luminosities versus temperature/spectral type in charts thereafter known as *Hertzsprung-Russell* or *H-R diagrams*. They discovered that most stars fall along the *main sequence*, a band on the H-R diagram which runs from hot, luminous O-type stars on the upper left to faint, cool M-type stars on the lower right. The Sun is situated in the lower middle of the main sequence. But several other regions of the H-R diagram contain concentrations of stars. In its lower left below the luminous O- and B-type upper end of the main sequence are the hot, but very faint, *white dwarfs*. Toward the diagram's upper right, above the *red dwarf* end of the main sequence, are the *red giants* (in which are included yellow G and orange K as well as red M stars). The red giant branch is rooted to the main sequence by the *subgiant branch*. And scattered along the upper edge of the diagram are the numerically few, but visually conspicuous, *supergiants*. Blue and white B and A, and red M, supergiants are much more numerous than yellow F, G, and K supergiants, hence there is a gap in the supergiant sequence. The reason for this gap is the rapidity with which massive stars expand from relatively compact B/A supergiants into bloated M supergiants.

Obviously spectral type (or temperature) alone does not uniquely determine a star's location on the H-R diagram: if it did, all stars would fall along the main sequence. So in the late 1940s W. W. Morgan and P. C. Keenan refined stellar spectral notation by adding luminosity classes to spectral types:

IaO or Ia+	= hyper-supergiants	II = bright giants
Ia	= luminous supergiants	III = normal giants
Iab	= moderately-luminous supergiants	IV = subgiants
Ib	= less-luminous supergiants	V = main sequence

Thus the Sun's complete MK spectral type is G2 V, Rigel's is B8 Ia, Procyon's F5 IV–V, Arcturus' K2p III, Betelgeuse's M2 Iab, and Zeta Puppis' O4fn Ia (the rather rare "n" suffix indicating the broadening of the star's spectral lines). Occasionally you will see the prefixes "c," "g," "d," and "D" in front of a star's spectral type: these mean "supergiant," "giant," "dwarf," and "white dwarf," respectively, and are also archaisms from 19th century astrophysics. (Occasionally one also sees "sd"–subdwarf.) The suffixes "e" and "p" mean that the star's spectrum has emission features or other peculiarities, "f" indicates strong emission bands of ionized helium and nitrogen (typical of the spectra of super-hot O-type supergiants), and "m" implies a stellar spectrum uncommonly abundant in *metals*–elements heavier than hydrogen, helium, and lithium.

I.4. Stellar Evolution

Stars condense within cool, dense, massive clouds of molecular hydrogen gas and interstellar dust. *Giant molecular clouds,* GMCs, produce not only more stars than *small molecular clouds,* SMCs, but also giant and supergiant stars, which SMCs do not.

Some dark clouds of interstellar dust and gas can be seen silhouetted against the star clouds of the Milky Way, the most conspicuous being the Great Rift (actually a chain of dust clouds) in the northern Milky Way and the Coalsack in the southern. Star formation is not occurring in every dark cloud because these clouds tend to be in pressure equilibrium: they resist gravitationally contracting into protostars because of a combination of their internal turbulence and their magnetic fields. Some trigger is necessary to overcome this pressure equilibrium and to initiate gravitational contraction. That trigger could be the shock wave from a supernova, or the radiation pressure from a nearby cluster of hot, luminous stars, or the cloud's passage through the gravitational enhancement of a spiral arm's density field.

Whatever the trigger, shortly after an interstellar cloud begins contracting, its internal gravitational field takes over and the contraction becomes a free-fall gravitational collapse. As the cloud collapses, it fragments first into cloudlets of several thousand Solar masses each–protoclusters–and then the cloudlets themselves fragment into individual *globules*–protostars. The reason for this fragmentation is not well understood, but it probably involves some combination of the original cloud's internal turbulence, magnetic field, and rotation. Anyway, gravitational collapse releases energy, which heats the interior of the globules. Some of this energy is used to dissociate the globule's molecular hydrogen, H_2, into individual hydrogen atoms, and then to ionize the hydrogen atoms. The globule's core eventually becomes sufficiently hot to nucleosynthesize hydrogen nuclei (protons) into deuterium (a hydrogen nucleus with one proton and one neutron) and lithium, with a release of more energy. Throughout its gravitational collapse, the globule/protostar has been self-luminous, radiating first at radio and microwave frequencies and later at infrared wavelengths. But not until the final phase of contraction does it really "shine" in visible wavelengths. At different phases of its gravitational collapse the protostar's outer layers become alternately opaque and transparent to the energy produced in its core: when opaque, the outer layers trap the energy within, heating the core and slowing gravitational collapse; when the outer layers are transparent, the core radiates energy and cools, which accelerates the collapse.

Finally the protostar's core temperature reaches 20 million °K, which initiates the fusion of four hydrogen nuclei (protons) into one helium nucleus (two protons plus two neutrons). This is the basic energy source of the vast majority of stars. The new star will still be concealed in remnants of its natal dust cloud–indeed, it will be observable only by the infrared radiation its visible light excites in the surrounding dust cloud. But this cloud will soon be dissipated by the star's stellar winds and radiation pressure. (M78 in Orion is an emission + reflection nebula around young stars which have just dispersed most of the obscuring dust in which they have hitherto been embedded; and numerous infrared sources – still-unveiled stars – have been identified within the large dust clouds which contain M78.) The new-born star's two energy sources–the last stages of gravitational collapse and the first stages of hydrogen burning–compete and conflict; thus the star will be erratically variable in brightness and even in surface temperature = spectral type = color. Such stars called *nebular variables,* because they are found only in complexes of bright and dark nebulae (such as Orion), or *T Tauri* or *RW Aurigae variables,* after two of the earliest-recognized nebular variables. Material falling back upon the new star's surface from the surrounding gas/dust cloud creates emission lines in the star's spectrum, enhances its luminosity, and aggravates its variations.

When hydrogen fusion begins, the star has reached the *zero-age main sequence.* Most of a star's lifetime is spent on the main sequence: its pre- and post-main sequence evolution are comparatively rapid. The Sun, for example, will linger some 10 billion years on the main sequence; but solar-mass protostars are thought to require only 75 million years to contract onto the zero-age main sequence; and the Sun's red giant lifetime, during which it will add the nucleosynthesis of helium into carbon and oxygen to its hydrogen-burning, will likewise be only a few ten million years. The reason for such a long main sequence lifetime is that the nucleosynthesis of hydrogen into helium in a stellar interior is a very controlled, stable, and fuel-efficient process.

Main sequence stars are in *hydrostatic equilibrium:* the inward-directed force of gravity is precisely counterbalanced by the outward-directed forces of gas and radiation pressure. The gas and radiation pressure depend upon the star's internal temperature: the hotter a star's core, the greater the gas pressure and the rate of hydrogen-burning. Thus if something heats up the core of a star, its outer layers will be driven outward by the increased gas and radiation pressure until these layers reach a new point of hydrostatic equilibrium. The wild card is *convection:* that is, under certain circumstances, cells of very hot gas will rise from the deep interior of a star toward its surface, bringing lots of energy, and exotic elements nucleosynthesized near the star's center, with them. The role of convection in stellar evolution is still very poorly understood; but it is not a factor in the physics of main sequence stars.

A star's place on the main sequence is determined by its mass: the more massive a protostar, the hotter and more luminous will be its main sequence descendant. The amount of energy released by a protostar's gravitational contraction is proportional to the amount of matter it contains; hence the greater mass of a larger protostar heats more of the protostar's interior to the critical 20 million °K level at which hydrogen-burning initiates and therefore more nuclear energy is generated in the cores of the more massive main sequence stars. A slight difference in mass makes a surprisingly large difference in luminosity. Along the main sequence, luminosity increases with the mass raised to the 3.5 power.

The range of stellar masses is actually not very large. The minimum mass required to generate sufficient energy from gravitational contraction to initiate hydrogen fusion in a protostar is only 8% the Solar mass. The result is an M7 or M8 red dwarf with an energy output of less than 1/1000 the Sun's. (Such stars radiate mostly in the invisible infrared; so their bolometric absolute magnitude is between +11 and +12, but their *visual* absolute magnitude is a mere +15.) On the other hand, the hottest and brightest main sequence stars, O2 and O3 objects, radiate roughly 2 million times more energy each second than the Sun but contain only about 120 Solar masses. The main sequence lifetime of the Sun is theoretically calculated to be 10 billion years: hence the main sequence lifetime of an O3 star of 120 Solar masses which is radiating energy 2 million times faster than the Sun is roughly 600,000 years (120 times 10 billion years divided by 2 million). By contrast, a red dwarf of 0.1 = 1/10 Solar mass radiating only 1/1000 as much energy each second as the Sun, will have a main sequence lifetime of one *trillion* years. These numbers, though crude approximations, dramatically demonstrate that the most luminous members of our Galaxy are also its most transient: they literally burn themselves out in what is but an instant of our Galaxy's lifetime. The red dwarfs, on the other hand, would, if such a thing was possible, outlive the universe itself.

M- and K-type dwarfs in effect have no evolutionary history: they will continue to glow feebly until the very end of time. With them will be the *brown dwarfs,* those bodies which did not have sufficient mass to ignite the hydrogen in their cores but weakly radiate in infrared light, microwaves, or radio waves the energy residual from their gravitational contraction. (Jupiter is a weak radio source and therefore technically a brown dwarf.) Solar-type and high-mass stars, by contrast, are evolving toward inevitable, and often showy, extinction.

The Evolution of Solar-Mass Stars. When the hydrogen is exhausted in the core of a Solar-mass star, hydrogen-burning continues in a shell around an inert helium core. The time required for core hydrogen exhaustion is, however, the bulk of the star's main sequence lifetime (in the case of the Sun, 9 billion years). As the hydrogen-burning shell slowly eats up toward the star's surface, it produces an increasingly massive helium core, which gravitationally contracts and therefore heats. The increase in energy from hydrogen shell-burning and from helium core-contraction increases the radiation pressure on the star's outer layers which, previously in hydrostatic equilibrium, expand. As the star's outer layers expand, they cool. Thus the star evolves off the main sequence *up* (more luminous) and to the *right* (cooler and redder) on the H-R diagram through the subgiant branch into the red giant region. In the case of a Solar-mass star, this requires about a billion years.

The density of the helium core–which is essentially helium nuclei in a sea of electrons–eventually becomes so great that it falls into *electron degeneracy*–that is, its electrons are as packed together as is possible. (One way of thinking about it is that all the boxes of "electron space" in the star's core become filled.) However, the helium nuclei themselves continue to behave according to the laws of conventional physics, absorbing more and more energy from the star's hydrogen-burning shell above. Finally, when the helium-nuclei component of the star's electron-degenerate core reaches 80 million °K, the helium begins nucleosynthesizing into carbon and oxygen with a net release of energy. But the degenerate-electron sea at first absorbs none of this new energy: it all goes back into the helium nuclei, more and more of which achieve the velocity necessary for their fusion into carbon and oxygen, generating yet *more* energy. The consequence is a runaway explosion called a *helium flash* which at its climax produces as much energy each second as our entire Galaxy! But *none* of this energy reaches the star's surface (it would blow the star apart): all the energy of the helium flash, which occurs when the star is at the very upper right tip of the red giant branch, is absorbed in lifting the star's core out of electron degeneracy.

Following the helium flash, the star's overall energy production decreases and it quickly evolves *down* (less luminous) and to the left (bluer) in the H-R diagram, ending up on the *horizontal branch* (a conspicuous feature of the H-R diagrams of globular clusters). It has a core in which helium is being nucleosynthesized (at a sustainable rate) into carbon and oxygen enclosed by a shell in which hydrogen is being nucleosynthesized into helium. The horizontal branch extends from about spectral type G0 left to the late-B range, the stars on it having absolute magnitudes around +1. Because of mass loss due to strong stellar winds while they were red giants, horizontal branch stars have masses of only about 70% to 80% of the Sun's.

Unlike the main sequence, the position of a star on the horizontal branch has nothing to do with its mass: no matter what the total mass of the red giant is, helium ignition occurs when the mass of its helium core is just under half the Sun's mass. The thing that determines a star's location on the horizontal branch is its chemical composition: stars richer in *metals* (elements heavier than hydrogen, helium, and lithium) end up on the redder (G-type) end of the horizontal branch and stars poorer in metals on the bluer (B-type) end. The reason is that atoms of heavier elements, when in a star's atmosphere, are efficient absorbers of the star's underlying blue light, re-radiating it at yellow and red wavelengths. Consequently a horizontal branch star with the Sun's metal-rich chemical composition will look comparatively redder than a metal-poor horizontal branch star both because its blue light is being absorbed by the heavy elements in its atmosphere and because that blue light is re-emitted in yellow and red, enhancing the star's underlying red and yellow radiation. Thus solar-abundance horizontal branch stars tend to congregate at the base of the red giant branch in the horizontal branch's *red clump*. In the middle of the horizontal branch is the *instability strip* of RR Lyrae variables, the pulsations of which are the consequence of their metal-poor atmospheres.

After some 70 or 80 million years, the helium at the center of a horizontal branch star is exhausted and helium burning continues in a shell around an inert, but now *extremely* hot, core of 50% oxygen and 50% carbon. As mass is added to it from the helium-burning shell, the star's oxygen/carbon core gravitationally contracts. The star now has a hydrogen-rich envelope, an outer shell in which hydrogen is being nucleosynthesized into helium, an inner shell where helium is being nucleosynthesized into oxygen and carbon, and an inert, but gravitationally-contracting, oxygen/carbon core. The shift from core- to shell-helium-burning increases the star's overall energy output. At first the extra energy simply brightens the star, increasing its luminosity but not changing its temperature (which means that it evolves *straight up* on the horizontal branch). But eventually its outer envelope, responding to the ever-greater energy flow from the interior, expands and cools and the star evolves back toward, and then up, the red giant region. This time it follows the *asymptotic giant branch,* which is slightly bluer than the post-main sequence red giant branch.

As the now extremely old, highly-evolved, asymptotic branch giant continues to expand and cool, approaching for the final time the extreme upper right of the red giant region, the conflict between its two interior energy-producing shells creates instabilities and the star suffers pulsations, appearing as a *long-period, semiregular, or irregular* M-type variable. Most of the time the hydrogen-burning shell dominates the star's energy production; but occasionally it will be temporarily extinguished by a flare-up in the helium-burning shell beneath. The star also once again suffers severe mass-loss from a strong stellar wind that blows material off its bloated atmosphere. This material sometimes cools into a envelope of dusty gas that absorbs the light of the underlying star, re-radiating it in the infrared wavelengths emitted by the hydroxyl radial (OH–the water molecule less one hydrogen atom). Such objects are called *OH/IR* (hydroxyl/infrared) *stars.* Moreover, some asymptotic branch red giants seem to experience such strong internal convection that core carbon is dredged up to the star's surface and they appear as intensely red, irregularly variable *carbon stars.* In some asymptotic branch giants (stars of spectral type S) such exotic elements as zirconium, barium, ytterbium, strontium, and technetium are mixed to the surface, implying that some very exotic nuclear reactions can occur in the oxygen/carbon cores of highly-evolved red giants.

However, an asymptotic branch red giant's core never reaches a sufficiently high temperature for the sustained nucleosynthesis of oxygen and carbon into heavier elements. Before that point the pulsations of the star's outer layers become so extreme that they reach escape velocity and are puffed out into space as a *planetary nebula,* which shines by the florescence induced in its gas by ultraviolet radiation from the hot stellar core left behind. The multiple shell structure of most planetaries suggests that an asymptotic branch red giant usually suffers several such shell ejection events. After only a few ten thousand years the expanding planetary will dissipate. Though the planetary nebula phase of a star's evolution is exceedingly brief, it occurs to so many stars that over 1,200 planetaries are known in our region of the Galaxy alone.

Meanwhile, the star's hydrogen-burning ends with the loss of its hydrogen envelope, and its helium-burning shell soon fizzles out as well. But now there is no internal energy source to withstand gravitation, so the ex-red-giant's core begins to contract, becoming even hotter. Thus the expanding planetary leaves behind an underluminous O-type subdwarf–very blue, very hot (sometimes over 100,000 °K), but very small and consequently very faint. This stellar object rapidly contracts, losing luminosity but not temperature, until it drops into electron degeneracy and becomes a *white dwarf.* The typical white dwarf contains half the mass of the Sun in a sphere the size of the Earth. The fact that all stars that begin with up to 8 Solar masses end up as white dwarfs with one Solar mass or less means that much of the interstellar gas and dust that originally condenses into stars is recycled into the interstellar medium–enrichened with heavy elements. A star loses most of this mass during its two red giant phases.

Any stellar remnant containing less than $1^1/_4$ Solar mass cannot contract further after dropping into electron degeneracy. (Stellar remnants with *more* than $1^1/_4$ Solar mass meet a rather different fate–as we shall see in the next section.) Consequently white dwarfs are simply dead objects radiating away residual heat. However, they have such small surface area that billions of years are required for a classic white dwarf with the surface temperature of an O, B, or A star to cool down into a "white" dwarf with the surface temperature of a yellow G-type star. Nevertheless several G- and K-type white dwarfs have been identified, the brightest of them the magnitude 12.4 Wolf 28, "Van Maanen's Star" in Pisces. Given sufficient time a white dwarf would cool all the way down to the 3 °K of outer space itself.

The Evolution of High-Mass Stars. The gravitational contraction of a high-mass protostar generates a great deal more energy than the gravitational contraction of a Solar-mass protostar. Thus a high-mass protostar reaches the critical 20 million °K central temperature at which hydrogen can be efficiently nucleosynthesized into helium long before its core is as dense as the core of a Solar-mass protostar. This dramatically alters its subsequent evolution. Initial hydrogen-burning in such stars does not take place in a compact core, but throughout an extended central region containing some 20% of the star's total mass. When the hydrogen fuel gives out in this extended core, it rapidly contracts, and hydrogen burning shifts into a thick shell. Core contraction and shell-burning both provide more energy, which causes the star to quickly (only 100,000 years, in the case of a 9 Solar-mass star) expand from a blue early-B subgiant into a red M-type supergiant. A huge amount of energy is required to drive the outer layers of a really massive star out a few hundred million miles; hence the overall luminosity of the star remains virtually constant as it expands.

Because of its initially lower central density, a massive star's contracting helium core will not drop into degeneracy before becoming sufficiently hot–80 million °K–to ignite its helium. In a 9 Solar mass star core helium ignition occurs after the star has been a red Ib supergiant for 70 or 80 thousand years; but in more massive stars the helium ignites while the star is still expanding toward the red supergiant region. The helium is nucleosynthesized into carbon-12 (^{12}C) by *the triple-alpha process:* three *alpha particles* (normal helium nuclei–two protons plus two neutrons) combine to create one ^{12}C nucleus (six protons plus six neutrons) with a net release of energy.

The onset of core helium-burning upsets the star's internal and external balance. The star's overall internal energy generation *decreases* because it no longer has energy from core contraction and because its hydrogen-burning shell thins. Thus its outer layers contract, and it evolves out of the M-supergiant region back left across the K, G, and F Ib supergiant sequence. (In extremely massive stars–those of 60 solar masses and more–core helium ignition and its attendant instabilities occur even before the star expands past the A Ia supergiant stage.) The contraction adds energy, preserving the star's overall luminosity. But the conflict between the star's hydrogen-burning shell and its helium-burning core, and the special properties of stellar atmospheres in this region of the H-R diagram, result in pulsations, and the star becomes a *Cepheid variable.* The clockwork pulsations of Cepheids are physically similar in cause and mechanics to those of the low-mass RR Lyrae variables on the horizontal branch, and both types of variable fall within a vertical zone in the H-R diagram called the *instability strip.*

In a fairly brief time (astronomically speaking) the helium in the star's core is exhausted and helium-burning continues in a shell around a gravitationally-contracting core of pure carbon. In stars of around 10 Solar masses the carbon core eventually becomes sufficiently dense to drop into electron degeneracy and the subsequent evolution of such stars is so complicated that it has defied theoretical modeling. However, in more massive stars, which started their main sequence lifetimes with less dense cores, the contracting inert carbon core will reach ignition temperature before dropping into degeneracy, nucleosynthesizing oxygen. If the star is sufficiently massive, the resulting oxygen core will not drop into degeneracy either, but in its turn ignite, nucleosynthesizing neon. This cycle–core exhaustion, core contraction, core ignition–will be repeated at the center of an extremely massive star until it has an "onion-shell" structure of superimposed layers nucleosynthesizing successively heavier elements (though not all the layers might be firing simultaneously).

But with the *iron-peak nuclides* the cycle ends. These nuclei, which include iron, copper, nickel, and cobalt, are very stable and therefore cannot be fused into yet heavier elements without *absorbing* more energy than they release. When this begins, the supergiant's core rapidly contracts, dramatically increasing in density and temperature. At a core temperature of 5 billion °K, the atomic nuclei first absorb huge quantities of free electrons, and then are shattered into alpha-particles (helium nuclei) by high-energy photons. This process absorbs so much of the core's energy so quickly that its pressure drops essentially to zero and it violently implodes.

This internal catastrophe is announced by an out-rush of neutrinos which, because the upper layers of the star are transparent to them, carries away most of the implosion's energy. (This preliminary neutrino outflow was actually observed from the 1987 supernova in the Large Magellanic Cloud.) The rest of the energy blasts the star's outer layers off in a *supernova*. The kinetic energy in the expanding supernova debris cloud is something like 2×10^{51} ergs. Because the energy production rate of the Sun is about 3.8×10^{33} ergs/second and there are 3.2×10^{7} seconds/year, the amount of time necessary for the Sun to radiate as much energy as is blown away in a supernova debris cloud is 15 billion years. But the Sun's entire main sequence lifetime is less than 10 billion years! (And this calculation is based only upon the amount of energy contained in the supernova's debris cloud: it does *not* include the energy carried out by the initial neutrino blast, which is probably 100 times greater.)

Hypermassive supergiants of more than 60 Solar masses evolve so rapidly that they neither become red supergiants nor begin synthesizing iron peak nuclei in their core before they go supernova. Precisely *what* happens to them is still uncertain; but at the moment of crisis their core seems to be largely oxygen. Apparently they undergo a few violent pulsations–sort of stellar grand mal seizures–during which the inevitable core collapse is delayed by episodes of core oxygen burning.

The fate of the supernova's collapsed core depends upon its mass. If it is less than $2^1/_4$ Solar masses, the core's collapse will end when its protons and electrons have been crushed into neutrons: it is a *neutron star*. Such objects are only about a dozen miles in diameter but have incredibly rapid rotational velocities because the gravitational collapse carried the core's rotational energy with it down into this tiny dozen-mile sphere. The original star's magnetic field has also been concentrated into this minute volume, and is so intense that the neutron star's radiation is channeled along the field's lines of force and beamed out its magnetic poles. The neutron star's rapid rotation sweeps its magnetic poles around much like a lighthouse beacon (unless the magnetic and rotational poles chance to coincide), and if we lie in the direction swept by the light beam, we see the star flash on and off as a *pulsar*. The Crab Pulsar at the center of the Crab Nebula flashes on and off in X-rays, radio waves, and visible light 33 times each second.

But a collapsed supernova remnant of more than $2^1/_4$ Solar masses has an even more drastic end. Gravity overwhelms even its degenerate neutrons, crushing them into a *Schwarzschild discontinuity*–a *black hole*–which is so dense that the escape velocity on its surface is greater than the speed of light. The state of the material in black holes is unknown because conventional physics cannot "see" into such objects.

II Stellar Groups

Stars are always formed in groups within interstellar clouds of gas and dust. Even the smallest molecular cloud produces dozens of Solar-mass stars. But how long a star group stays together depends upon how massive and concentrated it is: less massive, more scattered groups will soon disperse because of the Galaxy's tidal pull, or because of close encounters with giant molecular clouds or with other clusters. Consequently the most concentrated type of star group, globular clusters, have the oldest stars and the most scattered type of star group, stellar associations, have the youngest stars. In between are open clusters, the vast majority of which are indeed intermediate in age between globulars and associations.

II.1 Globular Clusters

Globular clusters are the most populous and visually impressive star group, for they contain hundreds of thousands of stars in spheres usually over 100 light-years in diameter. Unfortunately, the nearest of them are several thousand light-years away so moderate-aperture telescopes are necessary to see them at their best. Most globulars are very star-rich in their core, the star-density decreasing gradually through the cluster halo out to an ambiguous periphery: indeed, even on photos it is hard to tell where the globular ends and the surrounding star field begins. Astrophysicists have defined three different types of globular (and open) cluster radii: *core radius* is the distance from the center at which the cluster's brightness drops to half its central value; *median radius* is the radius within which half of the cluster's light is concentrated; and *tidal radius* is the distance from the center beyond which there are no more cluster members–the cluster's "true" radius in the sense that word is usually understood. However, the tidal radius of a cluster is always the most difficult of the three to determine.

The Shapley-Sawyer concentration classes rank globular cluster structure, class I globulars being the most star-dense toward the center and class XII the least. Some class XI and XII globulars–M71 in Sagitta and NGC 5897 in Libra, to name two–are hardly richer than the most populous open clusters.

Globular clusters have a large range of intrinsic luminosities (roughly, but not strictly, correlated with concentration class). Our Galaxy's most brilliant globular is Omega (ω) Centauri, which has an integrated absolute magnitude of –10.2, a luminosity of 1 million Suns. Second, with an absolute magnitude of –9.8, is NGC 6388 in the Tail of Scorpius. (The absolute magnitude of M54 in Sagittarius is –10.0; however, this globular has been found not to be a true Milky Way cluster but associated with a dwarf spheriodal galaxy just beyond our Galaxy's central bulge.) By contrast, the absolute magnitude of M71 in Sagitta is only –5.5, a luminosity of just 13,000 Suns; and even fainter is the magnitude 13.6 Palomar 1 in Cepheus, a loose class XII object that has an absolute magnitude of a mere –2. The average absolute magnitude for all the globulars in our Galaxy is around –7.3, and this seems to be about the same in other galaxies.

Our Galaxy's family of over 150 globular clusters is distributed in a sphere some 100,000 light-years in radius centered upon the Galactic Center. The number-density of globulars at the Sun's 30,000 light-years from the Galactic Center is only 1/300 of what it is on the outer edge of the Galaxy's bulge. Thus the constellations toward the bulge, Sagittarius, Scorpius, and Ophiuchus, are exceptionally rich in globular clusters. Only a few globulars (such as M79 in Lepus and NGC 1851 in Columba) can be found on the opposite side of the sky.

Our Galaxy's globular cluster family has two subgroups: *halo globulars* are distributed from the bulge all the way out to the edge of the Galaxy's halo and are distinguished by the extreme poverty of their stars in metals (elements heavier than hydrogen, helium, and lithium), a symptom of age; and *bulge globulars* are found only within the Sun's orbit in a somewhat flattened spheriod, and are only moderately metal-poor. A few globulars can be found more than 100,000 light-years from the Galactic Center, including NGC 2419 in Lynx, the famous "Intergalactic Wanderer." However, though these remote globulars are indeed gravitationally bound to our Galaxy, their original membership in our Galaxy's globular cluster family is questionable: they perhaps should be considered part of a intergalactic system that would include our Galaxy's dwarf spheriodal satellites.

The H-R diagrams of globular clusters all share certain special features. First, their main sequences are always very short, never extending further than type G. Thus all stars in them earlier than type G have already evolved off the main sequence up toward the red giant region. This by itself would prove that globulars are very old, because Solar-type early-G stars spend about 10 billion years on the main sequence. Globulars always have a long and well-populated arm of subgiants and red giants curving up and toward the right off the tip of their truncated main sequence. These subgiant and red giant branches are always very narrow, implying that the stars are of uniform chemical composition and were formed more-or-less simultaneously because any significant chemical and age differences would tend to smear them out. The narrowness of globular cluster red giant branches means that their asymptotic giant branches tend to be very conspicuous, well separated from the first giant branch.

Between a star's two periods as a red giant, it is on the horizontal branch, which extends to the left (that is, toward the blue side of the H-R diagram) from the base of the red giant branch. The HB lies between absolute magnitudes +0.5 and $+1^1/_2$, but the length to which the horizontal branch of any given globular reaches left (blue) is a measure of that cluster's age: disc globulars, some of which are a "mere" 9–10 billion years old, have only short stubs for horizontal branches; but the ancient halo globulars boast horizontal branches that cross the RR Lyrae variable instability strip and extend as far left as the original main sequence. Thus the more ancient globulars are both richer in RR Lyrae variables than younger globulars (disc globulars often do not contain even one single RR Lyrae star), and bluer in integrated color. The actual determining factor in the length of a globular's horizontal branch is less *age* than *chemical composition:* because the metals in stellar atmospheres absorb the underlying blue light of stars, re-radiating it in yellow or red wavelengths, the stars of metal-rich clusters are generally redder than the stars of metal-poor clusters, an effect that shows up most strikingly in the horizontal branch. But older globular clusters would be expected to be poor in metals anyway because they were formed early in our Galaxy's history before the interstellar medium was enriched by heavy elements from the first Galactic supernovae.

The ages of globular clusters are estimated from their *main sequence turn-off,* the point at which the main sequence curves up into the subgiant branch. The idea is that a cluster cannot be younger than the youngest stars of its main sequence, the ages of which can be computed by theoretical models of stellar evolution. The complicating factor is the *metallicity* of the cluster, which affects stellar colors: the main sequence of a population of metal-poor stars will lie to the left–bluer–of the main sequence of metal-rich stars. Thus the spectral type and absolute magnitude corresponding to a specific stellar age will be different along a metal-poor main sequence than along a metal-rich main sequence.

II.2 Open Clusters

The two major differences between globular clusters and open clusters are appearance and location. First, globular clusters are large, populous, more-or-less concentrated, and have a smooth distribution of stars. Open clusters, by contrast, are generally smaller, much less populous, and coarser-structured. Thus globular clusters have a certain uniformity (though not monotony) of appearance whereas open clusters present an almost bewildering array of appearances: some are as populous and concentrated as class XI and XII globulars, but others look like little more than enhancements of the background star field; some are swarms of similarly-bright stars, while others consist of only a half dozen bright stars with a few faint attendants; some are large but loose, others large but with evenly-distributed stars of comparable brightness, and still others mere knots that are hardly more than glorified multiple stars. Some open clusters have conspicuous star-chains, others remarkable star vacancies.

The second major difference between globular and open clusters is that globulars can be found in every direction in the sky (though they are more concentrated toward Sagittarius and the Galactic bulge) whereas open clusters are almost strictly Milky Way objects. Of the more than 1,000 open clusters that have been catalogued in our area of the

Galaxy, very few are found more than 25° off the galactic equator. (For that reason, open clusters are sometimes also called *galactic clusters.*) Some directions in the Milky Way are extraordinarily rich with open clusters, particularly Cassiopeia, Monoceros, Puppis with eastern Canis Major, and Crux with NE Carina. These are relatively dust-free "windows" through our Galaxy with views past accumulations of open clusters. However, open clusters are not especially concentrated within the spiral arms themselves–they are also abundant between as well as just above and just below the spiral arms. But few can be found more than 2,000 light-years above or below the plane of our Galaxy.

II.2.1 Open Cluster Properties

True sizes. Open cluster diameters range over a factor of about 10, the intrinsically smallest clusters being just a half dozen light-years across and the largest some 60 to 70 light-years in extent. Many open clusters have a core of bright stars several light-years wide embedded in something of a halo of fainter members roughly 20–30 light-years in diameter. Populous intermediate-age open clusters with a more-or-less even stellar distribution are typically 15–25 light-years across. However, open clusters, like globular clusters, merge imperceptibly into their stellar background, so these values are minima. A splendid example of the ambiguity of open cluster boundaries is the Perseus Double Cluster, each component of which melds seamlessly into the surrounding Perseus OB1 association.

Luminosities. Open cluster luminosities extend over about the same range as those of globular clusters, 8 to 9 magnitudes. But globulars all have the same types of stars, the brightest of which are modest-luminosity ($M_V \sim -2$ to $-2^1/_2$) K and M asymptotic branch giants, so differences between globular cluster absolute magnitudes are merely a matter of raw star numbers. But open cluster luminosities depend upon *both* numbers and cluster age–upon age particularly, because the younger a cluster, the more high-luminosity stars it is likely to contain: a young open cluster can be poor in numbers but very high in luminosity if it has a couple supergiants or a handful of bright giants. M29 in Cygnus, for instance, has little more than its five B0 giants; but these stars are sufficient to give the cluster an absolute magnitude of −8.2, almost a whole magnitude greater than that of the average globular cluster. Old open clusters, by contrast, boast only modest-luminosity red giants, and therefore must make up in star numbers for what they lack in single-star candle-power. But even the most star-rich evolved open clusters do not much exceed absolute magnitude −6, which seems to be a cap on the luminosity of old open clusters.

The most luminous open clusters are the rare supergiant-rich aggregations. In our part of the Galaxy the brightest open cluster seems to be NGC 6231 in Scorpius, which has a remarkable number of O-type giant and supergiant members and an absolute magnitude of −10.2–a luminosity of 1 million Suns. This is equal to the visual luminosity of our Galaxy's greatest globular cluster, Omega (ω) Centauri. But in terms of *total energy output* NGC 6231 is much brighter than Omega Centauri because its O-type giants and supergiants radiate chiefly in invisible ultraviolet wavelengths whereas the K giants which predominate Omega Centauri's light radiate almost exclusively in visible wavelengths. At the opposite extreme of open cluster luminosity is the Ursa Major Moving Group's absolute magnitude of −1.4, a luminosity of just 310 Suns. The scattered Coma Berenices Star Cluster has an absolute magnitude of only −2.0–which is, however, almost twice the luminosity of the Ursa Major Group. No doubt there are plenty of low-luminosity Ursa Major Group/Coma Star Cluster gatherings in our Galaxy; but they are difficult to recognize at any great distance from the Sun because they are so loose that they tend to be lost in the Milky Way's star fields.

II.2.2 Open Cluster Distance- and Age-Estimating

Except for those groups within the range of the Hipparcos satellite parallax measurements (some 600 light-years), the distances of open clusters are estimated from their H-R diagram, or, if the cluster's stars are plotted by color (more easily obtained than spectra) versus magnitude, from their *color-magnitude diagram* (CMD). Distance determination using the H-R or CM diagram is in principal straightforward. If the stars of two open clusters are plotted on the same H-R or CM diagram by apparent magnitude, the result is *two* main sequences, one below the other (unless both clusters chance to be at exactly the same distance from us). Obviously the lower main sequence will be that of the more remote cluster, which almost always will be the cluster with the unknown distance. The vertical magnitude difference between the two main sequences is then measured. Assume, for example, that it turns out to be exactly 3 magnitudes. Because 3 magnitudes corresponds to a brightness ratio of 16, the stars in the more remote cluster appear 1/16th as bright as those in the nearer cluster. Because brightness decreases with the *square* of the distance, and the square root of 16 is 4, the second cluster must be 4 times farther from us than the first. If you know the distance to the nearer cluster, you simply multiply by 4 to get the distance to the more remote cluster.

Of course things are not quite so simple in real astronomical life. For one thing, it is important to have as long a main sequence as possible plotted from the second cluster for comparison with the first cluster's, and for very remote clusters both magnitudes and colors–and more especially spectral types–can be difficult to obtain for faint K and M dwarfs. Moreover, interstellar absorption makes a cluster appear spuriously faint, hence spuriously far; and reddening by interstellar dust pushes the main sequence to the *right,* hence closer to the nearer, less-reddened, cluster's, thus reducing the measured magnitude difference between the two main sequences. These effects are usually worse the more distant a cluster actually is; but if the spectra of a sufficient number of cluster members can be obtained, both reddening and extinction can be approximately corrected. Finally, the difference in *metallicity*–chemical composition–between two clusters can affect the relative position of their main sequences: because metal-poor stars of the same mass as metal-rich stars are *bluer,* the main sequence of a metal-poor cluster will be farther *left* than it would be if its star were metal-rich. Once again, the effects of cluster metallicity can be corrected if the spectra of a sufficient number of cluster stars is obtained. But, particularly for remote clusters, obtaining stellar spectra can be very expensive of telescope time.

When the effects of reddening and absorption are taken into account, and when clusters are plotted on an H-R or CM diagram by *absolute* rather than by apparent magnitude, it is found that the main sequences of different clusters go up to different points. This is the cluster's *main sequence turn-off,* which is an index of its age. Clusters with O and B main sequence members are necessarily very young because O and B stars have very brief main sequence lifetimes. Clusters with main sequences up to type A must be considerably older than clusters with O or B main sequence members, but younger than clusters with only F-type main sequence stars. As a cluster ages, its main sequence gradually "peels away" to the right, its more massive stars first quickly evolving into red supergiants and destroying themselves as supernovae, and then its less massive stars gradually evolving into red giants and puffing their life away as planetary nebulae. Theoretical models of stellar evolution allow astrophysicists to correlate main sequence turn-off points with specific star, hence cluster, ages.

But once again, things are not quite so simple as they seem. First are the uncertainties in the theoretical models, particularly concerning how important convection might be in a main sequence star's internal structure. Furthermore, the assumed metallicity of the star models used in the theoretical computations affects the rapidity of their evolution, hence the derived age when these model stars leave the main sequence. And then there are the usual observational problems: because of the error-spread in magnitude estimates and colors (particularly for remote clusters with faint stars), the precise position of a cluster's main sequence turn-off is not always easy to fix. But even when observational errors are small (and the effect of binary stars has been compensated for), cluster main sequences perniciously persist in possessing width, and this smears out the upcurving arc of the turn-off. This width is largely caused by the slightly different chemical compositions in individual cluster stars. But differences in chemical composition–metallicity–can be much worse *between* rather than within clusters. Hence, even if the ages of two clusters are in reality identical, the position of the main sequence turn-off of a metal-poor cluster will be distinctly different from that of a metal-rich cluster.

II.3 Stellar Associations and Stellar Streams

Stellar associations are loose aggregations of recently-formed stars, often, but not always, still within or near the remnants of their birth-clouds of interstellar gas and dust. They frequently include one or two open clusters, are divided into two or three subgroups, and their involved clouds are usually still bringing forth new association members. There are three types of stellar association:

OB associations have O and B main sequence, giant, and supergiant stars.

B associations have B, but no O, main sequence and giant stars. They often are simply old OB groups.

T associations are aggregations of nebular T Tauri type variables still gravitationally contracting toward maturity as modest-mass, modest-luminosity A-, F-, and G-type main sequence stars. T associations are usually still partially embedded within dark dust clouds, their individual members still surrounded by patches of emission nebulae.

OB associations are often hundreds of light-years across because only huge giant molecular clouds can generate O and early-B type stars. T associations, by contrast, are always small, sometimes only a few light-years across. A T association can be part of an OB association complex; but the nearest T associations to the Solar System, only several hundred light-years away in small but dense dust clouds toward Corona Australis, Lupus, the south circumpolar constellation of Chameleon, and Taurus (just beyond the Pleiades), are isolated groups. Most B associations are probably just ex-OB associations: the huge Cassiopeia OB8, for example, would be a B association but for one O7.5 giant member. However, modest-sized associations that began with no main sequence O-type members are easily conceivable.

The Orion Association, centered about 1,600 light-years away, is the nearest major association to us. It includes all the bright stars in the constellation (except Betelgeuse and Gamma [γ]) as well as many, or most, of Orion's 4th, 5th, and 6th mag stars. The numerous bright and dark nebulae from the Sword up to Lambda (λ) Orionis are part of the complex. And star formation presently continues in the molecular cloud behind the Orion Nebula. The entire complex is probably 700–800 light-years wide, and as much deep.

The Belt and Sword of Orion illustrate *sequential association subgroup formation.* The supergiants of the Belt are several million years old and surrounded by a loose cluster, Collinder 70, of the B-type main sequence stars which formed with them. Stellar winds and radiation pressure from the Belt supergiants have cleared the region around them of most of the gas and dust that remained of the original giant molecular cloud in which they and Cr 70 formed. However, their radiation pressure and stellar winds–plus supernova shock waves from now-destroyed Belt supergiants–rammed into the giant molecular cloud of the Sword region and initiated gravitational contraction of the denser segments of those clouds. The result is the O and early-B stars of the Sword. But radiation pressure and stellar winds from the Sword stars (and perhaps a supernova shock wave or two) have in their turn initiated gravitational contraction further back in the Sword GMC, where a third generation of Orion Association stars is now coming into being.

Much nearer, even larger in apparent size, but older than the Orion Association is the Scorpio-Centaurus Association, which includes the majority of the 1st, 2nd, 3rd, and even 4th magnitude stars from Scorpius on the NE through Lupus and Centaurus to Crux on the SW. The Scorpio-Centaurus group technically is a B association, for it has no O-type stars: all its brightest members, except for the red M-type supergiant Antares and a couple rapidly-evolving late-B giants, are early-B main sequence, subgiant, and giant stars, many of them Beta Canis Majoris variables just beginning their evolutionary expansion off the main sequence. The association's lack of O-type stars might be the consequence of its age, already at least 20 million years; but it might well have lacked O-type members from the start because the main sequence progenitor of its most evolved star, Antares, probably was a B0 or B1 object rather than an O-type star.

The apparent size of the Scorpio-Centaurus Association is 70° × 25°. Because its center (located between Alpha [α] Lupi and Zeta [ζ] Centauri) is about 550 light-years distant, its true extent is 700 × 250 light-years. It is roughly 400 light-years deep. Its highly-elongated ellipsoidal shape is the consequence of *galactic rotational sheer:* an association might begin as a sphere, but because all its stars are orbiting the Galactic Center with the same velocity, those further from the Galactic Center, having longer orbits to cover at the same velocity, fall behind those nearer the Center. Hence with time an association elongates. Given that the Scorpio-Centaurus group is already about twice as long as it is deep but can have covered only about 10% of its full orbit around the Galaxy's Center (the "Galactic year" in our region being some 220 million years), the process of rotational sheer is obviously very rapid.

As the Scorpio-Centaurus Association illustrates, associations age rapidly: their most massive, luminous O-type members are soon lost to supernova events; and differential galactic rotation quickly sheers the rest of the group into an ever-lengthening ellipsoid. But even from birth, associations contain the seeds of their own destruction: they are so scattered that they lack the gravitational cohesion necessary to resist the strains of rotational sheer and of the tidal pull of the Galaxy's gravitational field; and, because of turbulence in the parent giant molecular cloud, individual association members have slightly idiosyncratic velocities and therefore are slowly drifting apart. The visual identity of an association will endure only so long as a significant number of its brighter members remain in the same general area of a spiral arm.

However, even as the association elongates and expands, the vast majority of its fainter stars "remember" in their approximately parallel space motions something of the original space motion through the Galaxy of the association's parent giant molecular cloud. The group has now degenerated into a *stellar stream.* The central five stars of the Big Dipper, though sometimes called an open cluster, are in reality merely the most concentrated part of the *Ursa Major Stream*–also known as the *Sirius Supercluster,* after its brightest member (in apparent magnitude). The Sun presently lies in the very midst of the Ursa Major Stream, for Stream members can be seen in every direction on the celestial sphere and include such widely-scattered stars (in addition to Sirius and β-γ-δ-ε-ζ Ursae Majoris) as Alpha (α) Coronae Borealis, Delta (δ) Leonis, Beta (β) Eridani, Delta Aquarius, and Beta Serpentis. Because the Ursa Major Stream's earliest stars are A0 and A1 main sequence objects, the group must be about 300 million years old.

The Ursa Major Stream (and the Sun) is within a still larger and older stellar stream. The Hyades in Taurus and the Praesepe, M44 in Cancer, though 400 light-years apart, are the dual open cluster cores of an extremely extensive and loose aggregation known as the *Hyades Stream* or *Taurus Stream.* This group, which includes among its brighter members Capella, Alpha Canum Venaticorum, Delta Cassiopeiae, and Lambda (λ) Ursae Majoris, extends over a region some 600 light-years long roughly centered on the Hyades. The Hyades Stream has a main sequence up only to the mid-A range and therefore is twice as old as the Ursa Major Stream.

III Nebulae

Nebulae are of two types, dark and bright. Bright nebulae glow either by florescence or by reflected star light (sometimes both). Dark nebulae, on the other hand, are "seen" only because they are silhouetted against a Milky Way star field or upon a bright nebula. In general, bright nebulae are merely the illuminated sections of much more extensive dark nebulae. Both bright and dark nebulae are part of the *interstellar medium,* which has four components: molecular clouds, neutral hydrogen gas, ionized hydrogen gas, and coronal gas (in order of increasing temperature and decreasing density).

1. Molecular clouds are composed mostly of molecular hydrogen, H_2, at temperatures of just 10 to 20 degrees above absolute zero. They are "seen" as dark nebulae because the tiny grains of carbon, silicon, and ice in them absorb and scatter starlight. Molecular clouds comprise only a few per cent of the volume of the interstellar medium but are extremely important because of their gravitational fields (they can contain hundreds of thousands of Solar masses of material) and because they are the sites of star formation. More will be said about molecular clouds in the section on dark nebulae, below.

2. Neutral hydrogen (H I) gas consists of individual hydrogen atoms (with the usual complement of helium and trace-element atoms), and comes in cold and warm components. The cold neutral hydrogen gas has temperatures up to only 100 °K, densities in the range of 10–20 atoms per cubic centimeter, and forms haloes around the dark dust clouds. However it is simply the coolest, densest part of the warm (6000 °K), neutral hydrogen gas, which is very low density (0.1 to 1 atom per cc) but occupies 20% of the volume of interstellar space. Both the cold and warm components of the H I gas can be traced by the 21-centimeter wavelength radiation emitted by the neutral hydrogen atom when the spin of its electron and proton becomes parallel. 21-cm radio observations have shown that the H I gas tends to be concentrated along our Galaxy's spiral arms (though it is reasonably dense between the arms as well), and that, in contrast to the molecular gas, 80% of the Galaxy's 4.3 billion solar masses of H I lies *outside* the Sun's orbit. The thickness of the neutral hydrogen disc (that is, its height above and below the Galactic plane) increases beyond the Sun's orbit; and at one side of its rim the neutral hydrogen disc curves down and at the diametrically opposed side it curves up. This *warp* in the neutral hydrogen disc is probably to be blamed upon the gravitational fields of the Magellanic Clouds.

3. Ionized hydrogen (H II) gas is simply the nebulosity, like M42 in Orion and M8 in Sagittarius, around young, hot O and early-B stars. These *H II regions,* which have temperatures around 8000 °K, occupy almost 10% of the interstellar medium and can be very dense (up to 10,000 atoms per cubic centimeter – like the molecular gas) but are often quite rarified (one atom per cubic centimeter). The Orion Nebula is a very dense H II region and therefore has high surface brightness. By contrast the huge (2°× 1°) IC 1848 in Cassiopeia is very thin and thus of very low surface brightness. H II regions are almost always in or around dark nebulae because they are fluoresced by the hot, young stars newly-formed from giant molecular clouds. The boundary between a cool, dense dark nebula and a hot, bright H II region is usually very active because the H II gas is expanding, sometimes almost explosively, into the cool material. One example of this kind of interstellar battle-front is the 1°-long north-south reef of very pale emission nebulosity, IC 434, upon which is superimposed the Horsehead Nebula, B33 in Orion. This reef is the zone of excitation along which the rarified H II gas fluoresced by the Belt supergiants, as well as the radiation pressure and stellar winds from those stars, is crashing into the dense, resistant molecular cloud east and SE of Zeta (ζ) Orionis. The bright seam right along the impact zone is called a *rim nebula.* More about H II regions shall be said in section on emission nebulae, below.

4. The fourth component of the interstellar medium, the *coronal gas,* is the largest in volume, occupying fully 70% of interstellar space. It is the extremely rarified (0.001–0.005 atoms per cc) but very hot (1 million °K) gas left by the passage of a supernova shock front. Coronal gas is so hot that it emits soft X-rays; and its pressure is so great that, despite its low density, long, thin channels of it called *worms* have worked their way well above or below the disc of the Galaxy out into the halo.

III.1 Dark Nebulae

Dark nebulae are clouds of cool, dense interstellar dust and gas silhouetted against bright nebulae or against the Milky Way itself. They are composed mostly of molecular hydrogen, H_2; but their light-dimming power derives from their tiny grains (some containing as few as 50, others in excess of 10,000, atoms) of graphite (carbon), silicates (silicon), ices (water), and possibly metals like iron and aluminum. These clouds obscure the stars behind them both by scattering, and by absorbing, the starlight. Both processes affect blue light more than red light (most terrestrial dust behaves the same way); hence stars seen through a foreground dust cloud are reddened as well as dimmed. Indeed, dark nebulae are transparent to long-wavelength infrared, microwave, and radio radiation. The process of scattering is in effect reflection: the photons of starlight are simply redirected. However, when a dust grain *absorbs* a blue or ultraviolet photon of starlight, it heats up and then re-radiates the photon's energy at infrared wavelengths. Because they are recently formed from giant molecular clouds of gas and dust, virtually all hot, highly-luminous O and B giant and supergiant stars are near dark nebulae, and so much of their blue and ultraviolet radiation is absorbed by interstellar dust and re-emitted in the infrared that the Milky Way (and other similar spiral galaxies) are actually brighter in infrared than they are in visible light!

Despite their seeming opacity, the density of dark nebulae is actually quite low, only a few thousand molecules of H_2 per cubic centimeter, less than that of the best terrestrial laboratory vacuum. The darkness of these clouds, then, is the consequence not of density but of *depth:* they are scores, sometimes hundreds, of light-years thick. Molecular clouds not silhouetted against bright backgrounds can be "seen" by the low-energy radio wavelengths emitted by their trace components–particularly by their carbon monoxide molecules, which radiate 1.3 and 2.6 millimeter-wavelength microwaves. What has been found from carbon monoxide radio observations is that three-fourths of our Galaxy's one billion solar masses of cool molecular gas is *within* the Sun's orbit around the Galactic Center, and that the bulk of this is between 12,000 and 17,000 light-years from the Galactic Center in what is called the *molecular ring.* Very little molecular matter is within the Galaxy's bulge, and very, very little much beyond the Sun.

These radio observations are confirmed by the unaided-eye appearance of the Milky Way. Except for a stretch in Perseus, where nearby (roughly 500 light-year distant) dust clouds narrow it, the galactic anticenter half of the Milky Way from Cassiopeia to Puppis is a fairly unbroken (though dim) band. However, along its Galactic Center half much of the Milky Way is bisected by the most conspicuous of all dark nebulae, the *Great Rift,* a series of more-or-less connected dust clouds from several hundred to several thousand light-years away that splits, like a long narrow island in the middle of a river, the Milky Way from Deneb in Cygnus SW all the way to Alpha (α) Centauri. (Because the Great Rift cloud chain is slightly tilted with respect to the plane of our Galaxy, it arcs out of the Milky Way in Ophiuchus, passes behind the Antares region of Scorpius, and rejoins the Milky Way in Ara.) The Great Rift defines the very inner edge of our Orion-Cygnus Spiral Arm. Similar dust lanes are often seen on the inner edges of spiral features in other galaxies (such as in M31).

Among the other conspicuous dust features along the Milky Way undoubtedly the most famous is the *Coalsack* in the far southern constellation Crux. The Coalsack is not as opaque as it looks on photos (or as its name sounds): something like 16% of the background starlight gets through. It is centered about 550 light-years away, measures 50–60 light-years across, and is simply a detached Great Rift feature. The so-called "northern Coalsack," centered midway between Deneb and Alpha Cephei, is neither as pronounced as the true Coalsack nor part of the Great Rift. However, the 7°-long *Pipe Nebula,* a splendid 7x binocular object that extends from SE to SW of Theta (θ) Ophiuchi, is another detached Great Rift feature. The Coalsack and the Pipe Nebula actuallly are not exceptionally large dust clouds: they look big simply because they are relatively nearby.

The shape of a dark cloud is conditioned by its environment: whatever predisposition it might have toward a Coalsack-like circularity is usually disappointed by radiation pressure and stellar winds from nearby O and early-B supergiant stars, by expanding supernova shells, by the gravitational effects of recently-passed globular clusters or other massive dark clouds, and probably by our Galaxy's magnetic field. And many dark nebulae have been, or are being, eaten into by star-formation regions. All these effects contribute to the interesting shapes of these objects.

Dark nebulae are not truly opaque: a little background light does leak through them (16% in the case of the Coalsack, as has been mentioned). Moreover, the relative darkness of a dust cloud can also be effected by how many foreground stars chance to lie between the cloud and ourselves. The number of foreground stars per unit sky area appearing on a dark nebula can give a crude idea of its distance; otherwise the distances of dust clouds that are not obviously involved with H II regions and stellar associations can be very hard to judge. In making such estimates, due account must be given to the star-richness in the direction of the dust cloud: the star-density toward the dust clouds around the North America Nebula in Cygnus, which is ahead of us in our own spiral arm, is greater than that toward, say, the conspicuous clouds B142 and B143 near Gamma (γ) Aquilae because toward Aquila we look lengthwise down a relatively luminous-star-poor interarm gap.

Dark nebulae are graded on a scale from 1, those that are little more than diminutions in the basic Milky Way background glow, to 6, those that are virtually black. The most frequent dark nebula designation is by Barnard number, "B," after the great Milky Way astrophotographer E. E. Barnard, who catalogued these objects around 1900. But LDN = Lynds Dark Nebula numbers are also often seen.

Star-dimming and Star-reddening by Interstellar Dust. The microscopic grains of dust that make dark nebula dark are spread generally, but lightly, through the neutral HI gas as well. Thus in every direction we look around the Milky Way, whether through a molecular cloud or not, the stars are to a greater or lesser degree dimmed and reddened. The amount of dimming and reddening can be expressed quantitatively. Dimming, technically called *extinction* by astrophysicists, is given in magnitudes, A_V ("A" for "absorption," "V" for "visible light"). It varies by distance and direction depending, of course, upon how much interstellar material we look through. For example, the stars of the Canis Major Association, though about 2,500 light-years distant, are practically extinction-free because toward southern Canis we are looking in a practically dust-free direction through the Milky Way. By contrast, the Garnet Star, Mu (μ) Cephei, though only slightly farther from us than the CMa Association, is dimmed $2^1/_2$ magnitudes (a factor of 10: that is, only 10% of its light gets through to us) by dust involved with the IC 1396 emission nebula.

It is also the fault of IC 1396 complex dust that Mu Cephei is so red–redder than even an M2 "red" supergiant like it should be. Interstellar reddening, like interstellar extinction, has been quantified by astronomers. The *color index* = (B – V) of a star or any other celestial object is, as was mentioned in Section I.3, the difference between its apparent magnitude as it appears on a blue-sensitive photographic plate and its apparent magnitude as it appears on a yellow-sensitive photographic plate:

$$B - V = m_B - m_V.$$

However, for a star behind any appreciable amount of interstellar dust, there is a difference between its *observed* color index and its *intrinsic* (or "real") color index: the star's observed color index will be numerically larger–that is, redder–than its intrinsic B – V, written $(B - V)_0$. The difference between the observed and intrinsic color indices for the same star (or other celestial object) is called its *color excess*, E (B – V):

$$E\,(B - V) = (B - V) - (B - V)_0.$$

For example, the observed color index of Mu Cephei is +2.4, but the intrinsic color index of M2 Ia supergiants is usually about +1.7; so the color excess of Mu Cephei is:

$$E\,(B - V) = +2.4 - (+1.7) = +0.7.$$

Color excess is an easily-obtained quantity because a star's observed color index is directly measured from photographic plates and its intrinsic color index can be inferred from its spectrum. Thus we do not need to know a star's distance to find out how much it has been reddened, E (B – V), by the interstellar medium.

III.2 Bright Nebulae

Bright nebulae are divided into four types based upon how and why they shine: reflection nebulae shine by reflected light, emission nebulae shine by the florescence induced in them by hot young stars, planetary nebulae shine by the florescence induced in them by a hot but tiny central stellar remnant, and supernova remnants shine sometimes by the exotic synchrotron process, but usually by the florescence induced by their explosive expansion into the surrounding interstellar gas and dust.

III.2.1 **Reflection nebulae** are exactly what the name suggests–nebulae that shine by light reflected from stars near, or embedded within, them. Most reflection nebulae are simply the illuminated parts of much larger interstellar molecular clouds, the minute grains of graphites, silicates, ices, and metals in the cloud scattering starlight in exactly the same way that dust in the Earth's atmosphere scatters sunlight. And, just as the Earth's atmospheric dust selectively scatters the Sun's blue light, producing our blue sky, so too the dust grains in the interstellar medium selectively scatters blue light. Hence color photos of the Pleiades reveal that the fine-textured filamentary reflection nebulosity in which the four stars of the Pleiades "dipper" is embedded is in fact *bluer* than these four blue giants themselves. And stars seen through dust clouds or reflection nebulae are reddened because of the scattering of their blue light. The Garnet Star, Mu Cephei, is so red because we view it through a dense peripheral dust cloud of the huge IC 1396 nebulosity. Sometimes the hydrogen gas through which the reflecting dust is spread is ionized and fluoresces as an emission nebula: M78 in Orion is an example of a nebula shining both from glowing gas and from reflecting dust. (The Pleiades' mid-B giants are too cool to fluoresce the hydrogen gas around them: stars later than type B1 do not produce sufficient ultraviolet radiation to produce emission nebulae.)

III.2.2 **Emission nebulae** are interstellar clouds of ionized gas glowing by basic florescence. These nebulae are always near or around hot O and early-B stars because it is the ultraviolet radiation from such stars that ionizes (that is, knocks the electrons off) the atoms in the gas, which is the first step in the mechanics of florescence. When the freed electrons recombine with the ionized atoms and cascade back down through the atoms' electron orbital shells, they emit a characteristic wavelength of radiation at every step. Hydrogen is far and away the most abundant element in the interstellar medium (including the dark nebulae, though it contributes nothing to their opacity): thus most emission nebulae shine with the light given off by electrons cascading down the orbital shells of hydrogen atoms. The beautiful red color seen in photos of many emission nebulae is radiated when an electron jumps from the third down to the second lowest orbital shell of hydrogen, a transition that yields the Hα line of red light at wavelength 6562 Å. However in dense emission nebulae, like M42 in Orion, radiation from doubly-ionized oxygen, the O III ion, at the blue-green wavelengths of 4959 and 5007 Å is especially strong. (In the early days of astronomical spectroscopy, the O III lines at 4959 and 5007 Å were a puzzle. At first they were thought to be from some unknown element dubbed "nebulium": only later was it realized that they came from electron transitions in O III ions that could not be observed in terrestrial laboratories simply because the density in even the *densest* emission nebula is less than that of the best terrestrial laboratory. Hence these two, and certain other, nebular spectral features were called "forbidden lines.") The surface brightness of the Orion Nebula is so great that its blue-green glow can be glimpsed even with only moderate-aperture optics.

The size of an emission nebula depends not only upon the actual dimensions of the gas cloud providing the hydrogen and oxygen to be ionized, but also upon the cloud's density, upon the number of O and early-B stars radiating the ultraviolet photons necessary for ionizing the gas, and especially upon the temperature of those stars. The denser the gas, the more quickly with distance from the central star or star cluster will the ultraviolet photons be "used up" (but also, from the standpoint of the observer, the higher will be the nebula's surface brightness). But the most important single factor in emission nebula size is star temperature, because very hot mid-O stars radiate chiefly ultraviolet light and therefore provide copious quantities of ultraviolet photons–much more than B0 or even O9 stars. The distance out to which a star of a given surface temperature can ionize a gas of given density is called the star's *Strömgren radius*. The difference between the ionizing power of a moderately-hot B0 star and that of an extremely hot O5 star is impressive. The Strömgren radius of an O5 V star is *nine times* greater than that of a B0 V star. Thus, because the volume of a sphere increases with the *cube* of its radius, the *volume* of interstellar gas ionized by an O5 star is $9 \times 9 \times 9 = 729$ times greater than that ionized by B0 star (assuming equal gas density). Consequently we find mid-O stars at the center of all really large emission nebulae (such as the Rosette in Monoceros, the Lagoon in Sagittarius, NGC 2174 in Orion, and IC 1396 in Cepheus); but the central stars of such compact nebulous patches as M78 in Orion and NGC 1333 in Perseus are always B0 objects.

If left to their own devices, emission nebulae would be simple spheres centered upon their hot illuminating star or star cluster. Among the few emission nebulae that appear circular and therefore must be more-or-less spherical in space are the Trifid Nebula in Sagittarius, the Rosette Nebula in Monoceros, NGC 2174 in Orion, and IC 1396 in Cepheus. In practice, however, the giant molecular cloud from which an H II region's fluorescing stars have just been born usually hem it in. The Orion Nebula, for example, though it fades out gradually toward the south, is bounded on the north by a very dense cloud. The Lagoon Nebula, M8 in Sagittarius, has an almost rectangular profile because of bordering dust clouds. The huge $2° \times 1°$ low-surface-brightness IC 1848 in Cassiopeia likewise is distinctly rectangular. And the peculiar shape which has given M17 in Sagittarius such flamboyant titles as the "Swan," the "Omega," and the "Horseshoe" is the consequence of a lobe of dust jutting over the nebula's central glow from the west coupled with a dark foreground cloud terminating it on the north.

The Orion Nebula, bright and impressive though it appears to us, is not especially large for an emission nebula, only about 30 light-years in extent. Its grandeur is in large part due to its nearness. By contrast, the Lagoon (three times farther from us) measures more than 60×40 light-years, the Rosette (likewise three times farther than M42) is over 100 light-years in diameter. But even these are not particularly large emission nebulae. The Eta (η) Carinae Nebula, NGC 3372 in the far southern Milky Way, is centered upon a very ultraviolet-photon-rich open cluster of more than a dozen O-type giants and supergiants and therefore is 200 light-years in extent. And larger emission complexes are known in other galaxies. The Tarantula Nebula, NGC 2070 in the Large Magellanic Cloud, is a truly gigantic 900 light-years across – 30 times the size of the Orion Nebula! – and so bright that, though 160,000 light-years from us, it is an unaided-eye object.

III.2.3 Planetary Nebulae are the death shrouds of highly-evolved asymptotic branch red giants in which the amplitude of the atmospheric pulsations reached escape velocity. The multiple rings of most planetaries implies that the ejection of the red giant's envelope usually required several such episodes. The resulting expansion velocities of the ejected rings are a rather gentle 10 to 20 kilometers per second.

The name "planetary" was given to these objects by William Herschel in the late 18th century because many of them resemble the small bluish-green disc of the planet Uranus (itself discovered by Herschel). However, planetary nebulae actually come in an impressive array of apparent (and presumably real) sizes, some so small that they look virtually stellar even at high powers in large telescopes, and others comparable in size to the Moon. The Helix Nebula, NGC 7293 in Aquarius, measures $16' \times 12'$–about half the Moon's apparent diameter!–and is so tenuous that it can be seen only at very low powers: any kind of magnification simply enlarges it right out of visual existence.

The blue-green color of planetary nebulae is from the 4959 and 5007 Å wavelength radiation of doubly-ionized oxygen, O III. The oxygen is ionized by ultraviolet photons from the planetary's central star, the core of the ex-red-giant, which no longer produces energy from nucleosynthesis but is still very hot. The nebula's hydrogen is likewise ionized and radiates rather strongly in the red Hα line at 6562 Å and the blue-green Hβ line at 4859 Å.

Planetary nebulae display a great variety of structures. One is the classic "ring" seen both in compact, bright planetaries like M57 itself as well as in the large, ghostly NGC 7293 in Aquarius, though entirely different optics are required to perceive the annularity of the two extremes: the "donut" of M57 is visible at 100× in small telescopes; but the "hole" in the glow of NGC 7293 requires extremely low powers with large RFTs. Some planetaries, such as the Eskimo Nebula (NGC 2392) in Gemini, NGC 3242 in Hydra, NGC 6543 in Draco, and the Saturn Nebula (NGC 7009) in Aquarius, have double or multiple rings. As visually intriguing as ring nebulae are such box-shaped, double-lobed planetaries like the Dumbbell in Vulpecula, the Little Dumbbell (M76) in Perseus, and the Bug Nebula (NGC 6302) in Scorpius. These objects apparently are true bipolar nebulae in which some physical process, probably the parent red giant's strong magnetic field, ejected the star's outer layers in opposite directions.

The central stars of planetary nebulae are highly compact bluish subdwarfs with colors equivalent to those of O and B main sequence stars but which, because they are no longer producing any energy from nucleosynthesis in their interiors, have gravitationally shrunk and are much smaller and fainter than main sequence stars of the same surface temperatures. Planetary central stars are in fact condensing into true white dwarfs: an object like the magnitude 13.6 sdO central star of the Helix Nebula, which must have an absolute magnitude of only around +10 and therefore is only 1/100th as luminous as the Sun, is probably close to true electron degeneracy. Because of their intrinsic faintness, the central stars of even the largest and therefore probably the nearest planetary nebulae are very difficult to discern even with large telescopes–particularly because the overlying nebula-glow competes for the eye's attention.

Planetary nebulae are estimated to last only some 30 or 40 thousand years. Nevertheless, as of 1991 1,340 planetary nebulae had been identified in our area of the Galaxy alone. (A very great many of our Galaxy's planetaries must be obscured by dust, or simply too small and remote to be discerned.) Their distances individually are very uncertain; but as a group they display a decided distribution in a thick disc around the Galaxy's bulge, their number-density decreasing from the bulge outward. Planetaries are significantly more numerous in such Galactic interior constellations as Sagittarius, Scorpius, and Aquila (and even in little dust-dimmed Serpens Cauda) than in the opposite direction toward Auriga, Gemini, and Monoceros. However, even toward the Galactic exterior planetary nebulae are not as rare as globular clusters, for planetaries are true intermediate-age thick disc members of our Galaxy, not ancient halo objects like the globulars, the density of which always declines precipitously outward from the bulge. A few planetary nebulae can be found well off the Milky Way, such as M97 in Ursa Major, NGC 246 in Cetus, and NGCs 7009 and 7293 in Aquarius; but these are relatively nearby objects, and consequently actually not all that far off the Galactic plane (because our Galaxy's thick disc population straggles out to more than 2,000 light-years off the Galactic plane). In general, planetary nebulae seem to have been produced by stars that had ages and chemical compositions similar to our Sun's (itself a member of the Galaxy's thick disc population), though there are some planetary progenitor stars that must have been much younger, and others much older, than the Sun.

Numerous planetary nebulae have been identified in neighboring galaxies, including the two Magellanic Clouds, the Andromeda and Triangulum spirals, and even in members of the Coma-Virgo Cluster. The brightest planetaries of M31 have an apparent magnitude of 19.8, corresponding to an absolute magnitude of –4.6. This seems to be a true physical cap on the potential luminosity of a planetary nebula, and is valuable because it provides a *standard candle* almost as good as Cepheid variables for estimating the distances to other galaxies. The brightest planetaries in Coma-Virgo galaxies suggest a distance to the center of that cluster of 15 million parsecs, nearly 50 million light-years. Because of the uncertainties in the distances of the Galactic planetaries, we cannot confirm that the $M_V \sim 4^1/_2$ luminosity cap on the M31 planetaries holds in our own Milky Way: differences in the chemical composition between M31 and our Galaxy could cause a difference in the planetary nebula luminosity cut-off.

III.2.4 Supernova remnants (SNRs) are the expanding debris clouds of supernova explosions. The most famous of them is also the most unusual: the Crab Nebula, M1 in Taurus, is the remnant of a supernova that occurred near Zeta (ζ) Tauri in 1054 A.D. and was reported by Chinese astronomers to equal Venus in brilliancy–which means that it must have reached an apparent magnitude of $-4^1/_2$ or −5. However, M1 is an easily-observed, high-surface-brightness object not because it is so young, but because it radiates by the exotic, and very powerful, *synchrotron process:* electrons streaming at velocities near the speed of light out from the Crab Pulsar, the collapsed neutron-star core of the star that exploded, emit highly-polarized radiation as they spiral around the lines of force in the nebula's powerful magnetic field.

Not all supernova remnants go through a Crab phase: all that Tycho's Star of 1572 in Cassiopeia, Kepler's Star of 1604 in Ophiuchus, the Cassiopeia supernova of about 1670 (which occurred deep within a dust cloud and is known only from the strong radio emission, the radio source Cassiopeia A, of its highly-obscured remnant), and the Lupus Supernova of 1006 A.D. (observed by Arabian astronomers) have left are nebulous shreds around the periphery of an expanding shock front. What these SNRs share with the Crab is strong emission in radio waves and X-rays–the consequence of the tremendous kinetic energy of the expanding debris cloud, which shocks the interstellar medium into which it plows, pushing some material ahead of it as it continues to expand and leaving the thin gas behind it with a temperature of 1 million °K.

Supernova remnants not only *heat* the interstellar medium, they enrichen it with the heavy elements produced both by the progenitor star and by the explosion itself–heavy elements that will end up in future generations of stars. Type I supernovae from Population II stars blow large amounts of iron into the interstellar medium. This iron was created both by an almost instantaneous nucleosynthesis when the carbon and oxygen white dwarf progenitors of this type of explosion collapsed into each other, and by the radioactive decay of nickel-56 into iron-56 in the debris cloud immediately after the supernova. (Indeed, a Type I SN's optical and ultraviolet emission comes exclusively from this radiative decay.) By contrast, a Type II supernova, the consequence of core collapse in a young, massive, rapidly-evolving supergiant, ejects a greater variety of heavy elements into the interstellar medium because its progenitor star had an "onion-shell" structure in which a series of successively heavier elements were being nucleosynthesized in concentric layers within the star. Hence a 100 Solar mass supergiant star that goes supernova will blast into the interstellar medium something like 2 Solar masses of carbon, 17 Solar masses of oxygen, $2^1/_2$ Solar masses of neon, and even one Solar mass of magnesium (among other elements). But in such a supernova event iron will also be produced in the expanding debris cloud by the radioactive decay of nickel-56 into iron-56, a process observed in the 1987 supernova in the Large Magellanic Cloud.

IV Galaxies

Galaxies are gravitationally-bound aggregations of millions, often billions, of stars, usually accompanied by massive clouds of gas and dust. They are the most numerous type of deep-sky object, but they are extremely distant – even the nearest large galaxy, M31 in Andromeda, is 2.4 million light-years from us. Consequently relatively few galaxies offer much visual detail even in large telescopes.

Of course, in reality galaxies are neither small nor faint and come in an impressive range of true sizes and luminosities. The most brilliant galaxies are the *supergiant ellipticals,* or *cD galaxies,* with absolute magnitudes of –23 to –24–luminosities approaching 1 *trillion* Suns!–and diameters of several hundred thousand light-years. (Most cD ellipticals are in extremely remote galaxy groups; but one visible with moderate-size telescopes is NGC 4889 of the Coma Galaxy Cluster.) At the opposite extreme are the *dwarf spheroidals,* probably the most numerous type of galaxy in the universe, but with absolute magnitudes down to –9, or less, and only a couple thousand light-years across, and therefore invisible much beyond the Local Galaxy Group. Our Milky Way has an absolute magnitude of –20.6 and a diameter approaching 100,000 light-years, and is significantly above average in size and luminosity even for a spiral system. But the Andromeda Galaxy is almost twice as large and bright as the Milky Way.

In the eyepiece the majority of moderately-bright galaxies display a tripartite structure: a stellar nucleus embedded within a small but moderately-bright core enveloped by a tenuous halo. But there are many variations on this basic theme, and the visual appearance of a galaxy in the eyepiece is not mere happenstance: it is the result of the galaxy's individual structure, which depends upon the galaxy's *morphological type.* The most commonly-used scheme of galaxian morphological classification was originated by Edwin Hubble around 1930. Hubble began with four basic galaxy types: *elliptical* (E), *spiral* (S), *barred spiral* (SB), and *irregular* (Irr or I). At first he regarded *lenticular* (S0 and SB0) as a "more or less hypothetical class" at the intersection of the most oval ellipticals (E6/7) with the most tightly-wound spirals (Sa and SBa); but later true lenticulars were identified. Unfortunately, Hubble termed the loosely-wound, spiral-arm-dominated Sc and SBc galaxies "late spirals" and the tightly-wound, bulge-dominated Sa and SBa systems "early spirals." The problem with this nomenclature is that his "early"-type Sa and SBa spirals are largely composed of *late*-type Population II stars (G, K, and M giants and dwarfs), and his "late"-type Sc and SBc spirals are dominated by *early*-type Population I O- and B-type giants and supergiants. Thus there is a *very* confusing contradiction between Hubble's nomenclature of galaxian morphology and the (still-used) nomenclature of stellar evolution.

Hubble's unfortunate use of "early" and "late" derived from his erronious speculations about galaxian evolution. He arranged his galaxy types into what became known as the *Hubble tuning-fork diagram,* and hypothesized that any individual galaxy begins as an E0 circular-shaped elliptical, evolves into an elongated E6/7 elliptical (the elliptical sequence thus forming the "handle" of his "tuning fork"), and then becomes either a barred SB0 or a non-barred S0 lenticular, from whence it follows either the barred or the non-barred spiral sequence (the two spiral sequences being the "prongs" of the tuning fork), SB0 → SBa → SBb → SBc or S0 → Sa → Sb → Sc. If this evolutionary theory had been correct, ellipticals would indeed be "early" and spirals "late" galaxies.

Subsequently Hubble's theory of galaxian evolution proved wrong, in part because of Walter Baade's discovery, a decade later, of stellar populations and their relationship to Hubble's galaxy types (that is, young *early*-type Population I stars appear in Hubble's "late"-type spiral galaxies, and old *late*-type Population II stars comprise Hubble's "early"-type elliptical galaxies). But Hubble's galaxy classification scheme also proved too simple to fully describe the reality about galaxian morphology. Later elaborations of the Hubble system were introduced by Allan Sandage and Gerard de Vaucouleurs. The Sandage/de Vaucouleurs extension of the Hubble system is sometimes supplemented with Sidney van den Bergh's DDO (David Dunlop Observatory) spiral galaxy luminosity classes.

The almost over-elaborate de Vaucouleurs system preserves Hubble's original division of galaxies into ellipticals, spirals, barred spirals, and irregulars – plus the afterthought of lenticulars and barred lenticulars – but adds notation for bars (B), for the lack of bars (A), for conspicuous inner and outer rings (r and R, respectively), and for a global "S"-design (s). Other details of de Vaucouleurs' designations include "E+" for supergiant ellipticals, "Sm" for extreme "late"-type spirals with incipient spiral arms like that of the Large Magellanic Cloud, "Im" for irregulars which resemble the completely non-structured Small Magellanic Cloud, and three divisions each for the barred and non-barred lenticulars (see Section IV.2 below). In this system the Andromeda Galaxy, usually simply designated as an Sb spiral ("b" in the original Hubble system suggesting that a spiral has a moderately-large bulge and moderately tightly-wound arms) is classified SA(s)b, the "A" indicating that M31 has no sign of a central bar and the "s" that its global structure is S-shaped. Something of the weakness of de Vaucouleurs' scheme is exposed in his classification M101 in Ursa Major, SAB(rs)cd, where "AB" means that M101 has an incipient bar, "cd" that its spiral arms are not quite as straggling as those of the loosest (Sd) spirals, and "rs" that the galaxy has an overall S-shape ("s") but with inner arms that close in a ring ("r") around the central bulge. Without a photo of M101, it would be very difficult to envision what an "SAB(rs)cd" galaxy might look like!

The DDO luminosity class applies only to spiral galaxies and is indicated by a Roman numeral following the spiral's de Vaucouleurs morphological type–SA(s)b I–II in the case of M31 and SAB(rs)cd I in the case of M101. Strictly speaking it is *only* a morphological description of a spiral's arms, class I systems having thick, well-developed arms and class V galaxies anaemic, poorly-developed arms. However, these classes did indeed prove to correlate with actual spiral galaxy luminosities. For example, the absolute magnitude of the class I-II M31 is –21.1, which is indeed a little lower than the –21.5 of the class I M101. However, the DDO luminosity class typology of spiral arms has been largely superceded by the 12 *spiral arm classes* of D. M. and D. G. Elmegreen (described in Section IV.3, below).

So how good a description of reality *is* the de Vaucouleurs/DDO galaxy classification scheme? Does it actually give a good basic insight into the true physical, possibly evolutionary, differences among galaxies? In 1995 A. Naim published an article in the *Monthly Notices of the Royal Astronomical Society* giving the results of a test of in which six astronomers had been asked to classify 831 galaxies. Frequently two astronomers placed the same galaxy in different classes, but the classes invariably were adjacent. Binney and Merrifield (*Galactic Astronomy*, p. 149) concluded, "While the classification scheme used is fundamentally sound, one should not take a galaxy's exact classification too seriously." Thus these classification systems should not intimidate the amateur. The thing to remember is that these systems are based upon galaxies' *photographic* appearances and therefore relate only in the most general way to what one actually sees in the eyepiece of even a very large telescope. And often even the photographic appearance of a galaxy can be rather different depending upon such factors as the length of the exposure and the precise color sensitivity of the film or receiver used. However, there really is a good basic correlation between the original Hubble classification scheme (excluding the bars, which are photographic features) and what a galaxy looks like in an amateur's eyepiece.

IV.1 Elliptical Galaxies

The main visual and photographic characteristic of elliptical galaxies is a smooth decline in brightness = star density from a bright nucleus out to an ambiguously-bordered periphery. Though they have no distinct boundaries, the profiles of elliptical galaxies are very distinct, ranging from circular E0 to highly elongated E6/7.

Elliptical galaxies are so-called simply because their appearance on the sky is elliptical. But their true 3-dimensional shapes seem to be more complicated. Some are *oblate spheroids*–that is, squashed spheres, where the axis of rotation is the system's short (minor) axis. Others are *prolate spheriods*–that is, torpedo-shaped, with the axis of rotation the system's long (major) axis. Still others are apparently true 3-axis ellipsoids (which is a rather difficult geometric form to visualize or draw. If you view an oblate or prolate spheroid along its axis of rotation it appears circular, and if you view it perpendicular to its axis of rotation it appears elliptical. But a 3-axis ellipsoid appears elliptical down whatever axis you view it.) In other ellipticals the *degree* of ellipticity changes from the galaxy's interior to its exterior. For example, the isophotes (lines of equal brightness) of supergiant cD systems are nearly circular toward their center but much elongated toward their periphery.

Traditionally, elliptical galaxies have been described as gas-and-dust free. That, however, is an over-simplification: 40% of all ellipticals have dust lanes (though these lanes are not nearly so thick and conspicuous as they are in spirals). And giant and supergiant ellipticals can be extremely rich in hot (1 million °K), low-density gas that emits X-rays, some systems containing as much as 100 *billion* solar masses of this gas. However, despite so much interstellar material, even giant and supergiant ellipticals do not seem to be experiencing significant star-formation at present, nor to have experienced it in the astronomically-recent past: the bright-star population of all ellipticals is the same as that of globular clusters–highly-evolved, subsolar-mass, Population II K- and M-type asymptotic branch giants. Nevertheless, ellipticals must have had a long *early* history of active star formation, because the K and M giants toward their centers are redder, hence more metal-rich, than the K and M giants toward their peripheries. Presumably the former were created in an interstellar medium enriched by the supernovae of the earliest generations of these galaxies' stars, the remnants of those generations being the metal-poor stars presently on their peripheries. The metal-richness of the interiors of elliptical galaxies implied by their redder color is confirmed by spectroscopic studies of some of these systems, which show both iron and magnesium to be more abundant in the stars near their centers than in the stars toward their perimeters.

Elliptical galaxies come in an impressive range of sizes and luminosities, from the cD supergiants hundreds of thousands of light-years across and nearly a trillion times brighter than the Sun, to dwarfs only a couple thousand light-years across and barely brighter than the average globular cluster. (Indeed, there is no reason morphologically or physically not to regard globular clusters simply as extreme low-mass, low-luminosity elliptical galaxies.) The cD systems are always at the center of rich, compact, galaxy clusters, and it is assumed that they have gotten as big and bright as they are simply by cannibalizing other cluster members. In fact, the ellipticity of a cD galaxy's halo always follows the ellipticity of the galaxy-distribution of the group around it. Presumably the large amounts of interstellar gas that are detected in cD systems come from the spirals they have "eaten."

Two varieties of dwarf elliptical galaxies exist (three, if you count globular clusters). The *dwarf ellipticals* proper are merely smaller versions of the giant ellipticals: they display the same gradual decline in star-density and brightness from a dense core out to a diffuse periphery, they have the same range of elongations, and they have the same bright star population–asymptotic branch giants with absolute magnitudes of just -2 or $-2^1/_2$. They are simply constructed on a Lilliputian scale. Some of them even contain a bit of interstellar gas and dust. The four satellites of the Andromeda Spiral–M32 and NGCs 147, 185, and 205–are typical dwarf ellipticals.

The *dwarf spheriodals* are either simply an extremely low-density, low-luminosity subclass of the dwarf ellipticals, or indeed a physically different type of galaxy. They have the bright-star population and stellar-distribution profile typical of ellipticals; but their star-density, from core to periphery, is a mere 1% of normal. Consequently, the average dwarf spheriodal in the vicinity of the Milky Way is about the same size as the normal dwarf elliptical–a few thousand light-years across–but has an integrated absolute magnitude of a mere -11. The dSph/E3 Draco System in fact has an absolute magnitude of a paltry -8.6. Dwarf spheroidals may well be the most numerous sort of galaxy in the universe. However, we cannot see them much beyond our Local Galaxy Group even with the Hubble Space Telescope because they are too spread out to produce a consolidated haze: they can be identified only by the loose gathering of pin-prick star-images made by their lucidae, the usual K and M asymptotic branch giants with absolute magnitudes of just -2 to $-2^1/_2$.

The nearest dwarf spheroidal to us is the dSph/E7 Sagittarius System, only about 80,000 light-years away. Its brightest members have apparent magnitudes approaching 15–within range of the very largest amateur instruments. But any chance of recognizing it is obviated by its location right behind the star-multitudes of our Galaxy's bulge. One of the evidences for the existence of the Sgr System is that counts of RR Lyrae variables in the "windows" through our Galaxy's dust here have *two* peaks: one, around apparent blue magnitude $15^1/_2$, is of the RR Lyrae stars in our Galaxy's bulge; and the other, at $m_B \sim 18$, is of the RR Lyrae variables in the Sgr System. However, the unusually luminous ($M_V = -10$) globular cluster M54 is a Sgr System member. It is not at all strange for a dwarf spheroidal galaxy to possess a globular cluster: the 420,000 light-year distant dSph/E3 Fornax System has five globular clusters. In terms of number of globular clusters as a function of galaxy luminosity, Fornax is the most globular-cluster-rich galaxy known in the universe!

After the Sagittarius System, the next nearest dwarf spheroidals to us are the Ursa Minor, Draco, and Sculptor Systems, which are from 230,000 to 260,000 light-years away and consequently have brightest stars in the magnitude 17–18 range. Obviously they are not targets for amateur telescopes, however large. Indeed, except for the four dwarf elliptical satellites of the Andromeda Spiral, all the elliptical galaxies visible in amateur instruments are giants or supergiants. The nearest giant ellipticals, M105 and NGC 3379 in the M96 Galaxy Group in Leo, are 37 million light-years distant but easily seen even with mere 10×50 binoculars. Several Messier-numbered giant ellipticals in the core of the Coma-Virgo Galaxy Cluster are also visible with 10×50 glasses: M49, M59, M60, M84, M86, M87, and M89.

The appearance of elliptical galaxies in the eyepiece is always the same: a bright stellar nucleus or (if magnification is sufficient) a bright core within a halo of ambiguous extent but (if aperture is sufficient) distinctly circular or elongated. Something to keep in mind while looking at these objects is that the night sky itself is far from truly *black:* it has its own intrinsic brightness from (1) natural air glow (due to aurora-like processes in the Earth's upper atmosphere), (2) zodiacal light (sunlight reflecting off solar system dust), (3) faint, unresolved stars in our Galaxy, and (4) faint, unresolved distant galaxies. In fact, if all the light from one square second of sky near the zenith was gathered into a single star-like point, it would have an apparent magnitude of about 22. However, almost *half* of an elliptical galaxy's light comes from areas of the system that have a surface brightness of mag 22 or less. Astronomers routinely trace galaxy brightness profiles down to mag 27–a brightness density only 1% of the night sky's (in other words, the galaxy adds 1% to the background sky brightness)–but the uncertainties at such light levels are rather large. In any case, what we see when we look at a dim galaxy is *its* light added to that of the night sky's, hence the galaxy's ambiguous boundaries.

But the nuclei of some elliptical galaxies–M87 is a good example–can be exceedingly bright. You might even say *suspiciously* bright: and in fact M87 itself is a radio galaxy with an exploding nucleus. Contrast the sharp brightness of the nucleus of M87 with the nuclei of the "quieter" giant ellipticals M84 and M86 just to its WNW.

The engine of M87's explosive activity is thought to be a black hole containing some 5 billion solar masses of material. It has been hypothesized that the star-dense nuclei of many elliptical galaxies harbor black holes (which do not necessarily wreck M87-type havoc in the host galaxy, even if present). The Hubble Space Telescope has allowed astronomers to trace the light intensity, hence (indirectly) the star density, of many elliptical galaxies to within a fraction of a second of arc of the galaxy's center. What has been found is that in many elliptical (and lenticular) systems the light-density profile levels off as the center of the galaxy is approached, but in other systems it continues to rise right to the limits of resolution. The latter group are the best candidates for nuclear black holes, the deep gravitational well of which would tend to attract stars toward the galaxy's center and would tend to increase the "turbulence" of the stars' motions around the center–that is, the scatter of individual orbital velocities around some average. A scattering of star velocities has in fact been observed in many ellipticals with star-density profiles that keep rising toward the nucleus.

IV.2 Lenticular Galaxies (S0 or SB0)

The main difference between elliptical and lenticular galaxies is that the latter contain discs which are extensions of the galaxy's bulge. These discs lack any hint of spiral arms (but do not necessarily lack dust), and often appear as nothing more than a collar around the equator of the galaxy's bulge. Morphologically lenticulars are such a smooth transition between E7 ellipticals and Sa spirals that it is difficult to discern the difference between an E7 elliptical and an $S0^-$ lenticular, or between an Sa spiral an an $S0^+$ lenticular: hence some galaxies are classified E7/S0, and other S0/a. The Sombrero Galaxy, M104 in Virgo, is a classic example of a transitional S0/a system.

Lenticulars can contain dust–even a complete dust lane–but any evidence of star formation in the dust (such as strong infrared radiation from the dust, the result of absorption and reradiation at IR wavelengths of ultraviolet photons from young, hot, O-type giants) marks a galaxy as an Sa or SBa system. In the Hubble/Sandage classification scheme nonbarred lenticulars are classified from $S0_1$ to $S0_3$, depending upon the amount of dust they contain: none in $S0_1$ systems, and a complete lane in $S0_3$. However, barred lenticulars are classified by the strength of the central bar: $SB0_1$ lenticulars have only little "knobs" at each end of the central bulge; but in $SB0_3$ systems the bar cuts boldly across, and beyond, the bulge. The de Vaucouleurs system uses the notation $SA0^-$, $SA0^o$, $SA0^+$ and $SB0^-$, $SB0^o$, $SB0^+$ for nonbarred and barred lenticulars respectively, the criterion of classification being the strength of the disc–inconspicuous in $SA0^-$ and $SB0^-$ galaxies but prominent in $SA0^+$ and $SB0^+$.

Like elliptical galaxies, lenticulars are composed of evolved Population II red and yellow giant and dwarf stars. About 30% of all lenticulars contain detectable quantities of neutral hydrogen, H I, gas–twice the percentage of the ellipticals with H I. In barred lenticulars the H I gas is generally in a ring well beyond the ends of the bar: apparently the rotation of the bar sweeps up and throws out all interior gas. In some lenticulars a gas-and-dust ring in fact encircles the galaxy around the *poles* of its bulge: in such *polar-ring galaxies* the rotation of the material in the ring is at right angles to the sense of rotation of the system's stars. Even more bizarre are lenticular galaxies with gas-and-dust rings that lie in the system's disc but rotate around the galaxy's center in a direction *opposite* to that of the disc's stars. We all know how strong a force angular momentum can be: it is the spin of a bicycle's wheels that keeps what is inherently a very unstable machine upright–and the faster the wheels spin, the more stable the bicycle. Consequently, a star system that contains two major components rotating in opposite directions could only have gotten that way from some cataclysm. Lenticulars with gas rings spinning thus *anti-parallel* to the stellar disc are thought to have gotten that gas from some external source: either from infalling intergalactic gas or, more likely, from other galaxies which happened to pass close to the lenticular and were too small to retain their interstellar medium against the big galaxy's gravitational pull.

That lenticulars might indeed have originally been normal spirals affected by galaxy collisions or near-misses is implied by the fact that they are very rare outside rich galaxy groups, but comprise fully one-half of the bright members of dense clusters in which collisions and encounters must be frequent. Indeed, there seem to be essentially *no* spiral galaxies near the center of some dense galaxy clusters; and those anywhere near the center of others prove deficient in neutral hydrogen gas. These observations suggest that lenticulars had been normal gas-rich spirals but, because of their location in dense galaxy clusters, the gas was swept out of them, and new-star formation ceased, billions of years ago. In subsequent galaxy-encounters some lenticulars reclaimed some material, observable visually as dust lanes or at radio wavelengths as neutral hydrogen gas rings. *For sure,* material in lenticular galaxy polar or counter-rotating rings could not have been shed from its own stars: it would have retained their sense of rotation around the galaxy's center.

Because of the tendency of lenticulars to segregate in dense galaxy clusters, few are found in the vicinity of the Milky Way and our Local Galaxy Group way out here on the relatively galaxy-thin perimeter of the Local Supercluster. The nearest lenticular to us apparently is NGC 3115, the Spindle Galaxy in Sextans, an edge-on lens-shaped S0⁻ system some 30 million light-years away with an absolute magnitude approaching −21. NGC 3115, however, seems out of the proper environment for a lenticular, for it is a long way from the rich galaxy clusters to its north in Leo and to its east in Virgo.

IV.3 Spiral Galaxies (S or SB)

Spiral galaxies are composed of a disc containing the spiral arms encircling a central bulge, the whole enveloped in a tenuous halo of stars interspersed with globular clusters. The disc is composed of Population I stars, the youngest of which–blue and red high-luminosity supergiants within star-formation regions rich with emission and dark nebulae–are concentrated along, and indeed define, the spiral arms. Older disc stars are in a more-or-less smooth distribution between, and above and below, the spiral arms (though they too have somewhat enhanced concentrations along very broad spiral features containing the optically-conspicuous supergiant-rich arms), with sharply decreasing density away from the galaxy's plane. The central bulge is usually a flattened sphere, sometimes at the center of a straight bar. The bulge is very dense toward its nucleus, but thins gradually outward, its chief stellar component being Population II red and yellow modest-luminosity giants. The halo begins in the bulge (the nature of the structural connection between the two, if any, is uncertain) and can extend out to twice the disc diameter in all directions. It is very thin even near the bulge, and decreases very rapidly in star-density outward. The halo's brightest members are also evolved Population II yellow and red K and M asymptotic branch giants; but its most conspicuous component are its galaxy's globular clusters.

Spiral galaxies are much richer in gas and dust than elliptical and lenticular systems. Most of their interstellar material is along the galaxy's plane, its greatest density in the spiral arms themselves. The dust is thickest in cool giant molecular clouds, which sometimes are in a ring just outside the bulge. Because a galaxy's giant molecular clouds are where its most vigorous star formation occurs, the inner spiral arms are almost always broader and brighter than the outer spiral arms. But the H I gas extends well out beyond the limits of the optical disc. Often the periphery of the gas disc is *warped* above or below the galaxy's plane, presumably because of the gravitational effects of the neighboring galaxies. Small discs of neutral hydrogen gas, of molecular gas, and of dust are often within a galaxy's bulge around its nucleus. If the galaxy is barred, conspicuous dust lanes often lie along the forward edges of the bar (the edges in the direction of galactic rotation) and curve into the bulge toward the galaxy's nucleus.

As Hubble originally conceived it, the Sa – Sb – Sc / SBa – SBb – SBc sequence of spiral galaxy types expressed (1) the degree of openness of the spiral arms, in Sa/SBa the arms being tightly wound and in Sc/SBc the arms loose and straggling; (2) the bulge-to-disc brightness ratio, greatest in Sa/SBa and least in Sc/SBc; and (3) the degree of resolution into stars of the spiral arms, least in Sa/SBa and greatest in Sc/SBc (because Sc spiral arms have more regions of luminous supergiant star formation). The de Vaucouleurs classifications preserved the spirit of these criteria but introduced transitional types (Sab, Sbc, etc.), accomodated a greater degree of spiral arm openness with a new type, "Sd," and gave "spirals" with barely detectable arm or arm-segments their own type, "Sm" (where m = "Magellanic," after the Large Magellanic Cloud, which has an incipient spiral feature).

The thing that makes spiral galaxies so distinctive is of course their spiral arms. In 1982 B. M. and B. G. Elmegreen introduced 12 spiral arm classes, class 1 spiral systems having chaotic, fragmented, *flocculent* arms and class 12 systems being the classic two-armed *grand-design spirals,* the archetypes of which are M51 in Canes Venatici and M101 in Ursa Major. Grand-design structure is thought to be caused either by a rotating bar in the galaxy's interior, or by the gravitational effect of a companion galaxy–in the case of M51, this would be the peculiar elliptical system NGC 5195 just to its north. But the mechanism by which flocculent arm-fragments are generated is not known.

One of the strangest things about spiral galaxies,

including our own Milky Way, is that the orbital velocity around the galaxian center of the stars and gas in the disc is more-or-less the same from the bulge all the way out to the disc's rim–and in fact *beyond* the disc's rim, because the neutral hydrogen gas that extends beyond the optical rim has the same rotational velocity as the optical disc itself. This is *very* puzzling: the laws of gravitation state that rotational velocity should *decrease* with distance from the center of mass (just as the orbital velocity decreases from planet to planet in the Solar System from Mercury out to Pluto). This conundrum has resulted in much speculation about a *dark population* composed of no one knows what lying out beyond the disks of spiral galaxies and influencing their rotational velocity. And the question of this *missing mass* has cosmological ramifications, for upon it hangs the fate of the Universe: if this dark population actually does *not* exist (and any of it around *any* galaxy – including our own – has yet to be observed), our Universe does not have sufficient density to reverse the cosmological expansion caused by the Big Bang, but will go on expanding – apparently forever. In this rather chilly scenario, all matter ends up either in black holes or in black dwarfs cooled to the ambient temperature of outer space itself.

As puzzling as this rotational velocity problem is how the spiral arms of spiral galaxies can maintain themselves in the face of it. Hypothetically the only way a spiral pattern could be *maintained* in a galaxy's disc is if the rotational velocity of the galaxy actually *increased* with distance from the galaxy's center: for the outermost arc of a spiral arm, having a longer orbit than the innermost arc, would have to travel faster to keep up. But a galaxy's spiral pattern is *not* sheered apart by its flat rotation curve from its bulge to the outer edge of its rim. Indeed, spiral arms seem to be very stable features: photographs of face-on spiral galaxies taken on red-sensitive plates and which therefore bring out the several-hundred-million year old thick disk population of evolved G, K, and M giants, shows that the density of this population is likewise enhanced along the galaxy's spiral arms in features that are smoother and broader arms than those traced by the blue O- and B-type giants and supergiants, but following the same curves. This implies that the spiral arms themselves are stable in a galaxy disc, no matter how transient their most conspicuous tracers (the short-lived O and B giants and supergiants). They are either a true spiral-form gravitational field in a spiral galaxy's disc, toward which gas and dust and stars are attracted, or the consequence of an initial spiral-form stellar density-distribution (perhaps set up by a gravitational interaction with another galaxy), which would have a spiral-form gravitational field. It is a question of what came first: the chicken or the egg.

The orbital motion of the stars, open clusters, giant molecular clouds, and neutral hydrogen gas in a spiral galaxy's disc around the galaxy's center is basically circular. But as a disc star ages, encounters with open and globular clusters and with giant molecular clouds give it an increasingly large component of velocity perpendicular to the galactic plane (called the *z-direction* by astrophysicists), and it therefore arcs above and below the galactic plane in a wave-like motion, the older disc stars (the *thick disc population*) arcing to greater distances above and below the galactic plane than the younger disc stars (the *thin disc population*). However, the stars and globular clusters in a spiral galaxy's halo follow long, thin comet-like orbits that, over vast amounts of time, carry them far into the galaxy's bulge near its center and then back far out toward the halo's perimeter. Spiral galaxy halos seem to be basically spherical, which means that halos as a whole do not rotate: their spherical form is said to be *pressure-supported* by the long in-and-out orbits of the halo stars. The shape of spiral galaxy discs, by contrast, are, because of the circular orbits of the disk stars, *rotationally-supported*. Spiral galaxy bulges are, for the most part, flattened oblate spheroids, suggesting that they too are largely rotationally-supported.

Spiral galaxies have an impressive range of luminosities (though not as great a range as elliptical systems). The most brilliant spirals, such as M58, M88, and M100 in the core of the Coma-Virgo Cluster, have absolute magnitudes around −22 (nearly 60 billion Suns). Toward the opposite extreme is the SBc NGC 5229 in the M101 Galaxy Group, which has an absolute magnitude around −16.5, a luminosity of just 330 million Suns. But even fainter spiral galaxies are known. And feeblest of all are the Sm systems: D127 in the NGC 4449 Galaxy Group in Canes Venatici, for example, has an absolute magnitude below −15.

IV.4 Irregular Galaxies (Irr or I)

Irregular galaxies are amorphous clouds of intergalactic gas and dust within which stars are presently forming. Thus they are practically pure extreme Population I systems, and not much different from our own Galaxy's complexes of associations, young open clusters, emission nebulae, and dense molecular clouds. The faintest dwarf irregular in our Local Galaxy Group, designated simply LGS ("Local Group System") 3, located 2.4 million light-years away, has an absolute magnitude of a mere −9.7, half a magnitude fainter than the Galactic open cluster NGC 6231, the core of the Scorpius OB1 association.

Indeed, compared to spirals, lenticulars, and normal ellipticals, irregular galaxies are all small and faint, similar in size and brightness to the dwarf ellipticals. The Small Magellanic Cloud, one of the largest and brightest Im systems, has an absolute magnitude of −16, just about the same as the M32 and NGC 205 dwarf elliptical satellites of the Andromeda Galaxy. The Local Galaxy Group contains a dozen dwarf irregulars. The Small Magellanic Cloud is a naked-eye object, of course. But at least two other Local Group irregulars, both of which have integrated absolute magnitudes around −14.4, are accessible to amateur instruments: NGC 6822 in Sagittarius can be glimpsed with just 10x50 binoculars; and IC 1613 in Cetus is visible in large amateur telescopes.

The origin of the dense intergalactic clouds from which dwarf irregular galaxies recently (by astronomical standards) formed is simply unknown. Could they be ejecta from the major spirals of the Local Group, the Milky Way, M31, and M33? Could they be actual remnants of the proto-galactic cloud from which the Local Group systems formed? (–probably not, though both the Large and the Small Magellanic Clouds are less rich in heavy elements than the Milky Way.) In any case, intergalactic material is very common in galaxy groups and clusters. The two Magellanic Clouds, for example, are enveloped in fairly dense neutral hydrogen (H I) gas; and a series of six increasingly small and lower-density H I clouds trail behind the Magellanic Systems in their orbit around the Milky Way. Eventually this *Magellanic Stream* gas might fall into the Milky Way and briefly enhance its luminosity by accelerated star-formation. (We then would be a *starburst galaxy.*)

A similar, but much larger, cloud of H I gas envelopes M81, M82, and NGC 3077 in the core of the M81 Galaxy Group. Most large galaxy clusters contain a vast, low-density medium of extremely hot (10 million °K) ionized gas detectable by its X-ray emission. Often this extensive intergalactic plasma contains *several times* more mass than the galaxies themselves! But this hot gas is not the remnant of the primordial cloud from which the galaxies formed: it always contains heavy elements that could only have been created in early generations of supergiant stars and subsequently blasted out into the intergalactic medium by supernova shock waves.

IV.5 Abnormal Galaxies

In Hubble's original galaxy classification scheme, abnormally-shaped systems were labeled *Type II Irregulars* (Irr II). In de Vaucouleurs' nomenclature they are simply called *peculiars* (pec).

IV.5.1 Peculiarly-Structured Galaxies

One group of abnormal galaxies are those which have conspicuous visual peculiarities: tails, jets, bright rings (sometimes within the bulge, sometimes well outside the disc), a rectangular profile, or exceptionally heavy dust lanes. M82 in Ursa Major and NGC 5253 in Centaurus both have conspicuously rectangular profiles. The very bright NGC 5128 in the Centaurus Galaxy Group has the star-distribution profile of a normal elliptical or S0 system, but possesses a very wide, thick dust lane easily visible in small telescopes. Another galaxy with an abnormally heavy dust lane is M64, the Black-Eyed Galaxy in Coma Berenices.

Photographs have found many strange galaxies in deep space. A few very remote galaxies appear as a thick ring around a small, bright central condensation. The Hubble Space Telescope image of the quasar PKS 2349–014 shows it as the bright "gemstone" in a ring galaxy. Long exposure photographs of many peculiarly-, and even some otherwise normally-, shaped galaxies reveal faint but very, very long tails extending for hundreds of thousands, sometimes even for millions, of light-years from the galaxy's body.

Most peculiarly-shaped galaxies seem to be interacting or colliding systems, or the wreckage from past galaxy-collisions. The abnormally thick or strangely-oriented dust rings in many elliptical or lenticular galaxies could be from the dust of spirals these systems have engorged. The long tails of many peculiar systems almost certainly were caused by the tidal stretching from a collision. And those "galaxies" which are virtually all ring probably became so when another system literally "bull's-eyed" them, carrying off their inner disc as well as most of their bulge: the exceptional brightness of these rings would result from enhanced star-formation induced by turbulence in that part of the galaxy-remnant's gas and dust (remember that much, or most, of a spiral galaxy's neutral hydrogen is *outside* its star-disk), and in gas and dust falling back from the collider as it passed away. Indeed, most peculiarly-shaped galaxies are bluer than normal, indicating that they have experienced enhanced star-formation over the past couple billion years – which would be the consequence of the turbulence induced in their gas and dust by the collision. Collision-wreckage systems tend to display prominent hydrogen-Balmer lines in their integrated spectrum and also to be infrared-luminous: the Balmer lines come from such galaxies' unusual numbers of recently-formed late-B/early-A stars (the main sequence lifetime of A0 stars, in which the H-Balmer lines are darkest, is less than 400 million years); and the infrared radiation is the result of absorption, and reradiation at infrared wavelengths, by dust of the ultraviolet photons from these galaxies' numerous hot, luminous O-type stars.

IV.5.2 "Active" Galaxies

A second sort of abnormal galaxy are those with *active galactic nuclei.* AGNs are extremely luminous, often in radio waves, X-rays, and visible light all at once. However, they are also dramatically variable, and usually on such short time scales that the variable source cannot be more than a few light-years across. Frequently a peculiarly-structured galaxy will have an AGN. The heavy dust lane of NGC 5128 in Centaurus seems somehow related to explosive behavior in its nucleus. M87 in Virgo is an extremely powerful radio source that displays in both radio and optical wavelengths a long jet projecting from its nucleus. In all cases the radio waves, X-rays, and even optical radiation from AGNs is largely generated by the *synchrotron process:* high-velocity electrons emitting photons as they spiral around strong magnetic lines of force.

Several varieties of AGNs have been identified over the past 60 years. The first was by the American Carl Seyfert in 1943. The *Seyfert galaxies* are otherwise normal Sc spirals which display very bright nuclei in which luminous, massive clouds of highly-ionized gas are turbulently moving with velocities of up to 5,000 kilometers per second. Seyferts are strong infrared sources, and strong and variable X-ray emitters. The brightest Seyfert is M77 in Cetus, an easy small telescope object displaying a very bright core within the face-on disc of a typical Sc system some 60 million light-years away. Most Sa and Sb spirals have nuclear gas clouds like those in the Seyfert Sc systems; but they are much lower in the ionization levels of their atoms and therefore have been labeled *low-ionization nuclear emission-line regions,* or *LINERS.*

Radio surveys around the sky have detected over 3,000 *radio galaxies,* more-or-less normal-looking giant ellipticals that are powerful radio emitters. The classic example of a radio galaxy is M87 in Virgo. At radio wavelengths these systems display a bright nucleus (most prominent at higher radio frequencies) flanked by two huge *radio lobes,* each centered sometimes millions of light-years from the radio nucleus (and therefore *way* outside the galaxy's optical image). The lobes are often connected to the nucleus by very long, very thin *radio jets.* Sometimes, as in the case of M87 itself, the jets are also optically visible. Apparently high-velocity material shoots out along the jets to the radio lobes, where the material's kinetic energy accelerates electrons to the near-light velocities necessary for synchrotron radiation. About 10% of all radio galaxies have optically-, as well as radio-, bright nuclei: they are termed *N galaxies.*

Up to the 1960s, the resolution of radio telescopes was so poor that usually it was difficult to identify radio sources with optical objects. But when resolution improved because of larger radio dishes and because of arrays of radio dishes that simulated extremely large radio apertures, it was found that literally hundreds of radio sources could be optically identified *not* with galaxies, but rather with very blue stellar images. However, the optical spectra of these *quasi-stellar radio sources,* or *quasars,* proved to have, superimposed upon a blue synchrotron background, Seyfert-galaxy type emission features that are so far redshifted it implies these objects are *extremely* remote. Thus quasars, because of their great distances and their intrinsically high luminosities (absolute magnitudes around -25 or -26–which means that at 32.6 light-years a quasar would be nearly as bright as the Sun!), are simply AGNs that swamp out the light of their host galaxy. Quasars photographed by the Hubble Space Telescope are invariably found to be within normal giant elliptical galaxies.

As intrinsically luminous as the quasars, and like them strong X-ray emitters, but strangely radio-quiet, are *quasi-stellar objects.* QSOs have Seyfert galaxy spectra and seem to be the distant, high-luminosity version of the classic Seyfert galaxy: but, like quasars, they swamp out the feebler glow of their host galaxy. However, complicating this picture is the fact that Hubble Space Telescope images have found some QSOs to be within elliptical systems rather than within Seyfert-like Sc spirals.

Some strong radio sources with the bluish star-like images and the broad, highly-redshifted optical emission lines of traditional quasars are visually variable. They are called *optically-violent variable quasars,* OVVs. Related to them, and the most difficult AGNs to detect, are the *BL Lacertae objects.* As the designation suggests, originally BL Lac, which can change in brightness by a couple magnitudes in less than a month, was thought to be some strange sort of variable star. The puzzling thing about it was that its spectrum is non-stellar: it has neither emission nor absorption lines; and the brightness profile of its continuum (that is, the variation in light intensity with wavelength) cannot be generated by a glowing gas (which of course is what stars are). Subsequently BL Lac was discovered to be a strong, rapidly-varying emitter of highly-polarized radio waves. Finally, in 1974, the faint glow of a normal elliptical galaxy was detected around the stellar point of BL Lac. OVVs and BL Lac objects are together termed *blazers* because of their strong variability. They are probably basically the same type of object: the lack of Seyfert-like emission features in the optical spectrum of BL Lac is doubtless simply because the bright background continuum swamps out the lines.

IV.6 Galaxy Groups and Galaxy Clusters

Just as for stars, clustering is the rule rather than the exception for galaxies. The Milky Way and the Andromeda Spiral are the two dominant members of a small aggregation of at least three dozen (mostly dwarf) systems called the Local Galaxy Group. The Local Group has a binary structure, with the Milky Way, its Magellanic satellites and a few dwarfs at one end and the Andromeda and Triangulum spirals, the Andromeda Spiral's dwarf elliptical satellites, and several more dwarfs at the other end. The Local Group is dominated by M31, which radiates almost as much light as all the rest of the Group's members together. The galaxy types in the Local Group–four loose-armed spirals (Sb M31, Sbc Milky Way, Sc M33, SBm LMC), with dwarf irregulars, ellipticals, and spheroidals–seems to be typical for our region of intergalactic space.

Indeed, the nearest galaxy group to ours, the 9 million light-year distant Sculptor Group in the southern heavens, is dominated by very loose-armed Sc and Sd spirals. Its brightest members, NGCs 45, 55, 247, 253, 300, and 7793, are distributed in a 20° area. Toward the Sculptor Group we look directly *away* from the interior of our Coma-Virgo

Supercluster: beyond it is relatively galaxy-empty space until, perhaps 70 million light-years away, we come to the near edge of the next galaxy supercluster.

On the opposite side of the sky, some 12 million light-years more-or-less toward the interior of our Coma-Virgo Supercluster, is the M81 Galaxy Group, the brightest members of which are the luminous Sab M81, the peculiar radio galaxy M82, and the Sc spiral NGC 2403 in Camelopardalis, somewhat removed from the M81 + M82 pair. Hence, like the Local Group, the M81 Galaxy Group has a binary structure and is dominated by fairly large, loose-armed spirals. Lesser members include NGCs 2366, 2976, 3077, and 4236. Sometimes the Sc spiral IC 342 and a few fainter galaxies near it in SW Camelopardalis are listed as M81 Group members.

A peculiarly-structured aggregation of peculiar galaxies is the 12–15 million light-year distant Centaurus Galaxy Group, strung out in a 30°, 8 million light-year, long chain that begins with NGC 5068 in Virgo on the north and ends at NGC 4945 in central Centaurus on the south. Along the chain are the three-armed, supernova-rich M83 in Hydra, the strange lenticular NGC 5102, the rectangular NGC 5253, and the giant radio galaxy NGC 5128. Why so much oddity should be concentrated in one group is not known.

The Sculptor, M81, and Local galaxy groups are out on the edge of the huge *Local Supercluster* (sometimes simply referred to as the *Virgo Cluster)*, which is centered some 50 or 60 million light-years away toward the M84-M86-M87 region in extreme NW Virgo. Because of our vantage on its perimeter, we see the Local Supercluster as a band of literally hundreds of galaxies of magnitude 13.0 and brighter extending from the bowl of the Big Dipper on the NW down through Canes Venatici, Coma Berenices, and Virgo into central Centaurus on the SE. The flattening of this galaxy band implies that the entire supercluster is rotating, the individual galaxies' orbital periods out on the rim being hundreds of millions of years. However, the galaxies of the Local Group are also falling in toward the core of the Coma-Virgo Cluster at a velocity of 250 kilometers per second. (Cosmologists call this the *Virgocentric infall.*) But our Local Group's galaxies are flowing at an even faster rate, 600 km/sec, toward a point behind the Centaurus Milky Way called the *Great Attractor.* Presumably this is a dense concentration of galaxies, but the dust of the interior of our Galaxy's spiral disc prevents a closer look.

The core of the Local Supercluster, the Coma-Virgo Galaxy Cluster, is in an area 10° in diameter centered midway between Denebola (Beta [β] Leonis) and Vindemiatrix (Epsilon [ε] Virginis). It is dominated by giant elliptical and lenticular galaxies, but also has some very luminous spirals. The preponderance of early-type, gas-poor, evolved-star-rich galaxies in Coma-Virgo and other supercluster cores is assumed to have something to do with galaxy encounters and near-misses in such dense environments, which long ago swept gas out of the interacting galaxies and ended star formation in them. And indeed, radio observations show that the galaxies in the Coma-Virgo Cluster are relatively poor in neutral hydrogen gas for their respective morphological types. And possibly some of the giant spirals that we see "in" the core of the Coma-Virgo Cluster are in fact somewhat foreground to it: hence Coma-Virgo could be even *more* spiral-galaxy-poor than it looks. Several spiral-dominated galaxy clusters, all roughly 15 to 30 million light-years from us, are spread between Coma-Virgo and ourselves from central Coma Berenices, through western Canes Venatici, and up into the Big Dipper; so there is good reason to think that some of the giant spirals that seem to be *in* the core of Coma-Virgo are actually on our side of it, (There is such a jumble of galaxies from Coma Berenices NW into the Big Dipper that different astronomers not only often assign different galaxies to different groups, but sometimes even see radically different galaxy groupings here.)

The distance to the Coma-Virgo Cluster is of extreme cosmological importance because it can be used as a measuring stick – a *standard candle* – to more remote galaxy clusters. However, different criteria yield a rather discouragingly broad range of distances to the cluster. The lowest distance estimate comes from the assumption that the planetary nebulae of Coma-Virgo galaxies have the same peak absolute magnitude, $-4^1/_2$, as the planetary population of M31: the brightest Coma-Virgo Cluster planetaries imply a distance of 15 megaparsecs, or just under 50 million light-years, to the cluster. The *highest* distance estimate to Coma-Virgo is from the Type I supernovae observed in it. The Type I SN explosions in our Galaxy have peak absolute magnitudes in blue light of $M_B = -19.7$. Because such supernovae in Coma-Virgo have peak apparent blue magnitudes of $m_B = 12.1$, they imply a distance modulus to the cluster of $m_B - M_B = 12.1 - (-19.7) = 31.8$ magnitudes, corresponding to 23 megaparsecs, or nearly 75 million light-years. But possibly the most reliable distant estimate to Coma-Virgo comes from Hubble Space Telescope observations of 20 Cepheid variables in the giant spiral M100: the Cepheid period-luminosity relationship applied to those stars results in a distance to M100 of roughly 17 megaparsecs ~ 55 million light-years.

Chapter 2
Open Clusters

I Summer Constellations

Literally hundreds of star clusters can be seen in large telescopes, and dozens are visible in even the smallest binoculars. Indeed, something like 20 star clusters can be identified with the unaided eye, a half dozen of which can even be resolved by the eye. These naked-eye clusters include the Ursa Major Group, the Hyades, Coma Berenices, the Pleiades, the Alpha Persei Group, and IC 2602 in the far southern heavens.

The intrinsic astronomical difference between open clusters and globular clusters is described in Section III of Chapter 1. From the observer's point of view, open clusters in general are wide-field, low-power objects often at their best in binoculars and small richest-field telescopes, but globular clusters are small, compact objects suitable only for moderate-to-large telescopes. The reason for this is simply that no globular clusters are near to us–the nearest globular, M4 in Scorpius, is about 7,200 light-years distant, and only a handful of globulars are within 10,000 light-years of the Sun. By contrast, there are a dozen open clusters less than 800 light-years away.

Thus globular clusters always require larger telescopes to be seen at their best. However, the nearer open clusters are at their best in just 10×50 glasses, and dozens of open clusters are splendid sights in giant binoculars. Indeed, open clusters along with emission nebulae and (surprisingly) galaxies are the best types of astronomical objects for giant binoculars. Therefore in this chapter a large number of open clusters will be discussed, many in significant detail. In addition to the basic attractiveness of their "clusterness," these groups often feature double stars, attractive star-colors and star-color contrasts, and sometimes even emission or reflection nebulae. In many open clusters the stars appear to be organized into aesthetically-attractive chains, streams, and subgroups. Some open clusters are the cores of large stellar associations which are themselves naked-eye or binocular objects. And many open clusters are astrophysically interesting, or at interesting locations within the spiral structure of our Galaxy.

In this chapter the clusters will be described by constellation, the constellations divided into Summer, Autumn, and Winter, depending upon the season of their best visibility in the early evenings (from mid-northern latitudes). The handful of open clusters in the Spring constellations are described in a brief section at the end of this chapter. Very few of the spring constellations have open clusters. The reason is that most open clusters lie in or very near the Milky Way, and during northern hemisphere spring the Milky Way lies low over the west, north, and east horizons and therefore, because of the thickness of the atmosphere over the horizon, is effectively invisible. Within the Summer, Autumn, and Winter sections the constellations are presented not alphabetically, but in their approximate west-to-east order along the Milky Way. This order is not only for the convenience of the observer, who would wish to move from one constellation to the neighboring one, but more particularly with the intention of presenting open clusters in an order that will make the astrophysical significance of their distribution along the Milky Way easier to understand. For example, nine open clusters will be described in the small constellation of Vulpecula but only three in the very much larger Aquila immediately to its SW along the Milky Way, the reason being that toward Aquila we look lengthwise down the open-cluster-poor gap between two spiral arms of our Galaxy whereas toward Vulpecula we look ahead into the forward are of our Orion-Cygnus Spiral Arm. This, and the reasons for the other peculiarities in the distribution of open clusters along the Milky Way, will be explained in detail in Chapter 3.

I.1 Sagittarius

Because it lies in the direction of the Center of the Galaxy, Sagittarius contains some of the brightest star clouds along the Milky Way, most notably the Great and Small Sagittarius star clouds, which are described in detail in Chapter 3. The constellation also is rich in open and globular clusters and has some of the best specimens of both cluster types for binoculars and small richest-field telescopes. Of Sagittarius' four Messier-numbered open clusters M23 and M25 are outstanding giant binocular groups. Its best small-instrument NGC open cluster is NGC 6530, a scattering of star-gems set in the haze of the Lagoon Nebula M8. NGC 6603 in the Small Sagittarius Star Cloud, though very distant and therefore not visually impressive, is astrophysically a very interesting cluster in a very interesting location within our Galaxy and consequently worth a look in giant binoculars.

Open Clusters in Sagittarius

Cluster	RA (2000.0)	Dec	m_V	Size	m_V *	*Spectrum	Distance	A_V	M_V	True Size	Age
M18	18^h20^m	−17°08′	6.9	10′	8.6	B2 III	4100 l-y	1.6	−5	12 l-y	50 mil
M21	18 04	−22 30	5.9	13	6.73	B0.5 III	5200	0.8	−5.9	20	
M23	17 57	−19 01	5.5	27	9.2		2100	1.1	−4.7	16	300
M25	18 32	−19 15	4.6	32	6.3v	F5-G1 Ib	2500	1.4	−6	20	30
N 6469	17 53	−22 21	8.2p	12	12						
N 6520	18 03	−27 54	7.6p	6	9.0p		5200	1.0		9	
N 6530	18 05	−24 20	4.6	14	6.87	06.5 V	5200	1.0	−7.4	21	
N 6603	18 18	−18 25	11.1p	5	14.0p		11,700	1.5		18	200
N 6645	18 33	−16 54	8.5p	10	12.0p						
N 6716	18 55	−19 53	6.9	10	8.28	B5	2000	0.4	−2	5.2	
Cr 394	18 53	−20 23	5.6	22	7.0		2000	1	−3.4	13	

M18: M18, easily found 1° due north of the Small Sagittarius Star Cloud, is one of the least-impressive of the Messier-numbered open clusters. Though a magnitude 6.9 object, it is quite compact (less than 10′ in diameter) and not very populous, with only about a dozen mag $8^{1}/_{2}$–11 members. In 10×50 glasses it appears merely as a knot of four or five stars embedded in a small shred of haze; but even 15× supergiant binoculars resolve only a half dozen M18 members. This cluster is some 4,100 light-years from us on the far side of the inter-arm gap between ourselves and the Sagittarius-Carina Spiral Arm, which contains the Lagoon, Trifid, Swan, and Star Queen nebulae (M8, M20, M17, and M16, respectively). M18 could be an outlying tracer of the Sgr-Car Arm because it is reasonably young, 50 million years old, and its brightest member is a B2 III blue giant.

M21 (Plate XI): This cluster, located only about 0.7° NE of the Trifid Nebula, is only slightly larger and richer than M18, though its stars are somewhat brighter. In 10×50 binoculars it appears as a knot of four or five stars, the two brightest of which are mag 6.7 and 7.3 objects. M21 resolves quite well in 15-20× giant binoculars; but its stars are scattered in ragged clumps, and are not very uniform in brightness. The "clusterness" of M21 is obvious; but it lacks the smooth elegance which makes M23 and NGC 6530 so beautiful. The group's lucida is immediately SE of the cluster center and has a faint companion just to its NW, with a third star visible somewhat farther to the ENE.

M21 is about 5,200 light-years away. Its two brightest stars are hot, luminous B0.5 III and B0 V objects, suggesting that the cluster is very young. Indeed, M21 seems to be a member of the large Sagittarius OB1 association, which features the B8 Ia supergiant Mu (μ) Sgr and the Lagoon Nebula with NGC 6530.

M23 (Plate XIV, Photo 3.4): The very large (27′ in diameter), very bright (mag 5.5), M23 can be spotted with the unaided eye (from a dark sky site) as a fuzzy dot 5° due west of the Small Sagittarius Star Cloud. The cluster's lucida is only a mag 9.2 star: the group is bright because it is very populous, with over 150 members. 10×50 glasses can only partially resolve it, revealing just a dozen faint stars twinkling in a granular haze.

In supergiant binoculars M23 truly comes into its own. With 15×100s it is resolved into dozens of stars scattered across a pale background haze. Most of the group's brightest members are in its concentrated central region; but the mag 9.2 lucida is out toward the cluster's NE edge. Because M23's brightest 20 or 30 stars are of similar brightness, the group has a very attractive swarm-like look. It perhaps is at its very best in 25×100 instruments, which have the light-gathering power to pull in many of the cluster's fainter stars and the magnification to separate them from each other. At least 50 members, excluding far outliers, can be counted. The beauty of this multitude of pin-prick star-sparks, the faintest ones twinkling in and out of resolution, is indescribable.

M23 is about 2,100 light-years away. This puts it in the inter-arm gap between our Orion-Cygnus Spiral Arm and the Sgr-Car Arm. Only intermediate-age and ancient open clusters are found in the spaces between spiral arms; and in fact M23 is estimated to be 300 million years old. Its brightest stars are blue-white B9 main sequence objects and evolved yellow G-type giants. Though it is populous, the cluster's total luminosity is only about 6,000 Suns. This is a little less than the much-poorer cluster M18. The difference is that M18 is still sufficiently young to have two or three very luminous blue stars.

M25 (Plate XIV, Photo 2.4): Only 2° due east of the Small Sagittarius Star Cloud is another naked-eye open cluster, M25, which is even larger and almost 1 magnitude brighter than M23. In contrast to M23, however, M25's brightness comes not from a swarm of faint stars, but from a handful of bright ones, mag 6½–8 objects. It easily resolves even in only 7× binocs. Supergiant glasses pull in at least three dozen cluster members (it depends upon how far out from the cluster core you count); but at 20× it is a little too loose and coarse for the best effect.

The brightest member of M25, situated squarely at the center of the cluster, is the Cepheid variable U Sagittarii, which ranges between magnitudes 6.3 and 7.1 in a period of 6.745 days. Its variations can be followed by comparing it to the group's second brightest star, a mag 6.8 object on the NW edge of the cluster. Open clusters with Cepheid variables are not really all that rare. NGC 7790 in Cassiopeia in fact has *four* Cepheids! And in Sagittarius itself the 7th mag Ceph BB Sgr is thought to be an outlying member of NGC 6716. Most moderately-massive stars are assumed to suffer Cepheid-type pulsations during their evolution. Thus Cepheids are found only in somewhat-evolved (but not ancient) open clusters. M25, for instance, is estimated to be at least 30 million years old.

M25 is located some 2,500 light-years away. Like the considerably-older M23, it is out in the interarm gap between the Ori-Cyg and Sgr-Car arms. It is a rather large cluster, 20 l-y across, and rather luminous, its integrated absolute magnitude of –6 corresponding to the brightness of around 22,000 Suns.

NGC 6469: Though it is a rather nondescript gathering of mag 12 and fainter stars, NGC 6469 is located only 2° WNW of the Trifid Nebula, M20, and therefore is of interest for its field. It requires telescopes to be seen at its best, but it can be glimpsed in 25×100 supergiant binoculars as a rather large, fairly rich field concentration of faint stars, irregular in outline and lacking background haze. NGC 6469 has the look of a fairly old, fairly distant, cluster which we see through Sgr-Car Arm dust.

NGC 6520: Like NGC 6469, NGC 6520 is a cluster just visible in large binoculars but of particular interest for its location. It is situated in the midst of the Great Sagittarius Star Cloud immediately east of the small, but conspicuous, dark nebula B86. Look for these two objects some 2½° north and slightly east of Gamma (γ) Sagittarii. Both nebula and cluster are only about 6' in diameter. They are best observed in conventional telescopes at about 70×; but NGC 6520 can be seen in 15×100 supergiant binoculars as a tight knot of four stars. The core of the cluster is a WNW-ESE pair of steadily-resolved stars in a tight patch of haze. A third star is just to the brighter pair's ESE; and a fourth star is a bit farther away to the WNW (in the direction of the bright field star on the far side of B86 from the cluster).

NGC 6520 is estimated to lie some 5,200 light-years from us. This is about the same distance as the Lagoon Nebula and the open cluster M21 to the north, which are major tracers of the Sgr-Car Spiral Arm along the Sagittarius Milky Way. However, the cluster is not a true Sgr-Car Arm group: it is too far from the core of the arm; and the fact that its brightest members are evolved orange giants of absolute magnitudes only around $-2^1/_2$ implies that NGC 6520 is too old to be a spiral arm open cluster. Nor is the cluster physically related to B86, which is so dark that it probably is much nearer to us.

NGC 6530 (Plate XIII, Photo 2.1): NGC 6530 is an interesting, and very beautiful, open cluster in an interesting and very beautiful setting. It is a fairly evenly distributed group of 7th, 8th, and 9th magnitude stars spread over the northern half of the eastern wing of the Lagoon Nebula, M8. The cluster's integrated apparent magnitude of 4.6 suggests that it could be seen with the unaided eye; but in fact the Lagoon itself is an easy unaided eye object. In 10×50 binoculars a half dozen NGC 6530 members, mag 6.9 and fainter, can be seen; but 25×100 supergiant glasses resolve at least fifteen stellar pinpricks, comparably bright and smoothly sprinkled over the Lagoon haze. The cluster's particular beauty comes from the uniform brightness and distribution of its two dozen stars. NGC 6530 is near its best at 25×, since it has no central condensation and higher powers begin to spread it out too much.

This cluster not only is visually, it is in fact *physically,* involved with the Lagoon Nebula. The group's lucida, a hot O6.5 main sequence object, is one of the mid-O stars providing the ultraviolet radiation fluorescing the Lagoon. Its second brightest member is a mag 7.5 B0 V object. NGC 6530 is about 21 light-years across and has an integrated absolute magnitude of –7.4, a very healthy luminosity of nearly 80,000 Suns.

NGC 6603: Yet another Sagittarius open cluster in a very interesting location is NGC 6603. This is a small (5' in diameter), faint group that can be just barely glimpsed with 25×100 supergiant glasses as a little patch of haze; but it is situated within the Small Sagittarius Star Cloud, M24. Look for it in the east-central part of M24, along the "waist" of the star cloud where it appears to "bend" a little toward the east.

NGC 6603 might not be visually impressive in wide-field, low-power instruments; but it is well worth a look because photometric studies have shown that it is physically involved with the M24 star cloud, perhaps just within M24's nearer edge. The cluster is, however, peculiarly evolved for a group that seems to be part of a spiral arm feature, for it is estimated to be 200 million years old. But the star cloud itself apparently is composed of stars up to 600 million years old. The true nature of M24 remains something of a puzzle.

NGC 6603 is a representative of a type of rich, but evolved, open cluster found toward the interior of the Galaxy. M11 in Scutum is another, and nearer, example of this type of cluster. The absorption of only 1.5 magnitudes to NGC 6603 is much less than half of what we would expect for a cluster at this distance and in this direction through our Galaxy and confirms that we view the group and its M24 star cloud in a "window" through the otherwise thick dust of the Sgr-Car Spiral Arm.

NGC 6645: This moderately large (10' diameter) but faint cluster is located a little more than 2° due north of M25 near the Sagittarius/Scutum border. Though at its best in conventional telescopes, it is an interesting sight in supergiant binoculars. 15×100s reveal an attractive scattering of delicate star-chips spread over a modest-sized area of very pale glow, the cluster's lucida on its NW edge and its second brightest member on the SE edge. A peculiar, almost perfectly straight, WNW-ESE line of one 7th, two 8th, and two 9th mag stars extends for over a half degree from just outside the cluster's NE edge to the line's 7th mag member at its ESE end.

NGC 6716: This group, located several degrees east and slightly south of M25, is a very large, but loose field concentration of three vague gatherings of six-to-eight 7th-to-10th mag stars, one gathering in the NE of the cluster area, the second in the SE, and the third in the west. The 10' diameter cited in the catalogues for the cluster suggests that only one of these three gatherings is considered to be NGC 6716 proper; but all three are so disorganized and random that none looks like a true open cluster. A 6th mag star is between the SE and west gatherings.

The catalogues give a distance of 2,000 light-years to NGC 6716, about the same as for M23. This would mean that NGC 6716, like that much more convincing cluster, would be an inter-arm group. However, this distance implies the abnormally small

Photo 2.1 Open cluster NGC 6530 in the Lagoon Nebula M8

integrated absolute magnitude of –2 for the cluster. The Coma Berenices Star Cluster also has an integrated absolute mag of just –2. Presumably NGC 6716, like that much nearer open cluster, is on the verge of disintegration.

Cr 394: Collinder 394 is a large field condensation, about 22' across, of seven or eight mag 7–10 stars, much like NGC 6716 in appearance and centered about $^2/_3$° SW of that cluster. This group's main claim to fame is that one of its members seems to be the Cepheid variable BB Sagittarii, located about 20' west of the cluster center halfway to the 5th mag star 29 Sgr. (The variable is the middle member of a short ESE-WNW line of three stars.) BB Sgr has a period of 6.6371 days, implying that its peak absolute magnitude is –3.4. The star's peak apparent mag is 6.7, so its distance (compensated for interstellar absorption) must be a little less than 2,000 light-years. This puts Cr 394 out in the inter-arm gap with another cluster with a short-period, low-luminosity Cepheid, M25.

Such Cepheids, and the intermediate-age open clusters in which they are sometimes found, are not rare in the Galaxy's inter-arm gaps.

I.2 Scorpius

Though the direction toward the Center of the Galaxy technically is in Sagittarius, $4^1/_2$° WNW of Gamma (γ) Sagittarii, it is not far NE of the Tail of the Scorpion. Thus SE Scorpius is, like Sagittarius, rich in both globular and open clusters. Indeed, the Tail of Scorpius boasts three of the brightest and largest open clusters in the heavens, M6, M7, and NGC 6321, and one huge star-group, Tr 24, that is at its best in binoculars. Among the constellation's many NGC open clusters is the one nearest to the direction toward the Galactic Center, NGC 6451. The NW part of Scorpius, dominated by the bright stars in the Head and Heart of the Scorpion, most of which are members of the nearby Scorpio-Centaurus Association, is rather far off the galactic equator and therefore lacks open clusters.

M6 (Plate VIII, Photo 2.2): Near the "Sting" of the Scorpion (Lambda [λ] + Upsilon [υ] Scorpii) are two of the largest and brightest open clusters in the sky, M6 and M7. They are easily visible to the unaided eye as two small patches of Milky Way haze. They were known to the ancient Greek and Roman astronomers. The Bedouin of the Arabian Desert called them *al-Humah al-'Akrab,* "the Venom of the Scorpion."

M6 is easily spotted about 5° north and slightly east of Lambda Sco. The core of the cluster is a conspicuous 20°-long NE-SW rectangle of about two dozen mag 6.8 to 9.5 stars. The cluster lucida, located at the NE end of the rectangle, is the semiregular orange supergiant variable BM Sco, which has a K2.5 Ib spectrum and a light range of mag 5.8–8.0 in a period of roughly 850 days. When it is near maximum, the star's ruddy color is obvious in supergiant binoculars and telescopes. In 25×100 instruments at least 46 M6 stars can be counted in and around the rectangle.

M6 is estimated to lie about 1500 light-years away. This means that it is located, like M23 and M25 in Sagittarius, within the inter-arm gap between our

Open Clusters in Scorpius

Cluster	RA (2000.0)	Dec	m_V	Size	m_V*	*Spectrum	Distance	A_V	M_V	True Size	Age
M6	17^h40^m	–32°15'	4.2	33'	5.8v	K2.5 Ib	1500 l-y	0.9	–5.0	14 l-y	100 mil
M7	17 54	–34 49	3.3	80	5.60	gG8	820		–3.7	20	220
N 6231	16 54	–41 48	2.6	14	4.77	B1.5 Ia+	6200	1.4	–10.2	25	3.6
Tr 24	16 57	–40 40		60	6.1v	B0 Ia	6200	1.1		110	3.6
N 6242	16 56	–39 30	6.4	9	7.3		3600	1.2	–5.0	9	50
N 6281	17 05	–37 54	5.4	8	7.9		1600	0.5	–3.6	4	220
N 6383	17 35	–32 34	5.5	20	5.70	O7V+O7V	5000	1	–6.5	30	10
N 6451	17 51	–30 13	8.2p	7	12.0p		6800	1.9		18	400

Photo 2.2 Open clusters M6 (left) and NGC 6383 (right) in Scorpius. The loose cluster of faint stars between them in the center of the field is Trumpler 28

Orion-Cygnus Spiral Arm and the Sagittarius-Carina Spiral Arm, the next spiral feature from us in toward the central bulge of our Galaxy. The bright-star population of the cluster is consistent with that of an intermediate-age inter-arm group: mid-to-late B-type giants and main sequence stars, supplemented with one Ib late-type supergiant. The cluster as a whole has a luminosity of just under 10,000 Suns. At maximum BM Sco reaches an absolute magnitude of -3.4, which corresponds to a luminosity of 1900 Suns.

M7 (Plate X, Photo 3.2): The huge open cluster M7, centered about 2° north and slightly east of the 3rd magnitude star G Scorpii, consists of 24 stars of mags 5.6 to 9.3 (as well as fainter members) spread over an 80' area–an area *seven times* larger than the full moon! The integrated apparent magnitude of the group is 3.3–about the brightness of G Sco itself. However, not all the cluster-glow visible to the unaided eye is from the group's stars. M7 happens to lie in front of a small but bright Milky Way star cloud. In fact, this star cloud, like the much larger Great Sagittarius Star Cloud to its north, is a section of the central bulge of our Galaxy.

M7 is so large and loose that it resolves extremely well even in mere 7x binoculars. In fact the cluster is so large and loose that powers of 25× or more are simply too much for it. The group makes its best aesthetic impression at about 15×. The cluster core is framed in a rough pentagon of mag 6–6½ stars. Conspicuous along the south side of the core is a shallow south-opening arc of a half dozen brighter cluster members. In the core itself is a short arc of four stars, at the SW end of which is an attractive double with mag 6.4 and 7.2 components.

M7 appears so large and bright partly simply because it is so close, only 820 light-years away. In terms of its location in our Galaxy, this distance places the cluster just beyond the Scorpio-Centaurus Association , the brightest members of which in this area include Kappa (κ), Lambda (λ), and Upsilon (υ) Scorpii. Neither the group's true size, 20 light-years, nor its true luminosity, 6250 Suns, are unusually large for an open cluster, though the former can be considered to be on the high end of average. M7's lucida is an orange G8 giant with an absolute magnitude of −1.4. Its other 23 brightest members are blue-white main sequence stars with spectra from B6 to A4. This suggests a cluster age of 220 million years.

NGC 6231(Plate VI, Photo 2.3): One of the most interesting and most beautiful open clusters in the sky is NGC 6231, in the Tail of the Scorpion $3^1/_2$° due south of the wide double Mu (μ) Scorpii. The core of the cluster is not especially large, just 14' in diameter; but it has an integrated apparent magnitude of a truly impressive 2.6, 1.1 magnitudes brighter than the famous Omega (ω) Centauri globular cluster. NGC 6231 therefore is an obvious naked-eye object. Unfortunately this group is almost 42° south of the celestial equator and consequently can be easily seen only from the southern half of the US.

Crowded into the 14' diameter core of NGC 6231 are two mag 6 and three mag $6^1/_2$ stars, as well as numerous lesser luminaries. The compactness of the cluster core means that it is an ideal object for higher-power and zoom binoculars (though it is at its best with at least 60× in telescopes). But almost all of the bright stars in the field around the core are also cluster members. In fact the brightest member of NGC 6231 is the mag 4.8 Zeta-one (ζ^1) Scorpii, located $^1/_2$° south of the cluster core.

Even if you live so far north that NGC 6231 appears only briefly in the haze over your southern horizon, it is well worth looking for because it is one of the richest concentrations of O and early-B supergiants known in our Galaxy. Zeta-one Sco is a B1.5 Ia^+ hyper-supergiant with an absolute magnitude of −8.8–a luminosity of 276,000 Suns! However, the star is so hot that most of its radiation is in the invisible ultraviolet and its bolometric absolute magnitude (its brightness to an "eye" sensitive to the entire electromagnetic spectrum) is −10.8. The Sun's *bolometric* absolute magnitude is +4.8: thus Zeta-one Sco radiates 1.7 *million* times as much energy each second as the Sun. Almost as luminous as Zeta-one Sco are the O8 Ia supergiants HD 151904, the mag 5.2 star almost 1° NW of the cluster core, and HD 151408, a mag 5.8 object 1° NNE of NGC 6231, which have bolometric absolute magnitudes of −10.5 and −10.2 respectively. The extremely hot (39,000 °K) mag 6.6 O6 giant on the NW edge of the cluster core has a bolometric absolute magnitude of −9.0.

Not only is NGC 6231 remarkable for its hyper-luminous O and early-B supergiants, it contains two of the rare Wolf-Rayet stars. Such objects are super-massive, super-hot stars suffering continuous and almost explosive mass loss, ejecting shells of turbulent gas at velocities of up to a nova-like 3000 kilometers per second. This instability is no doubt triggered by the first stages of these stars' evolutionary expansion toward the red supergiant stage. But they will never make it: their internal instabilities will probably cause them to go supernova first. One of the NGC 6231 Wolf-Rayets is in the dense cluster core. But the other, the mag 6.6 WN7 HD 151932, is easily identified $^1/_2$° due west of the cluster core.

NGC 6231 is estimated to lie about 6,200 light-years from us. Thus its integrated absolute magnitude, corrected for 1.4 magnitudes of absorption by interstellar dust, is −10.2, a visual luminosity of one million Suns. This is equal to the visual luminosity of our Galaxy's largest and brightest globular cluster, Omega Centauri. However, globular cluster stars, unlike the hot blue supergiants of NGC 6231, radiate most of their energy in visual wavelengths. Thus the total energy output of NGC 6231 is actually several times greater than that of Omega Cen. Moreover, the total mass of Omega Cen is in the hundreds of thousands of solar masses whereas that of NGC 6231 is only a few thousand Solar masses. Thus massive O- and B-type supergiants are extremely efficient energy-generators...

...but they are also very *short-lived* energy generators. The age of Omega Cen is probably at least 12 billion years whereas that of NGC 6231 is only 3.6 *million* years. Moreover, NGC 6231 will not look like it does much longer (in astronomical time-scales) because its supergiants will soon all go supernova. Young as the cluster is, however, its high-energy stars have already blown out of its vicinity most of the gas and dust that must have been around the group when it was born, for long-exposure red-sensitive photos reveal surprisingly little hydrogen emission nebulosity anywhere near

Photo 2.3 Open clusters in the Tail of Scorpius: the compact, supergiant-rich NGC 6231 is bottom center, the intermediate-age NGC 6242 top center, and the scattered Tr 24 in the middle of the field. The arc of emission nebula on the NE edge of Tr 24 is IC 4628

NGC 6231. Only extremely young open clusters (those less than a million years old) are embedded in large emission nebulae.

Tr 24 (Plate VI, Photo 3.1): NGC 6231 is the core cluster of a major Sagittarius-Carina Spiral Arm tracer, the Scorpius OB1 association. The richest part of the association outside the cluster is a 60' field of 6th, 7th, 8th, and 9th magnitude stars centered about 1° north and slightly east of the NGC 6231 core. This large, loose stellar aggregation has been designated as the open cluster Trumpler 24, though, unlike NGC 6231, its stars are too scattered for it to be a true gravitationally-bound group. In fact Tr 24 is so large and scattered that it is the perfect 7× binocular star field. The mag 5.8 O8 supergiant HD 151408 1° NNE of the core of NGC 6231 is just on the SSW edge of the Tr 24 star field. But the brightest star within Tr 24 is the small-amplitude variable V861 Sco, a B0 Ia supergiant that is an X-ray source and orbited by a massive, but apparently non-luminous, companion that could be a black hole.

The true size of Tr 24 is about 110 light-years. The full extent of Sco OB1 from Zeta-one Scorpii on the SW to the far NE edge of Tr 24 is about 2°~220 light-years. Farther NE is the very faint but large ($1^1/_2$°×1°) emission nebula IC 4628, strictly a photographic object. Sco OB1 is parallel to the galactic equator, probably because of the way in which it was formed.

NGC 6242 (Plate VI, Photo 3.2): About 1° due north of the center of Tr 24, and $1^1/_2$° SSE of the Mu (μ) Scorpii double, is the compact (9' diameter) but reasonably bright (mag 6.4) NGC 6242. In 10× binoculars this cluster appears merely as two or three mag $7^1/_2$–8 stars embedded in a patch of haze. It is reasonably populous and therefore a good sight in 25×100 supergiant binocs, which pull in many of the group's fainter members and spreads them out so they can be seen. The cluster is, however, at its aesthetic best at moderate powers in telescopes.

NGC 6242 is estimated to lie some 3600 light-years from us. Thus it is in the foreground of the Scorpius OB1 association and on the far side of the inter-arm gap between ourselves and the Sagittarius-Carina Spiral Arm. In terms of true size (12 l-y), luminosity (not quite 10,000 Suns), and age (50 million years), it is almost the twin of the comparably distant M18 in Sagittarius. Visually NGC 6242 is a half magnitude brighter than M18 and no doubt escaped inclusion in Messier's catalogue only because it is over 20° farther south.

NGC 6281 (Plate VI, Photo 3.2): 2° east of the Mu (μ) Scorpii double is another small (8') but moderately-bright (mag 5.4) open cluster, NGC 6281. This group is not as populous as NGC 6242, with only two 8th mag, four 9th mag, and a handful of fainter members. In standard binoculars it is a coarse, mottled, patch of haze, its two brightest members just visible. 25×100 glasses resolve ten or twelve stars in a loose N-S group.

Like NGC 6242, NGC 6281 is an inter-arm open cluster. But it is somewhat closer to us, 1600 light-years away, and somewhat older, 220 million years. The chief interest of these two groups is not in their appearance or in their intrinsic nature, for they are rather routine open clusters. They are interesting for their location in the sky and in our Galaxy: they enrichen the already visually-interesting NGC 6231 region; and, because they *lie in front* of the Scorpius OB1 association and the Sagittarius-Carina Spiral Arm, they can give the informed observer a sense of *depth* when scanning this region with wide-field instruments. This is a situation where one's observing experience is enhanced by looking with the *mind* as well as with the *eye*.

NGC 6383 (Plate VIII, Photo 2.2): Just $1^1/_4$° WSW of M6 is the mag 5.7 star h 4962, a spectroscopic binary with twin O7 V components. (h 4962 will be hard to mistake, for there is no comparably-bright star in its vicinity.) This star is the lucida of a large (20') but scattered cluster, NGC 6383, consisting of h 4962 plus some two dozen much fainter mag 8–11 associates. In 10×50 binocs a half-dozen 8th and 9th mag stars can be glimpsed just west, and barely out of the glare, of the lucida. 15×100 supergiant glasses reveal ten or twelve faint cluster members scattered from NW to SW of h 4962.

The combination of one or two bright O-type star with a couple dozen much more modest-luminosity associates seems to be a special variety of young open cluster, because several specimens of it can be found around the Milky Way. These include, in addition to NGC 6383 itself, NGC 2264 and NGC 2353 in Monoceros, and NGC 2362 in Canis Major. Presumably this type of cluster is generated within interstellar clouds of a certain specific initial size, mass, temperature, and density. NGC 6383 is not as aesthetically impressive as NGCs 2264, 2353, and 2362 (all of which will be described later in this chapter) because it is more distant than those groups and dimmed one whole magnitude by interstellar dust. It is situated squarely on the galactic equator some 5000 light-years away and is a tracer of the Sagittarius-Carina Spiral Arm.

NGC 6451: The nearest NGC-numbered open cluster to the direction toward the Center of the Galaxy is NGC 6451. The group is 1.7° SE of the direction of the Galactic Center, and can be found by scanning 4° due west from Gamma (γ) Sagittarii. NGC 6451 is neither large nor bright, and in 10×50 binoculars appears merely as a 6'-long, N-S patch of mottled haze in a rather rich star field. Its stars are so faint that it is definitely a telescope rather than even a supergiant binocular object. However, its location makes it worth looking for no matter how small your instrument.

Though NGC 6451 lies toward the Center of the Galaxy, in reality it is not anywhere near it. A recent study estimates the cluster to be 6,800 light-years from us, only one-fourth the way to the Galactic Center. The distance of NGC 6451 implies that it lies fairly in the midst of the Sagittarius-Carina Spiral Arm. However, it is not a true Sgr-Car Arm tracer because it is much too evolved, 400 million years old.

I.3 Serpens Cauda

Serpens Cauda is a Milky Way constellation only in the sense that it lies along the galactic equator: it does not have any bright Milky Way star clouds, and nowhere along it is there even any decently bright Milky Way background glow. The reason for this strange state of affairs is that the constellation is mostly covered by the Great Rift, which is a chain of dense, relatively-nearby dust clouds that block the light of the Milky Way beyond. Hence most of the open clusters we might have seen in the direction of Serpens Cauda are obscured. Only in the extreme SE corner of the constellation, and along its far northern edge, is the dust sufficiently thin for us to glimpse distant open clusters. Despite this, Serpens Cauda does in fact have two splendid groups for wide-field observers. The M16 open cluster, near the Serpens/Sagittarius border, is at its very best in giant binoculars and richest-field telescopes, for these instruments not only pull in faint cluster members, their low magnification allows the observer to see the nebula-glow in which the cluster is embedded. IC 4756 on the extreme northern edge of the constellation is extremely large and loose and thus at its best with the lowest power possible.

M16 (Plate XVI, Photo 2.4): The M16 emission nebula + open cluster complex is easily found 4° due north of the northern edge of the Small Sagittarius Star Cloud, M24. It is immediately apparent even in the smallest binoculars as a mag 6.0, 20' diameter glow over-spread with a few 8th and 9th mag stars. The only complication to identifying M16 is the equally-bright M17, the "Swan Nebula," which is 1½° NNE of the Small Sgr Star Cloud. M16 is about 2½° north and a little west of M17. All three objects, M16, M17, and at least the northern part of M24, fit in the same field of view of 7×50 and 10×50 binoculars. Robert Burnham, Jr. intro-

Serpens Open Clusters

Cluster	RA (2000.0)	Dec	m_V	Size	m_V*	*Spectrum	Distance	A_V	M_V	True Size	Age
M16	18^h19^m	−13°47'	6.0	21'	8.24	04 V	6500 l-y	3.1	−8.6	40 l-y	1 mil
N 6604	18 18	−12 14	6.5	4	7.50	08 Ibf	6500	3.2	−8.2	7.6	
I 4756	18 39	+05 27	4.6	52	8.7		1300	0.6	−4.0	20	830

Photo 2.4 The open cluster and emission nebula M16 in Serpens. North is to the right and west is up. The dark dust feature projecting over the nebula glow from the SE (Robert Burnham, Jr.'s "Star-Queen"), is bordered by extremely prominent rim nebulae: these are the zones of excitation between the denser parts of the dust residual from the formation of the M16 cluster and the stellar winds and radiation pressure of the cluster's hot, luminous O-type members

duced the splendid name "Star Queen" for the M16 nebula on the basis of the shape of the complicated dust feature that extends into the center of the nebula-glow from the SE. Historically M16 has been called the "Eagle Nebula."

The M16 emission nebula will be described in more detail in Chap. 3. The full extent of the M16 star cluster is about 15'. The core of the cluster is toward the northern edge of the nebula-glow. The cluster's stars are not especially bright, the mag 8.2 lucida being accompanied by one other 8th mag star and six each of the 9th and 10th mags. In 10×50 glasses only a half dozen cluster members can be counted. The group requires at least 15× and giant binoculars to begin to resolve well, but is virtually completely resolved in 25×100. In all these instruments the cluster star-chips are seen embedded in the indescribably soft haze of the nebula, and glitter like gems–a beautiful effect impossible to capture photographically, and lost in conventional telescopes, which simply spread the nebula-haze out to invisibility: magnification kills emission nebulae. The lucida of the M16 cluster is on the SW of the group's core. It is an attractive double, its slightly fainter companion just to its NNW.

The lucida of M16 is an extremely hot O5 main sequence star. In fact this cluster is one of the most impressive concentrations of O-type stars in our quadrant of the Galaxy. (Others include NGC 1893 in Auriga, the Rosette Cluster [NGC 2244] in Monoceros, IC 2948 in Centaurus, and NGC 6231 in Scorpius.) The group contains one other O5 V member, two O7 giants, and a half-dozen late-O main sequence stars. Its integrated absolute magnitude is an impressive −8.6, a luminosity of nearly a quarter-million Suns. The fact that the M16 cluster has two mid-O main sequence objects implies that it is very young, at most one million years old. Such young groups are almost always embedded in emission nebulae, for their stars, however powerful, have not had sufficient time to drive away the gas and dust left over from the original giant molecular cloud in which the group was born. And, like other young groups, M16 is the core of an association, Serpens OB1, a stellar aggregation which includes a hot O6 V star, two late-B Ia+ hyper-supergiants, and two Wolf-Rayets. M16 and Ser OB1 are about 6,500 light-years away and therefore major components of the Sagittarius-Carina Spiral Arm.

We see M16 through the fringes of the Great Rift, the dust clouds of which lie only several hundred light-years from us. Thus M16's stars are dimmed an average of 3.1 magnitudes. In other words, only 6% of the cluster's star-light gets to us! Members of Ser OB1 located west or NW of the cluster are behind even more Great Rift dust and consequently dimmed from 4$^1/_2$ to 6 magnitudes. Indeed, in a binocular scan the field west of M16 actually *looks* dusty: it has no background Milky Way glow, nor even the suggestion of a glitter of momentarily-resolved stars–just a poor scattering of field stars. A claustrophobic sight!

NGC 6604 (Plate XVI, Photo 2.4): Centered 1$^1/_2$° due north of M16 and a little deeper into (or, to be precise, a little more *behind*) the Great Rift is the open cluster NGC 6604. Like M16, this is the core cluster of a major Sagittarius-Carina Arm association, Serpens OB2, and is embedded in an emission nebula, Sh 2-54. NGC 6604 is at about the same distance from us as M16, 6500 light-years. However, because of the even thicker Great Rift dust in front of it, nothing of the nebula around NGC 6604 can be seen even in large telescopes, and most of the cluster stars are too faint to be discerned in apertures of less than 100 mm. With 10×50 binocs there is no sense "clusterness" at all in the location of NGC 6604, not even a pale unresolved haze. However, in 25×100 glasses the association itself is evident as a field concentration, about 1$^1/_2$° in diameter, of 8th and 9th mag stars: it is in fact quite obvious in contrast to the star-poor Great Rift around it. NGC 6604 is visible on the NW side of the association, appearing as a NW-SE partially-resolved lozenge of intermittently-visible star-specks. The long axis of the cluster points SE at the lucida of the association, a mag 8.2 O9 I-II near-supergiant. The lucida of the cluster is a mag 7.5 O8 Ib supergiant, and its second brightest member a very hot mag 8.5 O5 III giant.

IC 4756: One of the best big binocular open clusters in the summer Milky Way is IC 4756, easily found 4° WNW of Theta (θ) Serpentis, the 4th magnitude star at the tail-tip of Serpens. This group is almost a full degree (52') across. Its integrated apparent magnitude of 4.6 suggests that it can be seen with the unaided eye from dark sky sites as a small smudge of pale haze. 10×50 binoculars resolve IC 4756 into an impressive swarm of 9th and 10th magnitude stars spread over an area framed by a trapezoid of mag 6–6$^1/_2$ stars (which, however, are not true cluster members).

But it is with 15×100 instruments that the group really begins to come into its own. It is extremely large but looks very rich because of how evenly spread and evenly bright its 80 mag 8.7–11 members are. (They are hard to count, however, because the faintest are just above the threshold of resolution and keep winking in and out of sight.) Resolution is complete: there is no background haze. The main body of the cluster is enclosed within an equilateral triangle formed by its three brightest stars; and the main cluster concentration is toward the NNE side of the large trapezoid that encloses the group. IC 4756 might be even a better sight in 25×100 than in 15×100 glasses, for the cluster does not loose the sense of richness with the greater magnification.

IC 4756 has the appearance typical of the more populous of the evolved open clusters: an evenly-distributed swarm of comparably-bright stars. Other old groups with this look include NGC 752 in Andromeda, NGC 2477 in Puppis, and NGC 7789 in Cassiopeia. And indeed IC 4756 is estimated to be near a billion years old. It is some 1300 light-years from us and therefore about 20 l-y across. Its integrated absolute magnitude (corrected for $^1/_2$ magnitude of dimming by Great Rift dust) is –4.0, a luminosity near 3500 Suns. Its brightest members, however, have absolute magnitudes only between 0 and +0.5.

I.4 Scutum

Scutum is one of the smallest constellations along the Milky Way, covering an area only 9½° E-W by 12° N-S. However, it has a couple bright Milky Way star clouds (described in Chap. 3's opening section) and a surprisingly large number of open clusters. There are two reasons for Scutum's star-cloud and star-cluster richness: first, it is SE of the core of the Great Rift, and therefore we have good visibility through the Milky Way toward it; and second, when we look toward Scutum we look toward the star- and cluster-rich interior of our Galaxy.

Scutum possesses two Messier-numbered open clusters, M11 and M26. M11 is one of the best open clusters in the heavens for moderate-sized telescopes, but too compact for richest-field instruments (especially binoculars). Three of Scutum's NGC open clusters can be seen in larger binoculars, but only NGC 6664 is aesthetically interesting. Two of the constellation's faint open clusters, NGC 6704 and Tr 35, though not easy even in supergiant glasses, are attractively situated on the edge of the Scutum Star Cloud.

All six of the Scutum open clusters that will be described lie between 5000 and 6500 light-years from us. This is exactly the distance of the major Sagittarius-Carina Spiral Arm clusters and nebulae in Sagittarius (M8, M17, M20, and M21). However, toward Scutum the Sgr-Car Arm is arcing in toward the interior of the Galaxy. Consequently these six Scu open clusters are not Sgr-Car Arm tracers, but evolved inter-arm gap clusters. The Scutum Star Cloud itself is part of the distant in-curving arc of the Sgr-Car Arm.

M11: M11, popularly known as the "Wild Duck" Cluster, is a bright (mag 5.8) but very compact (13' diameter) group located just east of the center of the Scutum Star Cloud. It is on the south end of a "bay" of obscuring matter that bulges south down out of the Great Rift of Serpens. The best way of finding the cluster is to start at Lambda (λ) Aquilae and "star-hop": first 1½° SW to 12 Aql, then 1° west to the mag 5.0 Eta (η) Scuti; and finally 1½° WSW from Eta Scu to M11.

M11 is extremely populous for an open cluster, with 870 members from magnitude 8.0 to mag 16.5. However, it is rather far away, 6200 light-years, and almost as star-dense as a class XI globular cluster. Thus it does not resolve very well in low-power instruments. In fact, in 10×50 glasses it is indistinguishable from a globular cluster. In 15×100 binoculars M11 is larger, but still very globular in appearance. A single bright star is conspicuous near the cluster center, and a somewhat fainter star can be made out WSW of the center; but there is no other good evidence of resolution. Like those of globular cluster images, M11's edges are very hazy and ill-defined, though its over-all profile is unambiguously circular.

However, with the extra magnification of 25×100 supergiant binoculars M11 begins to look impressive. It remains highly concentrated, but dozens of cluster members are resolved. The mag 8.0 lucida can be seen to mark the east corner of a triangular mass of stars occupying the greater part of the interior of the cluster. The NE and SE edges of this triangle are especially star-dense (they are the "wings" of the "Wild Duck"), but the west side of the triangle is more open, the star-density here declining gradually out to the cluster's edge.

M11 is as dense as it looks, for its central star density of 2.4 per cubic light-year is comparable to that of a class X or XI globular. At the very center of M11, a sphere with a radius of 4.3 light-years (the

Scutum Open Clusters

Cluster	RA (2000.0)	Dec	m_V	Size	m_V*	Distance	A_V	M_V	True Size	Age
M11	18^h51^m	−06°16'	5.8	13'	8.0	6200 l-y	1.3	−6.9	23 l-y	250 mil
M26	18 45	−09 24	8.0	14	10.3	5200	1.8	−4.7	21	90
N 6649	18 33	−10 24	8.9	6.6	10.5	5200	3	−5.1	10	
N 6664	18 37	−08 13	7.8	16	9.98	6200	2	−5.6	29	
N 6704	18 51	−05 12	9.2	6	12.2	6300	2.3	−4.5	11	35
Tr 35	18 43	−04 48	9.2	9	11.4	6000	3.2	−5.3	16	

distance between the Sun and Alpha [α] Centauri) would contain over 800 stars! The cluster's total mass is estimated to be 2900 times the Sun's. M11 is about 250 million years old, but, because of its richness, has an integrated absolute magnitude of –6.9 (a luminosity of 20,000 Suns), which is rather high for an intermediate-age inter-arm open cluster.

M26: Scutum's other Messier open cluster, M26, is easily located 1° ESE of Delta (δ) Scuti. But it is rather small (9' in diameter) and only magnitude 8.0. Thus in 7×35 and even 10×50 binoculars it appears only as a small hazy spot. The cluster begins to resolve in 15×100 supergiant instruments, which show it as a small, low-surface-brightness patch of haze with its mag 10.3 lucida on the SW edge, its second brightest member on the south edge (east of the lucida), and a third star visible on the west edge. The extra power of 25×100 glasses resolves M26 into a compact grouping of a half-dozen faint stars with no background haze.

M26 is not much more interesting astrophysically than it is visually. The cluster is some 5,200 light-years from us, and dimmed a rather heavy 1.8 magnitudes by Great Rift dust. Its integrated absolute magnitude, –4.7 (a luminosity of around 7,000 Suns), is fairly "normal" for an intermediate-age inter-arm open cluster.

NGC 6649: NGC 6649, located 4° north and slightly west of Gamma (γ) Scuti (the cluster is at the NE corner of an equilateral triangle with two 6th magnitude stars), is so small (7') and so faint (mag 8.9) that it is definitely more a group for moderate powers in telescopes than for binoculars, however large, or even RFTs. However, it can be discerned in 15×100 glasses as a field condensation of three or four faint stars on a small, low surface brightness patch of haze, its mag 10.5 lucida on its south edge. NGC 6649 is the same distance from us as M26, 5200 light-years, and intrinsically slightly brighter, with an integrated absolute magnitude of –5.1. However, it appears a full magnitude fainter than M26 because it is dimmed 3 mags by Great Rift dust. The cluster's main point of interest is that it contains a Cepheid variable, V367 Scuti, which has a period of 6.29 days and a median apparent mag of 11.6. The star's median absolute mag is –3.6 (2500 Suns) but it is dimmed a full 4 mags by dust.

NGC 6664: Though NGC 6664 is the second brightest open cluster in Scutum after M11 (it is 0.2 magnitude brighter than M26), and in fact *larger* than M11, it is not necessarily easy to see, particularly in wide-field instruments. The problem is that the cluster is located only $^1/_2$° due east of the mag 4.2 Alpha (α) Scuti, and the glare of that bright star tends to overpower the group's members, which are only of mag 10.0 and fainter. In 10×50 glasses the cluster appears merely as an amorphous patch of very pale haze. 15×100 binoculars gets the group sufficiently far from Alpha Scu that it can be seen as a N-S oval of delicate star-specks, most of them in a concentration ENE of Alpha. Because of the generally smooth distribution of its comparably-bright stars, NGC 6664 has a fine fragile beauty. This beauty is not lost at 25×.

NGC 6664 seems to be about the same distance from us as M11, 6200 light-years. Its integrated absolute magnitude of –5.6 is respectably luminous for an evolved inter-arm open cluster. It is also rather large, 30 l-y across. However, the group's real claim to fame is that contains two Cepheid variables. One of these is the cluster lucida, EV Scuti, a short-period (3.09 days), small-amplitude (mags 10.0–10.3) star with a peak absolute magnitude of a modest –3.4. Inter-arm open clusters with short-period, modest-luminosity Cepheids are not rare: two such groups are in Sagittarius (M25 and Cr 394), and two in Scutum (NGC's 6649 and 6664). Short-period, modest-luminosity Cephs are modest-mass objects that evolve more slowly than high-mass stars and therefore would be expected to be found in the somewhat evolved open clusters that are common in the inter-arm gaps.

NGC 6704: Both NGC 6704 and Tr 35 (discussed below) are very faint magnitude 9.2 objects more suitable for moderate and large telescopes than for binoculars or richest-field telescopes. Indeed, they cannot even be glimpsed in anything less than supergiant binoculars. However, they are worth looking for if you are observing with such an instrument because both clusters are situated in interesting spots on the fringes of the Scutum Star Cloud.

NGC 6704 is faint and small (only 6' across), and has no stars brighter than mag 12.2. Even in 15×100 glasses it appears merely as a granular patch of haze. But it is located in an arm of the Scu Star Cloud that extends ESE toward Beta (β) Scuti across B111, the dark "bay" down from the Great Rift on the south shore of which lies M11. Moreover, the cluster is easy to find, because it lies exactly 1° due north of M11. NGC 6704 is around 6,300 light-years away, about the same distance as M11. It appears faint in part because it is dimmed 2.3 magnitudes by B111 dust.

Tr 35: Trumpler 35 is situated near the NE end of the more northwesterly of the two semi-detached star cloudlets that lie in the Great Rift just NW of the Scutum Star Cloud. (See Chapter 3 for a detailed description of the Scu Star Cloud.) This cluster is as faint (magnitude 9.2) as NGC 6704 and not much-larger (9' across). However, it has somewhat brighter stars, up to mag 11.4. Thus it can be partially resolved in 15×100 binoculars, though it is so scattered that it appears merely as a field condensation in the star cloudlet rather than as a more conventional open cluster haze with embedded star-sparks. Like NGC 6649 and NGC 6664, Tr 35 seems to contain a Cepheid variable, the nearby RU Scu, which has the rather long period of 19.7 days and an average apparent mag of 9.3. The star's period implies its median absolute mag is –5.2 (a luminosity of 10,000 Suns). Given that RU Scu is dimmed 3.2 mags by dust, its distance–and presumably that of Tr 35–is around 6,000 light-years.

I.5 Aquila

In terms of area, Aquila has to be considered one of the major Milky Way constellations. Moreover, its Milky Way glow, though split by the Great Rift, is very bright, particularly just NW of Gamma (γ) Aquilae. However, the Aquila Milky Way is astonishingly poor in open clusters. Aquila occupies six times more area than Scutum to its SW and twice as much as Vulpecula on its north; but in this chapter we shall discuss six open clusters in Scutum, nine in Vulpecula, but only *three* in Aquila. The reason for Aquila's open-cluster poverty is simply that toward it we are looking not *into* or *along* a spiral arm of our Galaxy, but *down* the length of an open-cluster-poor inter-arm gap–in this case the gap between the Sagittarius-Carina Arm, the outer edge of which is in Scutum, and our Orion-Cygnus Arm, the inner edge of which is toward Vulpecula.

NGC 6709: Aesthetically the best open cluster in Aquila is NGC 6709, located 5° SW of Zeta (ζ) Aquilae. In 15×100 supergiant binoculars it appears moderately large and well-resolved, its pale background glow scattered with lots of faint but distinct stars. The two brightest cluster members are on the east and SW edges of the group. Both stars are E-W doubles, the former a closer, more attractive, pair than the latter. A third bright cluster star, toward the NW edge of the group, gives NGC 6709 a triangular look. But cluster members straggle outside the triangle, particularly to the south and SSE. To a long, careful look this is an attractive group.

NGC 6709 is a run-of-the-mill inter-arm open cluster some 3900 light-years from us. Its true diameter is a modest 15 l-y and its true luminosity an equally modest 6600 Suns.

NGC 6755 + NGC 6756: The open cluster pair NGC 6755 + NGC 6756 is a visually interesting "double cluster" 5° WNW of Delta (δ) Aquilae in the middle of the Great Rift. However, unlike the true Double Cluster in Perseus to be described later in this chapter, NGC 6755 and NGC 6756 do not look anything like each other (but this contrast in appearance is part of the visual interest of the pair), and they are not physically related.

The larger and brighter of the two is NGC 6755, which is the same size as NGC 6709, Aquila's best open cluster, though 0.8 magnitude fainter than that group. It can be glimpsed in 10×50 binoculars as a patch of haze, but cannot be resolved in such modest instruments because its brightest stars are only mag $10^1/_2$ objects. The cluster begins to resolve with 15×100 supergiant glasses, in which it appears as a rather large granularity on the basic background sky-glow, over which is scattered a loose field condensation of resolved and momentar-

Aquila Open Clusters											
Cluster	RA (2000.0)	Dec	m_V	Size	m_V*	*Spectrum	Distance	A_V	M_V	True Size	Age
N 6709	18h51m	+10°21'	6.7	13'	9.1	K4 III	3900 l-y	1.0	−4.7	15 l-y	315 mil
N 6755	19 08	+04 14	7.5	14	10.35	B2 III	5200	3.2	−6.6	21	30
N 6756	19 09	+04 41	10.6p	4	13.0p		4900	3.8	−4	6	50

ily-resolved stars. In such supergiant instruments NGC 6755 is attractive in a subtle, delicate sort of way. Three of the brighter cluster stars form a triangle in the south central part of the group; and the other bright cluster stars can be discerned on or near the group's east, west, and north edges.

A WSW-ENE line of three brighter field stars north of NGC 6755 points toward NGC 6756. Such help is necessary for finding this cluster because it is very small (4'), very faint (10th magnitude), and comprised of extremely faint stars. Even in 25×100 instruments it is only an intermittently visible shred of paleness, with no resolution nor even any granularity from partial resolution.

NGC 6755 and NGC 6756 are both around 5,000 light-years from us. However, they are physically unrelated, NGC 6756 being perhaps twice as old as NGC 6755. The lucida of the latter is a B2 III giant, suggesting a cluster age of 30 million years–rather young for an inter-arm open cluster. NGC 6755 seems to be not only rather large, over 20 l-y across, but also rather luminous, its integrated absolute magnitude of −6.6 being definitely high for an inter-arm group. However, all these numbers are rather uncertain because both clusters are heavily dimmed by Great Rift dust, NGC 6755 by 3.2 and NGC 6756 by 3.8 magnitudes.

I.5 Ophiuchus

Most of the very large constellation of Ophiuchus lies well outside the Milky Way. The galactic equator just nips the extreme SE corner of the constellation SE of Theta (θ) Ophiuchi. This area of Ophiuchus is toward the central bulge of our Galaxy and consequently rich in globular clusters. However, it is remarkably free of open clusters. In the NE part of Ophiuchus, east and NE of Beta (β) and Gamma (γ) Oph (the two stars marking the right shoulder of the Serpent-Wrestler), a bright Milky Way stream begins behind the conspicuous V-shaped asterism called "Taurus Poniatovii" (66-67-68-70 Oph) and extends NE into the Epsilon (ε) + Zeta (ζ) area of Aquila. Two open clusters can be found in this region of Ophiuchus: IC 4665, so large and loose that it is the perfect low-power binocular cluster; and NGC 6633, a group at its best in supergiant instruments.

NGC 6633: 3° NW of the large binocular open cluster IC 4756 in Serpens (described earlier) is the 27' long NGC 6633. The best way to get to this group, located in a bright-star-poor region of the Milky Way, is from (δ) Aquilae in the center of the body of the Eagle. From Delta Aql go 6½° WNW to the 4th magnitude Theta (θ) Serpentis. Then from Theta Ser go 4° NW to IC 4756. Another 3° NW from the center of IC 4756 (which is about 1° in diameter) brings you to NGC 6633. A 6th mag star is just ¼° SE of the cluster.

NGC 6633 is very bright, its integrated apparent magnitude of 4.6 suggesting that it should be visible to the unaided eye on moonless nights from dark-sky sites. It is, however, not very populous or concentrated. In standard binoculars it appears merely as a condensation of mag 7½-9 stars in this region's already-rich Milky Way field: it can be easy to scan right over. But it has an interesting structure: its SW half is a 10'-long triangular gathering of some eight or nine mag 7½-8 stars from which extends 10' to the NE a rectangular scattering of several additional cluster members.

In 15×100 and 25×100 supergiant instruments more NGC 6633 members are resolved and it therefore has more of a sense of "clusterness." At 15× the arrowhead shape of the SW half of the group is more obvious. Two dozen stars can be counted in the arrowhead, which can be seen to be pointing NW. There is no background haze, just a hint of a few momentarily-resolved stars. Ten or twelve stars appear in the rectangular extension NE from the arrowhead. How-

Ophiuchus Open Clusters											
Cluster	RA (2000.0)	Dec	m_V	Size	m_V*	*Spectrum	Distance	A_V	M_V	True Size	Age
N 6633	18^{h}28^m	+06°34'	4.6	27'	7.57	A0 III	1030 l-y	0.5	-3.4	8 l-y	630 mil
I 4665	17 46	+05 43	4.2	40	6.87	B4 V	1140	0.5	-4.0	13	80

ever, the cluster is so loose that at 25× the triangular shape of the arrowhead is more ambiguous.

NGC 6633 is about 1030 light-years away. It must be a rather evolved cluster because its lucida has an absolute magnitude of just -0.4. The integrated absolute magnitude of the group as a whole, -3.4, corresponds to a modest luminosity of just 2000 Suns. NGC 6633 is therefore a fairly average evolved inter-arm open cluster.

IC 4665: The extremely large IC 4665 is centered only 1° NE of Beta (β) Ophiuchi. It is not populous, and very loose, but a fine sight in low-power binoculars. 7×50 glasses show 10 or 12 7th and 8th magnitude stars distributed around a ragged 40' diameter ring. In 15×100 supergiant instruments IC 4665 is still an attractive cluster – *not* from any virtue of concentration, which it most certainly lacks, but from the uniform brilliance of its brightest stars, which with this aperture really *are* bright. In form the group appears as a ring with a short "handle" on its NW side and a single bright star almost at its center. The "handle" consists of four progressively fainter stars arcing NW and then west away from the ring. Some other stars NW of the circlet look like a subgroup of the cluster because of their fairly uniform brightness and by contrast with the otherwise poor star-field around IC 4665. At least two dozen cluster stars can be counted.

Studies have found that IC 4665 members are scattered over a 2° area. Because the group is about 1140 light-years distant, its full diameter must therefore be around 40 l-y. The ring of the cluster is about a dozen light-years across. IC 4665 is only 80 million years old, but it is already so loose that it will soon disintegrate.

I.6 Sagitta

The tiny, but visually attractive, star-pattern of Sagitta occupies a tiny area of the summer Milky Way north of Aquila. It has some beautiful star-fields for binoculars (described in Chap. 3), but very little else either for binoculars or telescopes. Its "show-piece" object is the rather small and faint globular cluster M71, suitable only for moderate-to-large telescopes. Sagitta's one open cluster was not even listed in the NGC or IC.

H 20: Harvard 20, located immediately NE of M71, has respectable numbers on paper: its integrated apparent magnitude of 7.7 and visual diameter of 9' imply that it should be visible in 10×50 binoculars. However, it is in fact difficult with anything less than giant binoculars. In super-giant instruments it is easy but not impressive. 15×100's show the cluster as a hazy little knot of two or three stars just ESE of a ENE-WSW pair of 9th mag stars. Even in 25×100 glasses the group appears only as a small gathering of faint stars, looking more like a Milky Way field condensation than a true cluster.

H 20 is faint and small mostly because it is so far from us, 9000 light-years. In reality it is on the high side of average in both size (24 l-y) and luminosity (11,000 Suns). In fact in size and luminosity it is almost the equal of its globular neighbor, M71, which is 25 l-y across and has a luminosity of about 13,000 Suns. However, in M71 the luminosity is a result of sheer star-numbers whereas in H 20 the cluster brightness is from a few luminous giants.

Sagitta Open Cluster									
Cluster	RA (2000.0)	Dec	m_V	Size	m_V*	Distance	A_V	M_V	True Size
H 20	19^{h}53^m	+18°20'	7.7	9'	8.9	9000 l-y	0.8	-5.3	24 l-y

I.7 Lyra

The small but conspicuous star-pattern of Lyra lies on the NW fringes of the Cygnus Milky Way. Though small in area, it is an outstanding constellation for binocular observers, with the blue-white Vega, several binocular doubles, and even a globular cluster (M56) and a planetary nebula (M57, the famous Ring Nebula) suitable for giant binoculars. The outer billows of the Cygnus Star Cloud reach as far as the Eta (η) + Theta (θ) Lyrae region, which is a fine star-rich binocular field. Lyra also has two interesting open clusters very different in appearance and true nature: Stephenson 1, the Delta (δ) Lyrae cluster, is a small, sparse, group dominated by two bright members, and can be seen even in mere 7×35 glasses; and NGC 6791 is large and rich, but populated by such faint stars that it can be seen only with supergiant binoculars and telescopes.

NGC 6791: The most interesting thing about NGC 6791, located just under 10' ENE of the 4th magnitude Theta (θ) Lyrae, is that it is one of the most ancient open clusters known in our Galaxy, perhaps 9 billion years old (almost twice the age of the Sun–and just about the age of the youngest globular clusters). Like other highly-evolved open clusters, it no longer has any highly-luminous stars, its brightest members being of absolute magnitudes 0 to –0.5 and therefore only around a hundred times as luminous as the Sun. Unfortunately, because the cluster is so distant, 12,400 light-years, its brightest stars have apparent magnitudes (corrected for 0.7 mag of interstellar absorption) of 13.0 and fainter. Thus the cluster can be resolved only in 8-inch and larger telescopes. The group's integrated apparent magnitude is 9.5; but because it is so large, 15' in diameter, its surface brightness is extremely low. The cluster is invisible in 10×50 binoculars. In 15×100 supergiant glasses NGC 6791 is very difficult, visible only with averted vision as a rather large, but ambiguously-shaped and very low-surface-brightness patch with a few faint (no doubt foreground) stars sprinkled over it. The increased magnification of 25×100 supergiant instruments sufficiently darkens the sky background that the cluster can be partially-resolved into a fine-textured granularity, appearing as a very, very low-surface-brightness glow with knots of haze in a very rich Milky Way star field. NGC 6791 gives the peculiar impression of being *behind* the hundreds of resolved stars in the field–which it is, because our view toward it is at an angle up through the disk of our Galaxy and therefore there are many more stars in front of it than behind it. The cluster's true size is quite large, 36 l-y, and its true luminosity, about 3300 Suns, is definitely high for such an ancient open cluster.

Ste 1: The 10'-wide naked-eye double Delta (δ) Lyrae is a true physically-involved binary about 1020 light-years away. The two stars, separated by about 3 l-y, are a splendid color-contrast pair in binoculars: the magnitude 5.5 Delta-1 (δ^1) is a bluish B3 V object, and the ruddy-orange Delta-2 (δ^2) is an M4 II variable that irregularly ranges between mags 4.5 and 5.0 (barely discernable, given the lack of any nearby similarly-bright stars for comparison).

The Delta Lyrae pair are the lucidae of a rather scattered (20' in diameter) and relatively poor open cluster catalogued as Stephenson 1. The core of the cluster is a loose gathering of four or five mag $8^1/_2$–10 stars centered SW of Delta-2. The southern member of this gathering is a close N-S double. Ste 1 is at its best in the higher powers of zoom binoculars, or in. RFTs. The true size of the cluster is only about 6 l-y and its integrated absolute magnitude only –3.6. It seems to share the true space motion of the Pleiades and therefore to be a member of the Pleiades Stream.

Lyra Open Clusters

Cluster	RA (2000.0)	Dec	m_V	Size	m_V*	*Spectrum	Distance	A_V	M_V	True Size	Age
N 6791	19^h21^m	+37°51'	9.5	15'	13.0		12,400 l-y	0.7	–4.0	36 l-y	9 bil
Ste 1	18 53	+36 55	3.8	20	4.5v	M4 II	1020		–3.6	5.7	

I.8 VULPECULA

Though Vulpecula has no bright stars, it occupies a slice of the summer Milky Way about 10° high and a full $2^1/_2$ hours of right ascension long. Its stretch of the Milky Way is bisected by the Great Rift; but here the Great Rift can be seen in binoculars and RFTs to be more mottled and blotchy than it is to the SW in Aquila and Serpens Cauda, where it is more of a blank opacity. Thus the transparency through the Galaxy is better toward Vulpecula. Moreover, in this direction our view across the Galaxy is along the inner edge of our own Orion-Cygnus Spiral Arm rather than down an inter-arm gap, as toward Sagitta and Aquila to the SW. For both reasons Vulpecula is much richer in open clusters than Sagitta and Aquila. Its best group aesthetically is NGC 6940, a rich, populous cluster. The peculiarly-shaped Cr 399, the "Coat Hanger," is splendid for small binoculars. The ring of NGC 6885 and the large star-field St 1 are good in moderate-aperture glasses. Astrophysically the most interesting open cluster in Vulpecula is NGC 6823, a tight clump of hot O-type stars fluorescing the NGC 6820 emission nebula. NGC 6823, its surrounding emission nebula, and the NGC 6800, NGC 6802, and NGC 6830 groups in Vulpecula are giant binocular and RFT objects.

NGC 6800: Conveniently located just $^1/_2$° NW of Alpha (α) Vulpeculae is the reasonably large (15' diameter) but rather loose aggregation of faint stars, NGC 6800. It can be glimpsed with 8×40 binoculars as a patch of faint haze. With 10×50s that haze begins to look granular from the partial resolution of the cluster's brightest members, 10th magnitude objects. 15×100 supergiant instruments resolve NGC 6800 into a moderate-sized gathering of delicate star-specks with no background haze, very attractive in a subtlety delicate way because of how tiny these star-specks are and of how evenly they are distributed. At higher powers the real looseness of the cluster begins to become obvious.

NGC 6802: The very small (3.2' diameter), very faint (magnitude 8.8) NGC 6802 is located $^1/_3$° east of the easternmost star (a 7th mag object) of the large straggling group Cr 399, which is described below. Cr 399 is an easy asterism even for the smallest binoculars; but NGC 6802 is difficult for anything less than moderate-aperture telescopes. In 15×100 supergiant glasses the cluster is barely visible with averted vision as a little smudge of haze within an attractive trapezoid of mag 9–10 stars–a situation reminiscent of M44, the Praesepe open cluster in Cancer, which is within a trapezoid of 4th-5th mag stars. The NW member of the NGC 6802 trapezoid is a 1'-wide E-W optical double with mag 9.1 and 9.9 components. With 25×100 instruments the cluster enlarges to a tiny patch of fine-textured granularity, though even with the increased power it is not immediately visible at first glance. Because its brightest stars are 13th mag objects, NGC 6802 requires at least 8-inch telescopes to be resolved.

NGC 6802 is about 3300 light-years distant and we view it through the dust clouds of the Great Rift, which dim it by 2.6 magnitudes. The cluster's integrated absolute magnitude is –4.0, a luminosity of

Vulpecula Open Clusters

Cluster	RA (2000.0)	Dec	m_V	Size	m_V*	*Spectrum	Distance	A_V	M_V	True Size	Age
N 6800	19h27m	+25°08'		15'	10.0p						
N 6802	19 31	+20 16	8.8	3.2	12.9		3300 l-y	2.6	–3.8	3.1 l-y	1700 mil
N 6823	19 43	+23 18	7.1	12	8.73	B0.5 Ib	8200	2.5	–7.4	29	5–6
N 6830	19 51	+23 04	7.9	12	9.9		8200	1.8	–5.9	29	~100
N 6882	20 12	+26 33	8.1	10	9.9						
N 6885	20 12	+26 29		22	5.92	B8 V	410?				
N 6940	20 35	+28 18	6.3	31	9.0–9.5	M5 II	2700	0.8	–4.0	24	800
Cr 399	19 25	+20 11	3.6	60	5.2	K0 III	420		–2.0	7.3	
St 1	19 36	+25 13	5.3	60	7.0		1000	0.5	-2.7	19	300

Plate I

Plate II

Plate III

NGC 7000
Deneb
Cygnus
Star
Cloud
Vega
Altair
Scutum
Star
Cloud
M24
Great
Sgr
Star
Cloud
Mars
Antares
Norma
Star
Cloud

Plate IV

Plate V

Plate VI

Plate VII

Plate VIII

Plate IX

Plate X

Plate XI

Plate XII

Plate XIII

Plate XIV

Plate XV

Plate XVI

3300 Suns. The most interesting thing about this otherwise unpretentious star-group is its great age, 1.7 billion years, for not many open clusters are known to be older.

NGC 6823: Astrophysically the most interesting open cluster in Vulpecula is NGC 6823, a tight clump of a half-dozen magnitude 8.7–10.4 stars, with another half-dozen or so 11th mag attendants, 3½° SE of Alpha (α) Vulpeculae. This group is so tight, however, that it cannot be recognized as a cluster at powers of less than 15×. 15×100 supergiant binoculars reveal a knot of three stars, with a fourth just to the NW and a fifth a bit further away to the south. Star-poor though it is, at 15–20× NGC 6823 definitely looks like a cluster. Due east of the group is a peculiar field condensation of five approximately mag 10 stars arranged in a figure like the "five" on dice. Due west of the cluster (or to be more precise, extending from SW to WNW of the cluster) is a shallow sine-wave curve of six virtually equally-bright mag 9–9½ stars, a 10th mag star following the curve of the wave at its NE end. These visually intriguing asterisms are probably, however, merely chance alignments of stars.

NGC 6823 is embedded in a rather large (40'×30') but very low-surface-brightness nebula. The nebula-glow cannot be seen in standard 10×50 binoculars (too little light-gathering power) or in 25×100 supergiants (too much magnification) but *is* visible in 15×100s. In fact, in such instruments the cluster can be found with the help of the nebula-glow, which is quite conspicuous.

The NGC 6823 + NGC 6820 complex is around 8,200 light-years away. Thus the true size of the nebula is about 100×70 l-y, and that of the cluster nearly 30 l-y (though the group's central knot is more compact). The hydrogen-florescence of the nebula is stimulated by the ultraviolet radiation from two very hot O7 V stars at the center of the cluster, mag 9.34 and 9.97 objects only 37" apart in position angle 308°. The lucida of the cluster, however, is the mag 8.8 star nearby, a B0.5 Ib supergiant with an absolute magnitude of -6.0 (a luminosity of 22,000 Suns). The cluster and nebula would look a good deal more impressive than they do if they were not obscured 2½ magnitudes by interstellar dust (most of it from the Great Rift).

NGC 6823 is the open cluster core of a major association, Vulpecula OB1, whose members are distributed over a 3½°×1½° area along the galactic equator (a true extent, assuming 8200 l-y to the association, of about 500×215 l-y). The brightest member of Vul OB1 is a mag 7.0 star with a B8 Ia spectrum and the unusually high absolute mag (for a star of that spectral type) of -8.1, a luminosity of 145,000 Suns. (Rigel, also a B8 Ia supergiant, has an absolute mag of just -7.1.) NGC 6823 and its hot O-type members are on the SW end of the association. But on the NE end are slightly older types of luminous stars, a G1 Iab-Ib yellow supergiant and an M1 Ia red supergiant. This suggests that the NE end of the association formed first and then the radiation pressure, stellar winds, and supernova shock fronts from its stars induced the gravitational contraction of NGC 6823's progenitor giant molecular cloud. The M1 Ia supergiant member of Vul OB1, the mag 9.0 BD+24°3902, is one of the most luminous red supergiants known in our Galaxy, with a visual absolute mag of -7.5 and a bolometric absolute mag (total energy output) of -8.8, 290,000 times that of the Sun. This star is obscured 5 mags by dust; but a considerable amount of that dust is probably around the star itself, material that has condensed from the outer layers of this cool supergiant's extensive atmosphere.

Vul OB1 has an unusually large number of highly-luminous stars with bolometric absolute magnitudes of -9.0 and greater. Two are in the cluster (the two O7 V stars) and the other seven in the rest of the association. The most brilliant of them is an O8.5 Ib supergiant with a bolometric absolute mag of -10.3, meaning that its total energy output is equivalent to that of 1.1 million Suns! Another notable Vul OB1 member is the extremely long-period Cepheid variable SV Vul, located 4° east and slightly south of Beta (β) Cygni. This star has a range of mags 6.7–7.8 in a period of 45.0 days. Its peak absolute mag, according to the Cepheid period-luminosity law, is -6.5.

NGC 6830: Only 1.8° due east of NGC 6823 is the moderately-large (12' diameter) and rather faint (magnitude 7.9) NGC 6830. This group can be found by looking 1° SW of the 5th mag 13 Vulpeculae. NGC 6830 is definitely more of a conventional-telescope than a binocular cluster because even 25×100 supergiant glasses show it only as a knot of a half-dozen stars on a pale background haze. The group is of interest basically because it is near NGC 6823 not only in the sky, but also in space, for it too is estimated to be 8,200 light-years distant. However, though this means it might be as little as 260 l-y from NGC 6823, it is definitely not physically related to that cluster: the main sequence of NGC 6830 goes up only to spectral type B6 whereas that of NGC 6823 goes up to O7. Thus NGC 6823 is at most a few million, whereas NGC 6830 must be around *100* million, years old.

NGC 6885/6882: There has been some controversy in both professional and amateur literature about the two clusters NGC 6885 and NGC 6882. The former seems to be a 20′ diameter ring, the mag 5.9 20 Vulpeculae on its east side but the other members considerably fainter. This ring is quite conspicuous in both standard and giant binoculars, some 15 stars being visible around it with larger glasses. The interior of the ring is peculiarly star-poor. To its north is a conspicuous N-S line of three brighter stars, the mag 5.5 19 Vul in the middle of the line and mag 5.5 18 Vul just to the line's west. The physical reality of the NGC 6885 ring has not been established; nor do 18, 19, and 20 Vul seem to be physically related.

The other cluster, NGC 6882, apparently is a handful of mag 10 and fainter stars a few minutes of arc NW of 20 Vul. It is a mere field concentration not evident even in supergiant binoculars. Its physically reality too is open to question.

The NGC 6885 ring can be tricky to find, even sign-posted as it is by the mag 5.9 20 Vul. The problem is that this region of the Vulpecula Milky Way is confusingly-rich in mags 5½, 6, and 7 stars, but has no really bright object to serve as a celestial landmark. To find NGC 6885 start at the 3rd mag Epsilon (ε) Cygni and first go 3° due south to the 4th mag 52 Cyg and then 3½° SW to the conspicuous Vulpecula open cluster NGC 6940 (described immediately below). From NGC 6940, go 4° due west to the 1°-wide N-S pair of 5th mag stars 21 and 23 Vul. 1° SW of the southern star of this pair, 23 Vul, is NGC 6885.

NGC 6940: The finest open cluster in Vulpecula aesthetically is NGC 6940, which is located a little over 2° SSE of the 4th magnitude 41 Cygni and can be found by "star-hopping" from Epsilon (ε) Cyg in the manner described in the final paragraph of the NGC 6885/6882 section above. NGC 6940 has literally hundreds of members of mag 9.0 and fainter in a ½°-long N-S oval. Its integrated apparent magnitude is 6.3, and therefore the cluster can be seen even in small binoculars as an elliptical haze grainy with partial resolution. In 10×50 glasses the group's background glow is sprinkled with sparkling star-specks. 15×100 supergiant instruments resolve NGC 6940 into scores of faint or momentarily-visible stars and star-sparks, with no hint of background haze. Its multitude of sharp little star-points are more-or-less evenly distributed across its oval area, though there is a definite concentration of cluster members just south of its geometrical center. The cluster's lucida, a mag 9.0 M5 II red giant, is on the NW edge of this concentration. A lesser concentration of cluster stars is toward the group's NW edge. The cluster is over-scattered by quite a few bright non-member field stars. A trapezoid, oriented NW-SE, of brighter field stars is just south of the cluster's central concentration.

NGC 6940 is estimated to lie about 2,700 light-years away. Thus the oval of the cluster is around 24 l-y long and the group's integrated absolute magnitude (corrected 0.8 mag for interstellar absorption) is a rather modest −4.0. That such a populous cluster has such a modest total luminosity is a consequence of its great age, 800 million years: NGC 6940 has lost all its bright upper main sequence members. The cluster's lucida, the variable FG Vulpeculae, is a mid-M luminosity class II red giant very similar to Alpha (α) Herculis. It ranges between mags 9.0 and 9.5 in a period averaging roughly 80 days. The mag 8.6 star on the cluster's NE edge, and the mag 9.1 star on its SW edge, are blue giants with spectra of B8 III and A0 III, respectively, and are too young to be true NGC 6940 members. They are both probably well beyond the cluster.

Cr 399: Collinder 399, easily-spotted 5° south and slightly west of Alpha (α) Vulpeculae, is a very large but very loose field concentration of ten 5th to 7th magnitude stars arranged in a very distinctive asterism called the "Coat Hanger." It is best viewed in 7× or 10× binoculars. It consists of a 100'-long, almost perfectly straight, E-W line of six bright stars, from the central two of which hangs "down" (south) a 45'-diameter circle of four more bright stars. (Technically four stars can form only a quadrilateral; but something about this group says "circle" to the eye.) Even at 15–20× Cr 399 has a sense of clusterness, especially as you scan from east to west over it. This sense no doubt is helped by the group's star-poor Great Rift background. The Great Rift here is interesting to sweep, for it consists of lanes and patches of relative darkness superimposed upon a very dim Milky Way background glow. Look at it when Vulpecula is as near the zenith as it gets for your latitude.

If Cr 399 is a true physical open cluster (which is open to some doubt), it is only a few hundred light-years from us on the near side of the Great Rift dust clouds. Its integrated absolute magnitude would be, however, a very modest −2 or −2½, a luminosity of just several hundred Suns. Like the Coma Berenices Star Cluster, which also has an absolute mag of just −2, Cr 399, if real, is on the edge of total disintegration from our Galaxy's tidal pull.

St 1: Centered just 1½° ENE of Alpha (α) Vulpeculae is a very large (1° in diameter) field concentration of nearly 20 7th and 8th magnitude stars that has been catalogued as the open cluster Stock 1. The brighter stars of the group are distributed in two large clumps, the western clump the richer. St 1 is in truth a very scattered aggregation even at only 15×, but it stands out well because it has little background competition from its Great Rift field.

1.9 Cygnus

Cygnus spans a stretch of the summer Milky Way 35° long. The constellation contains two of the brightest star clouds north of the celestial equator: the Cygnus Star Cloud, a 20°-long oval between Beta (β) and Gamma (γ) Cygni; and the North America star cloud north of the North America Nebula (NGC 7000) and NE of Deneb (Alpha [α] Cyg). These two star clouds, and the other splendid Milky Way fields in Cygnus, are described in detail in Chapter 3.

There are two reasons for the brilliancy of the Cygnus Milky Way. First, though the Great Rift does extend NE into Cygnus as far as Deneb, it is not as wide as it was in Aquila or Serpens Cauda. Thus to the NW of the Great Rift in Cygnus, and beyond its NE end at Deneb, the interstellar medium is quite thin, permitting us to see thousands of light-years across our Galaxy's spiral plane. Second, toward Cygnus we are looking along the forward length of our Orion-Cygnus Spiral Arm, hence among the myriads of stars that lie along the Arm (which in this direction seems to be especially star-dense). Our view here is *ahead* both in the sense that this is the direction toward which we are going in our orbit around the Center of the Galaxy, and in the sense that this is the direction toward which the spiral arms of our Galaxy wind up. The direction of Galactic rotation of the solar neighborhood is toward Deneb. Therefore, because our spiral arm winds up toward Cygnus, we find that the brightest part of the Cygnus Milky Way, the Cygnus Star Cloud (which corresponds to our longest view down the length of our Ori-Cyg Spiral Arm), lies on the *Galactic interior* side of Deneb.

Because toward Cygnus we look down the length of a spiral feature of our Galaxy, it is not surprising that the constellation not only has exceptionally bright Milky Way star fields but is also rich in bright nebulae and open clusters. The several emission nebulae Cygnus offers the wide field observer, which include the North America Nebula and the Veil Nebula, will be described in Chap. 3. Here no less than 20 Cygnus open clusters will be discussed. Strange to say, none of them is an outstanding object aesthetically. The constellation does have two Messier-numbered open clusters, M29 and M39, but they are among the least populous of the Messier groups. Moreover, M29 is so compact that it requires large telescopes to be seen at its best and M39 is so loose that it looks scattered even in small binoculars.

Cygnus Open Clusters

Cluster	RA (2000.0)	Dec	m_V	Size	m_V*	*Spectrum	Distance	A_V	M_V	True Size	Age
Cyg OB2	20^h33^m	+40°12′		2°	9.06	O5.5 If	5700 l-y	5–20	~−11	200 l-y	<1 mil
M29	20 24	+38 32	6.6	6′	8.88	B0.5 II	6000	3.5	−8.2		11
M39	21 32	+48 26	4.6	31	6.8	A0 III-IV	830		−2.5	7.5	200
N 6811	19 38	+46 34	6.8	20	9.9		3300	0.5	−3.7	20	
N 6819	19 41	+40 11	7.3	9	11.5		7600	0.8	−5.3	21	3100
N 6834	19 52	+29 25	7.8	5	9.6		7000	2.1	−6.2	10	50
N 6866	20 04	+44 00	7.6	6	10.7		4100	0.5	−3.4	7.1	
N 6871	20 06	+35 41	5.2	20	6.80	O9.5I+WN	5700	1.5	−7.5	33	1
N 6883	20 11	+35 51	8.3	14	8.84	B1 IVpe	4500	1.0	−3.4	18	50
N 6910	20 23	+40 47	6.6	7	7.01	B2 Ia	4000	3.6	−7.4	8	
N 6991	20 57	+47 25		5	5.69	B8 Ia	5700	1.5		8.3	
N 6997	20 57	+44 38	10.0	8	11...						
N 7039a	21 11	+45 39		25	7.11	B1.5 V	2200	0.25		23	30
N 7039b	21 11	+45 39			11.3		5000				1000
N 7062	21 23	+46 23	8.3	5			5600	1.3	−4.2	8	
N 7063	21 24	+36 30	7.0	9	8.88	B8 V	2050	0.3	−2.4	5.4	130
N 7082	21 29	+47 05	7.2	25			4300	0.9	−4.2	33	
N 7086	21 30	+51 35	8.4	12			3800	2.2	−4.2	14	
I 1311	20 11	+41 11	13.0p	5	17.0p						
I 4996	20 16	+37 38	6.8	5	7.61	B0.5 III	6000	2.0	−6.5	9	7.5
Be 86	20 20	+38 42	7.9	5	9.72	O9 V	6000	3.2	−6.6	9	

Cygnus' best open cluster aesthetically is NGC 6871 in the Cygnus Star Cloud, which is a star chain rather than a conventional roundish star cluster. The populous NGC 6811, located outside the NW edge of the Cygnus Star Cloud, is a fine sight in supergiant glasses. Many of the Cygnus open clusters are astrophysically interesting as very young groups with some highly-luminous members–M29, NGC 6910, NGC 6991, IC 4996, Be 86. But all these are poorly-populated, and often rather tight, and therefore are not outstanding binocular or even telescope sights.

M29 (Plate XVIII, Photo 3.10): About 1.7° SSE of Gamma (γ) Cygni is a small (6′) knot of a half dozen 9th and 10th magnitude stars, the open cluster M29. It is just recognizable as a cluster in 10×50 binoculars. With 15×100 glasses the group begins to bear a remarkable resemblance to its photographic appearance, consisting of two short arcs of three or four stars each curved in opposite directions. (The curves are cupped to the NW and SE.) It is very distinctive in appearance, though, as photographs reveal, not rich and therefore without any haze from unresolved members.

M29 requires 100× in moderate-to-large telescopes to be seen at its best. But it is worth looking for in low-power rich-field instruments because its brightest members are all highly-luminous hot B0 and B1 giants. The cluster's integrated absolute magnitude is an extremely high −8.2, a luminosity of 400,000 Suns. It is fairly distant from us, 6000 light-years, but would be much more impressive visually if it was not dimmed $3^1/_2$ magnitudes by interstellar dust. In the absence of dust, M29 would be a 3rd magnitude object!

M29 is one of the three Cygnus open clusters in the large stellar association Cygnus OB1. The other two, IC 4996 and Be 86, resemble M29 in consisting of a handful of luminous blue giants and not much else–except those groups are even less populous than M29 and therefore have integrated absolute magnitudes of $-6^1/_2$ (which, in all truth, is high enough!) The brightest member of the association is the peculiar eruptive blue supergiant variable P Cygni, a hot B1 star with an apparent mag presently around 4.9 and an absolute mag that must be near $-8^1/_2$. The second brightest Cyg OB1 member is the mag 6.19 F5 Iab yellow supergiant 44 Cyg, which has an absolute mag of about −7.0 The association also includes three Wolf-Rayets, four M3 Ia and Iab red supergiants, and five mid-to-late O-type giants and supergiants with bolometric absolute magnitudes around −10.0 and therefore total energy outputs approaching one million Suns. (In fact Cyg OB1 seems to be unusually rich in such rare stars.) The one emission nebula involved with the association is NGC 6888, the Crescent Nebula (described in Chapter 3).

M39: The large (31') open cluster M39 is easily spotted in small binoculars about 9° ENE of Deneb. (The nearest reasonably-bright stars to the cluster are Rho [ρ] Cygni 3° to the south and Pi-one [π^1] Cyg $2^1/_2$° to the ENE.) Its integrated apparent magnitude is 4.6, and in fact the cluster is not difficult to spot with the unaided eye on moonless nights as a small patch of enhanced Milky Way glow. M39 consists of two dozen mag 6.8 to 10.5 stars distributed in a $^1/_2$°-wide triangle, the corners of which, marked by three of the brighter cluster members, point toward the SW, SE, and north. A fourth bright star is near the midpoint of the triangle's NE side.

M39 is only 830 light-years from us. Thus its integrated absolute mag is a mere $-2^1/_2$, a luminosity of just 830 Suns. Its true size is also very modest, $7^1/_2$ l-y: this cluster looks big only because it is relatively nearby. M39 is a rather evolved group, its age estimated to be 200 million years because its main sequence goes up to a mag 7.35 B9 star. In addition to the A0 III-IV lucida, the cluster has two other early-A subgiants, mag 7.65 and 7.85 stars with A1 IV spectra.

NGC 6811: One of the aesthetically best open clusters in Cygnus is NGC 6811, located well out of the Cygnus Milky Way 2° NE of Delta (δ) Cygni. This is a large (20'), bright (magnitude 6.8), and well-populated group. However, its brightest members are only 10th mag objects, so it is at its best in moderate-aperture telescopes at 70× or so. NGC 6811 can be seen in 10× binoculars as a hazy patch. In 15×100 supergiant glasses it begins to resolve into a crowding of tiny star-specks on a very dim background haze–not spectacular, but pretty. The extra magnification of 25×100 instruments improves the cluster's resolution to over two dozen stars in a compact but evenly-distributed mass; a fine object to careful looking.

NGC 6811 is around 3300 light-years away, so its true size is about 20 l-y and its true luminosity some 2100 Suns. Its main sequence goes up to spectral type A2, so it must be around the age of the Hyades, 660 million years.

NGC 6819: Another rich but evolved open cluster in the Delta (δ) Cygni region is the moderately-bright (mag 7.3) but rather small (9') NGC 6819, located 5° south of Delta Cyg and $4^1/_2$° ENE of Eta (η) Lyrae. This cluster is even more populous than NGC 6811; but because its brightest members are only mag 11.5 objects it requires larger telescopes to be resolved, and it is so concentrated that higher magnifications are necessary. In 10×50 binoculars NGC 6819 appears only as a hazy dot. 15×100 instruments begin to suggest resolution, the cluster appearing as a fairly conspicuous little glowing patch brighter toward the center and overspread with intermittently-visible star-sparks. Its surface brightness is similar to that of such loose globular clusters as M71 in Sagitta. The extra power of 25×100s unambiguously resolves a small group of six or eight cluster stars in the core of an amorphous, moderately-bright patch of haze.

NGC 6819 looks so small, and its stars appear so faint, in part simply because the cluster is so far from us, 6,800 light-years–about twice the distance of NGC 6811. Its true size, 18 l-y, is about the same as that of NGC 6811; but its greater population gives it a luminosity, 10,000 Suns, almost 5 times higher. And NGC 6819 is about 4 times older than NGC 6811, with the very advanced age (for an open cluster) of 2.7 billion years. Only initially highly populous open clusters have the gravitational cohesion to survive for so long.

NGC 6834: The very small (5'), rather faint (magnitude 7.8) open cluster NGC 6834 is located

$4^1/_2°$ ENE of Beta (β) Cygni on the SE margins of the brightest part of the Cygnus Star Cloud. Despite the transparency of the Galaxy in this direction, NGC 6834 is the only conspicuous open cluster in the SW half of the Cyg Star Cloud. Why there are not more open clusters here is a very good question. NGC 6834 is too small to be recognized as a cluster with standard binoculars, appearing merely as an 8th mag "star." 15×, however, makes the cluster evident as an intensification of the Milky Way background glow. In reality the star-group is quite populous; but moderate-sized telescopes are necessary to see it at its best.

NGC 6834 is quite distant, 7000 light-years. Because it lies behind the fringes of the Cygnus Great Rift, it is dimmed 2.1 magnitudes by interstellar dust. The cluster actually is rather luminous, then, its integrated absolute mag of -6.2 corresponding to a brightness of 25,000 Suns and definitely above average for an open cluster. The main sequence of NGC 6834 goes up to spectral type B3, implying that the group is only 50 million years old–possibly sufficiently young to be considered a true tracer of our Orion-Cygnus Spiral Arm.

NGC 6866: 3° SW of 31 Cygni is a small (6') but moderately bright (magnitude 7.6) cluster, NGC 6866, the third evolved open cluster in the Delta (δ) Cyg region, the other two being NGC 6811 and NGC 6819. NGC 6866, like NGC 6819, is more a telescope than a binocular or an RFT cluster, for it is compact and its brightest members are only mag $10^1/_2$ and less. In 15×100 supergiant glasses it appears as a little patch of glow around a N-S line of four resolved stars, the most northerly being the brightest. The pale haze around the line calls attention to it. This star line is reminiscent of the N-S line of bright cluster members that is so conspicuous a feature of NGC 2301 in Monoceros. Except for this line, NGC 6866 would be visually very uninteresting. In 25×100 binoculars the line resolves into a half dozen stars in the center of a coarse, loose gathering of fainter cluster members. NGC 6866 is some 4,100 light-years distant, a small (7 l-y) and intrinsically rather faint (absolute mag -3.4) group. It seems to be about the same age as NGC 6811, 600 or 700 million years old. Such evolved clusters are often found this far off the galactic equator.

NGC 6871 (Photo 3.9): The open cluster–or rather star-chain, for that is how it appears–NGC 6871, located $2^1/_2°$ ENE of Eta (η) Cygni in the heart of the Cygnus Star Cloud, is described in some detail in the section on Cygnus in Chapter 3. It is the ideal wide-field instrument object and completely destroyed aesthetically by the magnifying powers of telescopes. This region of the Cygnus Star Cloud has several striking chains and groupings of 6th to 9th magnitude stars, many of which are intrinsically luminous members of the Cygnus OB3 association. Cyg OB3, a major tracer of the Orion-Cygnus Spiral Arm ahead of us, includes NGC 6871, covers a $3^1/_2°\times1^1/_2°$ area of the Cyg Star Cloud, and is notable for its abundance of exotic stars. The lucida of NGC 6871, for instance, is the mag 6.8 multiple star ADS 13374, the primary of which is a spectroscopic binary with O9.5 Ib supergiant and WN4.5 Wolf-Rayet components. Cyg OB3 also has two other WN stars. The intrinsically most luminous member of the association is the primary of the multiple Σ2624, an O4 If^+ extreme supergiant with a bolometric absolute mag of −10.2 (a total energy output of one million Suns). And the X-ray source Cyg X-1 is a Cyg OB3 object: it consists of a mag 8.9 O9.7 Iab supergiant orbited by what seems to be a black hole into which is vortexing superheated (hence X-ray emitting) material from the blue supergiant.

NGC 6871 and Cyg OB3 are some 5,700 light-years away. Thus the true size of the association is 350×150 l-y. Its stars are dimmed an average of only about $1^1/_2$ magnitudes by interstellar dust–a far cry from the 2 to 4 magnitudes of dimming suffered by the stars of the cluster M29 and its association Cyg OB1 further NW along the Cyg Star Cloud. The M29 association is not only somewhat more behind the Great Rift than NGC 6871 and Cyg OB3, it is behind outlying dust of the IC 1318 emission nebula complex (described in the section on Cygnus in Chapter 3).

NGC 6883: Centered about 1° due east of the brightest star in the southern section of the NGC 6871 star chain, 27 Cygni, is the conventionally-shaped open cluster NGC 6883. This is a mere field condensation, some 14′ across, of about a dozen 9th and 10th magnitude stars, easily overlooked in the rich star field around it. 15×100 supergiant glasses show the group as a vague gathering of faint stars supported by just a hint of sparkling background half-resolution. The mag 7.8 star on the cluster's NW edge is not a true NGC 6883 member because it is a B1 III giant that must lie well beyond the group, which is 4,500 light-years from us. The true lucidae of the cluster are the E-W pair of mag 8.8 and 9.3 stars on its western edge, which have spectra of B2 IVpe and B2 III-IV, respectively. The main sequence of NGC 6883 goes up to spectral type B3 (its mag 8.8 and 9.3 members are subgiants that have already begun their evolutionary expansion), which implies a cluster age of 50 million years. Its star-poverty is proven by its modest integrated absolute mag, a mere –3.4, low for a cluster of its age.

NGC 6910 (Plate XVII, Photo 3.18): The very compact (7′ in diameter) open cluster NGC 6910 is a tight knot of one mag 7.0, one mag 8.1, one mag 8.5, and two mag 10.3 stars only $^{1}/_{2}$° ENE of Gamma (γ) Cygni. It is too crowded a group to be resolved in low-power binoculars. 15×100 instruments reveal that the mag 7.0 lucida is on the SE edge of the group, that the mag 8.1 star is on its NW edge, and that the mag 8.5 star, visible only with averted vision, is slightly SW of the line joining them. There is a provocative, but very subtle, sense of haze between the two brighter stars.

Aesthetically NGC 6910 may not be much even in telescopes, but it is worth finding for the sake of its mag 7.0 and 8.5 stars alone. The former is a brilliant B2 Ia blue supergiant with a total energy output of over 190,000 Suns, and the latter an ultrahot O6 III giant with the even more astonishing total energy output of 575,000 Suns. Half of NGC 6910's visual luminosity is due to the B2 supergiant; but over half of its total energy output is from the O6 giant alone. The two mag 10.3 cluster members are hot O9.5 and B0.5 main sequence objects; but they are far and away out-shown by the cluster's O6 giant and B2 supergiant. The mag 8.1 star on the NW edge of the group is not an actual cluster member: it is somewhat far from the core of the cluster, has an anomalous B1.5 IV subgiant spectrum, and is dimmed only 0.9 mag by interstellar dust-much, *much* less than the 3.6 to 3.9 mags of obscuration suffered by the true cluster members. This subgiant must be in the foreground of the group.

NGC 6910 is around 4,000 light-years away. Thus its true diameter is only 8 l-y, but its integrated absolute mag is a very respectable –7.4–mostly courtesy the B2 supergiant. Its heavy obscuration is caused by three dust features between ourselves and the cluster: (1) the Great Rift, have around 2000 l-y from us; (2) dust near the cluster, which photos reveal to be embedded in a pale emission nebula and thus within a significant amount of interstellar material; and (3) dust involved with the IC 1318 complex of emission nebulae, which is either in the foreground of NGC 6910, or possibly physically related to it.

NGC 6910 is the open cluster core of yet another major Orion-Cygnus Spiral Arm association in this direction, Cygnus OB9. The group covers a $2^{1}/_{2}$°×$1^{1}/_{2}$° (175×105 l-y) area and has 16 mag 7.5 to 10.0 members, including seven O7 and O8 stars, two additional early-B supergiants, and the mag 8.1 M3-4 Ia-Iab red supergiant variable RW Cygni. The absorption suffered by some Cyg OB9 stars is in excess of 4 magnitudes.

In addition to Cyg OB1, Cyg OB3, and Cyg OB9, the Gamma Cygni region has two other major associations, Cygnus OB8 and Cygnus OB2. Cyg OB8 is the most remote of these young-star groups, some 7,400 light-years away. Its members are spread over an area about 3° in diameter centered about 2° WNW of Gamma Cyg. The two brightest Cyg OB8 objects are mag 7.5 stars with B0 Ia and O5 Vnp spectra.

But the single most important and interesting of all the Cygnus clusters and associations is Cygnus OB2, a remarkable stellar aggregation some 2° in extent centered about $2^{1}/_{2}$° ENE of Gamma Cyg out in the Great Rift (see Plate XVIII.) Cyg OB2 is the largest, most populous, and most massive concentration of O-type main sequence, giant, and supergiant stars yet discovered in our Galaxy. It is estimated to have over 2000 O and B type stars, of which no less than 100 are O-type objects, including *at least* one O3 supergiant, one O4 III giant, and *four* O5 supergiants. More than a half dozen Cyg OB2 members have bolometric absolute magnitudes around –11, corresponding to total energy outputs of 2.1 million Suns, each. In visual light the most luminous star in the association is a B5 Ia+ extreme supergiant with an absolute mag of –10.0–practically equal to that of the most luminous globular cluster in our Galaxy, Omega (ω) Centauri.

Unfortunately all this luminosity is obscured by 5 to 20 magnitudes of interstellar dust. Much less than 1% of the visual light of the Cyg OB2 stars gets through to us. The lucida of the group in terms of apparent brightness is a mag 9.06 O5.5 If supergiant which has a visual absolute mag near –7.0 but is dimmed 4.8 magnitudes. The association is some 5,700 light-years away, so in the absence of dust this star would be a 4th mag object. The B5 Ia+ supergiant member of the association is obscured 10.3 mags (some of which seems to be due to circumstellar material). In the clear, this star would have an apparent mag of 1.2–the same as Deneb's!

But is Cyg OB2 really a simple OB association? Its total mass is calculated to be somewhere between 40,000 and 100,000 times the Sun's, and its central density estimated to be about 100 solar masses per cubic parsec. (The total mass density in stars and interstellar material in the Sun's neighborhood, by contrast, is 0.2 solar mass per cubic parsec!) These are values more characteristic of globular clusters than of stellar associations, or even of populous open clusters. And in fact that is exactly what Cyg OB2 seems to be: a newly born globular cluster, the first object of its type to be discovered in our Galaxy. Several such globulars have long been known to exist in the Large Magellanic Cloud; but it was thought that perhaps the conditions for the formation of globular clusters no longer obtain in our Milky Way.

NGC 6991: Due north of the North America Nebula (NGC 7000) and 3¼° NE of Deneb is the magnitude 5.69 star HD 199478, a Rigel-type B8 Ia supergiant. In 15×100 binoculars and other large wide-field instruments, the star can be seen to be accompanied by a handful of very faint attendants, most of them gathered to its east. HD 199478 and its companions comprise the open cluster NGC 6991 (incorrectly classified as "non-existent" in the Revised NGC). This group is embedded in a small (9'×7') bluish glow, the reflection nebula IC 5076, which is blue because it is reflecting the light of the blue-white HD 199478. Like most reflection nebulae, IC 5076 is a very low surface brightness object. NGC 6991 and IC 5076 are some 5,700 light-years distant, thus the cluster is about 6 l-y across and the nebula measures 15×12 l-y. They lie on the western edge of a thin, roughly N-S dust lane that separates the very bright star cloud NE of the North America Nebula from the slightly fainter star cloud to its NW.

NGC 6997 (Plate XIX, Photo 3.11): NGC 6997 is neither large (only 8' in diameter) nor bright (officially magnitude 10.0–though this is probably an underestimate). However, it is in a very intriguing and beautiful location: the west central part of the North America Nebula, NGC 7000. (The cluster is approximately at the location of "West Virginia.") Though it consists only of a scattering of about a dozen 11th and 12th mag stars, NGC 6997 is easily visible in 10×50 glasses as a small hazy spot in the center of the triangle formed by the three brightest stars in the western half of NGC 7000. It resolves rather well in 100mm supergiant binoculars, but because it is so scattered it is better at 15× than at 25×. It stands out well from the background star field and gives the illusion of richness because of how comparably-bright and evenly-distributed its stars are. And of course the aesthetic effect of the cluster is enhanced by the nebula-glow in which it seems to be embedded. Unfortunately, the cluster is not physically involved with the nebula. A star that probably not only *is* involved with NGC 7000, but in fact must be providing much of the ultraviolet light fluorescing it is the mag 5.96 HD 199579, an O6 Vf object located just ⅓° north of NGC 6997 at the northern corner of the triangle enclosing the cluster.

NGC 7039: Though far from one of Cygnus' more impressive star groups, NGC 7039, located 2°

NNE of the magnitude 3.9 Xi (ξ) Cygni, is of interest as a possible "double" cluster–a "double" in the sense that it seems to be the line-of-sight combination of a nearer star group, catalogued as NGC 7039a, with a more distant group, NGC 7039b. The same situation appears to be the case with another Cygnus open cluster, IC 1311, discussed below. Given that our view in this direction is several thousand light-years down an open-cluster-rich spiral feature, and given how large in apparent size many even distant open clusters really are, it is not surprising to find a couple such line-of-sight coincidences in Cygnus.

In 10×50 binoculars NGC 7039 appears as a large (25' diameter) group of a dozen mag 6.6 to 10.0 stars. However, photometric studies have revealed that only half of these stars belong to NGC 7039a. The mag 6.6 star is not even one of them: the true cluster lucida is a mag 7.1 star whose B1.5 main sequence spectrum implies that this is a rather young group. NGC 7039a, if it exists, would be about 2,200 light-years distant.

The much more distant NGC 7039b can be glimpsed in 15×100 supergiant glasses as a small area of haze between a NE-SW pair of mag $6^1/_2$ stars (the NW star the pseudo-lucida of NGC 7039a). An 11th mag star is on the east edge of the haze, a 10th mag star on the west edge, and a 9th mag star on its south edge; but only the first of these, the cluster's mag 11.3 lucida, is a true NGC 7039b member. The group is estimated to be 5,000 l-y away–over twice the distance of NGC 7039a–and to be perhaps 1 billion years old.

NGC 7062: The tiny (5' long) and faint (magnitude 8.3) NGC 7062 is located $1^1/_2°$ west of 71 Cygni and a half degree SW of a rich triangular star field that has 71 Cyg on its SE corner and the open cluster NGC 7082 (described below) on its north corner. Thus, though it is not itself a very conspicuous sight, NGC 7062 enhances the interest of a fine Cygnus Milky Way field. However, it is really more of a conventional telescope than a wide-field instrument cluster, for even in 25×100 supergiant binoculars it appears only as a small oval of faint, partially-resolved haze with faint stars at either end, the brighter of the two stars on the eastern end of the oval. NGC 7062 is rather distant, 5000 light-years away, and rather evolved, a few hundred million years old.

NGC 7063: The rather bright (mag 7.0) but compact (9' across) NGC 7063 is located immediately SW of 69 Cygni. In 15×100 binoculars it is a well-resolved little cluster of eight or ten mag 9 to 11 stars. It is very eye-catching during an innocent field scan, and aesthetically pleasing because of the uniform brightness of its stars. NGC 7063 is not distant, about 2050 light-years away; but intrinsically it is both small, only $5^1/_2$ l-y across, and very faint, its absolute mag of –2.4 corresponding to a luminosity of just 700 Suns.

NGC 7082: The open cluster NGC 7082 is a very large (25' long) oval of eight or ten moderately-faint stars located NW of 71 Cygni on the north edge of a fine triangular star field. Its peculiar shape makes it easy to spot. The oval's NE and SW sides are curved rows of three stars each. A pair of 9th magnitude stars are on the figure's east end. The NW end of the oval is open except for a single faint mag 10 star between the end stars of the two curved rows. NGC 7082 is roughly 4,300 light-years distant, which implies that its oval is a healthy 33 light-years long.

NGC 7086: The faint (magnitude 8.4) and rather small open cluster NGC 7086 lies a couple degrees NW of Pi-one (π^1) Cygni on the west edge of a narrow but bright and conspicuously star-strewn Milky Way stream. However, to the cluster's NW, west, and SW there stretches a several-degree diameter, and impressively murky, dust cloud. Thus the situation of NGC 7086 is visually very interesting, but the group requires supergiant binoculars to even begin to resolve. In 15×100s the cluster is a small granular patch of haze, the mag 10.2 lucida, located at its center, the only star visible. But the extra power of 25×100s resolves a sprinkling of faint stars across the cluster's pale background glow. NGC 7086 is some 3,800 light-years from us. It is dimmed a very respectable 2.2 mags by the outer fringes of the big dust cloud to its west, which is centered about 2500 l-y from us and therefore well in the foreground of the group.

IC 1311: IC 1311, located 2 ° WNW of Gamma (γ) Cygni, seems to be the same sort of problem as the earlier-discussed NGC 7039: the superimposition of a field condensation of brighter foreground stars that may or may not be a true physical group upon a compact and much more distant true open cluster. In the case of IC 1311 the foreground stars, 7th to 9th magnitude objects, are distributed in a conspicuous ring visible without any problem in standard binoculars. The small, distant cluster, located within, and toward the west side of the ring, requires 15×100 supergiant glasses to be glimpsed as a tiny patch of haze.

IC 4996: Located only $^1/_2$° SW of P Cygni, the brightest star (in apparent and real terms) of Cygnus OB1, is one of the three open clusters in that scattered association, IC 4996. (The other two Cyg OB1 clusters are M29 and Be 86.) This group is only 0.2 apparent magnitude fainter, and one minute of arc smaller, than M29, but sufficiently more compact that it requires moderate powers in telescopes to be recognized as a cluster: even in supergiant binoculars all that is seen is the magnitude 7.6 cluster lucida with a mag 8.0 star just to its north. Neither IC 4996 nor M29 seems to have many faint stars.

But the few bright stars of IC 4996 are, like those of M29, hot, luminous early-B giants. The cluster lucida is a B0.5 III giant; and the star to its north is a multiple with a B1 II primary and a mag 9.0 B0 V secondary 32" distant in position angle 77° (and resolvable at 25×). Because of the absence of fainter cluster members, IC 4996 looks more like a Sigma (σ) Orionis-type multiple star than a true open cluster. That two such similar groups as M29 and IC 4996 are in the same association is probably no accident but a consequence of the physical conditions (density, temperature, and perhaps chemical composition) in the interstellar material in this part of the Orion-Cygnus Spiral Arm.

Be 86: Berkeley 86, located about $^2/_3$° WNW of M29 and $^3/_4$° NE of P Cygni, is not really a wide-field observer's object. Indeed, it offers very little even to the telescope, just a handful of 10th magnitude and fainter stars in a small (5') area. In 15×100 supergiant binoculars Be 86 appears only as a tiny field condensation (not even as a real "knot") of a half dozen faint stars and is not obvious unless you already know it is there. However, it is one of the three open clusters in the 6,000 light-years distant Cygnus OB1 association (with M29 and IC 4996), and therefore worth mentioning–and looking for–on that account alone. Be 86 is the same size as its also hard-to-resolve Cyg OB1 colleague, IC 4996, but a whole magnitude fainter. And its lucida, though a hot O9 main sequence star, appears 2.1 mags dimmer than the lucida of IC 4996. However, much of the difference in appearance between the two clusters is due to the fact that Be 86 is obscured 1.2 mags more by interstellar dust than IC 4996. In reality both clusters have the same integrated absolute mag (–6$^1/_2$, a luminosity of 80,000 Suns) and are the same size (9 l-y).

II Autumn Constellations

II.1 Cepheus

Northeast from Cygnus the Milky Way divides into two streams. The brighter stream heads ENE across the northern half of the small constellation of Lacerta into Cassiopeia. Here we have a clear, relatively dust-free, view through the Milky Way. Hence Lacerta, and the adjoining Delta (δ) + Epsilon (ε) region of Cepheus, is rich in open clusters, some very distant. In this direction our view is generally ahead down the forward length of our Orion-Cygnus Spiral Arm. But the view angles across the Ori-Cyg Arm out toward the Galaxy's rim; consequently some of the distant open clusters we see in the N Lacerta/SE Cepheus Milky Way are in the Perseus Arm, the next spiral feature out toward our Galaxy's rim.

The other Milky Way stream from Cygnus heads straight north into the large rhombus formed by Alpha (α) + Beta (β) + Iota (ι) + Zeta (ζ) Cephei. In binoculars this is a region rich in field stars but lacking a Milky Way background glow. The lack of background glow is due to dust clouds some 2000–3000 light-years distant in our Ori-Cyg Spiral Arm. Embedded in these dust clouds is Cepheus' most outstanding wide-field object, the huge ($2^1/_2$° diameter) emission nebula IC 1396. This nebula is involved with the major Ori-Cyg Arm association Cepheus OB2, to which belongs many of the binocular field stars in the Cepheus Milky Way stream.

The one open cluster in the Cep OB2 association, NGC 7160, is neither large nor populous. In fact, Cepheus, despite being in such a promising direction of the Milky Way, does not have any open clusters that are outstanding in anything less than moderate-aperture telescopes. However, most of the constellation's clusters are worth looking for even in binoculars because of what or where they are: NGC 188, NGC 6939, NGC 7142, and NGC 7762 are evolved groups up to 9 billion years old; and NGC 7235, NGC 7380, and NGC 7510 are very young clusters in the distant Perseus Spiral Arm. Moreover, some of Cepheus' clusters are paired with other objects in the same binocular fields: NGC 6939 is less than a degree from the nearly face-on spiral galaxy NGC 6946; NGC 7380 is embedded in an emission nebula visible in binoculars and low-power telescopes; and NGC 7762 can be seen in supergiant binoculars in the same field with the difficult but interesting emission nebula NGC 7822.

NGC 188: Any open cluster that is found well away from the Milky Way is sure to be old because only the very oldest open clusters have had sufficient time to work themselves out of our Galaxy's spiral disc, in which all clusters are born. Thus it is no surprise that NGC 188, located only 4° from Polaris (Alpha [α] Ursae Minoris) 22.3° from the galactic equator, is computed to be 7.2 *billion* years old and thus one of the most ancient open clusters known. Its 5,000 light-year distance implies that it is almost 2,000 l-y off the Galactic plane well out in the Galaxy's thick disc population.

Cepheus Open Clusters

Cluster	RA (2000.0)	Dec	m_V	Size	m_V*	*Spectrum	Distance	A_V	M_V	True Size	Age
N 188	00^h40^m	+85°20′	8.1	13′	12.1		5000 l-y	0.4	−3.3	19 l-y	7.2 bil
N 6939	20 31	+60 38	7.8	7	11.9		4100	1.6	−4.2	8.4	2.2 bil
N 7129	21 43	+66 06	11.5p	7			2500			5.1	
N 7142	21 46	+65 48	9.3	4.3	12.1		9700	1.3	−4.4	10	4.4 bil
N 7160	21 54	+62 36	6.1	7	6.64	B1 III	3100	1.1	−4.9	6.3	
N 7235	22 13	+57 17	7.7	4	8.84	B9 Iab	13,000	2.5	−7.8	15	4–5 mil
N 7380	22 47	+58 06	7.2	20	8.6v	06 Vn	11,400	2.0	−7.5	66	~4 mil
N 7419	22 54	+60 50		2	10.0p		19,600			11	
N 7510	23 11	+60 34	7.9	7	9.68	B1.5 II	10,200	3.6	−8.2	21	~4 mil
N 7762	23 50	+68 02	10.0p	11	11.0p		2700	3.2		8.6	
Be 94	22 23	+55 51	8.7	2.3	9.65	06	11,400	2	−6.0	7.6	~4 mil

Like other evolved open clusters, NGC 188 has no really luminous stars, its most brilliant members being G8 to K4 class III orange giants with absolute magnitudes between +0.7 and +2 (luminosities of from 44 down to 13 Suns). Thus, given the group's distance, its brightest stars are only mag 12.1 and fainter and the cluster can be fully resolved only in moderately-large telescopes. However, it can be partially resolved in 100mm supergiant binoculars. In 15×100s NGC 188 appears as a large, circular, but very, *very* pale patch of haze behind two faint field stars, with just a hint of tiny star-sparks embedded in the haze. In 25×100s the patch becomes mottled, and under good conditions can be resolved into a scattering of very faint stars. The cluster's integrated apparent magnitude is 8.1; but because this is spread over a 13' area, the group's surface brightness is too low for it to be detected in 10×50 binoculars. And NGC 188 can be difficult to find because of its bright-star-poor location, and because it is so near the celestial pole, where directions are confusing. Look for it 1° SSW of the 5th mag star 2 Ursae Minoris, and keep in mind that on the celestial sphere all directions away from Polaris are *south,* that *west* is always the direction the stars appear to move on the celestial sphere, and that, as you look at Polaris, the stars appear to move *counterclockwise* around it.

NGC 6939: In extreme SW Cepheus about 2° SW of Eta (η) Cephei 10×50 binoculars show two patches of haze about $^2/_3$° apart NW-SE. A pair of 6th magnitude stars less than $^1/_2$° apart $1^1/_2$° SSW of Eta Ceph point directly at the SE patch of haze, which is the spiral galaxy NGC 6946, a mag 8.8 object 13' in diameter. The other patch, mag 7.8 but only 7' across, is the open cluster NGC 6939. The cluster can be glimpsed even in 7×35 binoculars (it has the slightly higher surface brightness of the two objects); but the galaxy requires at least 50mm aperture. Something of the true nature of these two objects becomes evident from their contrasting images in supergiant binocs. With 15×100s they are easily seen as two low-surface brightness patches conveniently close for comparison: however, NGC 6946 will appear as a diffuse, even glow whereas NGC 6939 is distinctly brighter toward its center and contains a couple faint stars (12th mag objects). The extra magnification of 25×100s partially resolves NGC 6939 into a mottled haze, and NGC 6946 can be seen to be slightly brighter toward its center–the region of its bulge and inner spiral arms. On the southern edge of the galaxy is a small triangle of field stars, the faintest and most northerly of them just within the galaxy glow.

The low surface brightness of NGC 6946 is the consequence of the fact that it is a loose-armed Sc galaxy which we see virtually face-on. The low surface brightness of NGC 6939 is the consequence of the fact that it is a loose open cluster and so evolved that it has no members brighter than absolute mag –0.1. Indeed, NGC 6939 is estimated to be 1.6 billion years old. Like the even more ancient NGC 188, the high age of NGC 6939 is implied by its large angular distance off the galactic equator, 12°. At the cluster's estimated 4,100 light-years, this angular distance corresponds to 800 l-y off the Galactic plane. The cluster is dimmed a healthy 1.6 mags by outlying dust of the large cloud of interstellar matter that covers much of north central Cygnus. This dust also makes the distance of NGC 6946 somewhat uncertain; but recent estimates place the galaxy 18 million l-y away.

NGC 7129: About 4.7° SSE of Beta (β) Cephei and a little more than $2^1/_2$° WNW of the double star Xi (ξ) Ceph is the compact (7') open cluster NGC 7129, a gathering of only a half dozen magnitude 10, 11, and 12 stars embedded in a very faint reflection nebula. With 15×100 supergiant binoculars the cluster appears as a little field concentration of three approximately mag 10 stars with a mag 11 star just to the west of the concentration and a mag $9^1/_2$ star the same distance further west of the mag 11 object. NGC 7129 *looks* like a cluster–in part because of the comparable brightness of the three stars in the central concentration. However, nothing of the reflection nebula can be seen. The dust of the NGC 7129 reflection nebula is within a chain of relatively nearby, 2500 to 3000 l-y distant, dust and gas clouds that extends entirely across central Cepheus from near (β) Cep on the west to past Iota (ι) and Omicron (o) on the east. Embedded in these clouds is not only NGC 7129, but also the NGC 7023 reflection nebula and the NGC 7822 emission nebula which are described among Cepheus' bright nebulae in Chapter 3. The NGC 7129 cluster is a small group of modestly-luminous early-B stars that has just recently formed in the Cepheus dust clouds.

NGC 7142: Located just $^1/_2$° SE of NGC 7129 (and 1° due west of a mag $6^1/_2$ field star) is the very small (4.3' in diameter) and very faint (mag 9.3) open cluster NGC 7142. This is visually and astrophysically an entirely different kind of open cluster from NGC 7129. In 15×100 binoculars it appears as a very pale amorphous patch with a 10th mag star on its SE edge and mag 11 stars on its east and NNW edges. These, however, are mere field stars, since the cluster's lucida is a mag 12.1 object. Moderately-large telescopes are necessary to resolve the group, and in them NGC 7142 will be seen to resemble NGC 188 and NGC 6939: a swarm of evenly-distributed, similarly-bright stars. This is the signature of highly-evolved open clusters, and in fact NGC 7142 is estimated to be 4.4 billion years old. And like those other two evolved Cepheus clusters, NGC 7142 is well off the Galactic plane: because it is at the large distance of 9,700 light-years from us, its 10° from the galactic equator converts to 1500 l-y off the Galactic plane. (The young NGC 7129 is even farther from the galactic equator than NGC 7142; but because it is so much nearer to us–less than 30% the distance of its neighbor–it is in fact much nearer the plane of the Galaxy.) NGC 7142 is beyond the dust clouds in which the NGC 7129 open cluster + reflection nebula complex is embedded: we see it through a "window" in those clouds. Even so it is dimmed a respectable 1.3 mags by dust.

NGC 7160: About 4° due east of Alpha (α) Cephei and 1° SSW of the beautiful red 5th magnitude variable VV Cep, is the compact (7' long), poorly-populated, but young and interesting open cluster NGC 7160. This group consists of only a handful of stars brighter than the 10th mag, but its two brightest members are mags 6.6 and 7.0 and separated only by 1' NW-SE. About $1^1/_2$' to their SW is a 1'-wide SW-NE pair of mag 7.9 and 8.9 stars. These two star-pairs give the cluster the attractive appearance of a "double-double." But not even in supergiant binoculars is there any indication of background haze in support of this "double-double," just the tantalizing suggestion of two or three fainter cluster members, one NE of the brighter double.

NGC 7160 is a young open cluster, for its mag 6.6 lucida is a blue B1 III giant and its mag 7.0 and 7.9 members are B1 main sequence objects, implying a cluster age of less than 20 million years. It might be involved with the huge Cepheus OB2 association, which also includes the famous "Garnet Star," Mu (μ) Cephei, and the large emission nebula IC 1396 several degrees SSW of the cluster, as well as VV Cep near the cluster. If NGC 7160 is indeed an Cep OB2 cluster and therefore about 3,100 light-years from us, its integrated absolute mag (corrected for 1.1 mags of interstellar absorption) is just –4.9, a luminosity of 80,000 Suns and rather low for such a young group.

NGC 7235: NGC 7235 is a small (4'), rather faint (magnitude 7.7), and poorly-populated open cluster, but it is easily found just $^1/_2$° NW of Epsilon (ε) Cephei. In 15×100 binoculars it is a tight, partially-resolved haze around its mag 8.8 lucida, the cluster's second and third brightest members visible just north and a few minutes of arc west of the mag 8.8 object. Moderate powers in telescopes reveal a dozen or so stars distributed in a tiny but loose ENE-WSW oval, the lucida at the oval's SW end.

The mag 8.8 member of NGC 7325 is a blue B9 Iab supergiant with an absolute mag of –6.8 (a luminosity of 44,000 Suns). This suggests that the cluster is quite young. Its 13,000 light-years distance implies that it is in the Perseus Spiral Arm, the next spiral feature out toward the Galaxy's rim from our Orion-Cygnus Arm. The cluster seems to be the core of a small association that includes the mag 8.36 HD 239895, a B8 Ia Rigel-type supergiant only 20' NE of the cluster (and easily identified since it is the brightest star in the area). The main sequence of NGC 7235 goes up to spectral type B0.5, confirming that this is indeed a young group. Its integrated absolute mag, corrected for 2.5 mags of absorption, is a robust –7.8, corresponding to a luminosity of 110,000 Suns.

Photo 2.5 A field of open clusters in the Cepheus Milky Way. This is a black-and-white print of Plate XXIV, to bring out the star fields and star clusters. The loose cluster in the emission glow toward the lower right (SW) corner is the 11,700 light-year distant NGC 7380, a major tracer of the Perseus Spiral Arm. To its NNE in the right center of the field is the comet Hönig. About the same distance further NNE is the small NW-SE gathering of faint stars NGC 7419. This cluster is perhaps 19,000 l-y away in the +II, or Outer, Spiral Arm of the Galaxy: its brightest stars are hot blue OB giants, but on Plate XXIV it can be seen that they appear yellow because of interstellar dust. The bright disc of the emission nebula NGC 7538, another Perseus Arm tracer, is toward the upper left. To its SSW is a very short NE-SW dash of stars, the young Perseus Arm open cluster NGC 7510

NGC 7380 (Plate XXIV, Photo 2.5): Just under $2^1/_2°$ east and slightly south of the famous variable Delta (δ) Cephei is the large (20') and rather bright (magnitude 7.2) open cluster NGC 7380. This group is best observed with telescopes because its stars are mag 9 and less. However it is embedded in a large, low-surface-brightness emission nebula requiring wide-field instruments to be seen: any kind of power simply magnifies the nebula-glow out of existence. The mingled nebula and cluster light can be seen in 10×50 binoculars as a faint Milky Way patch, two or three resolved stars glittering in it. With 100mm supergiant glasses NGC 7380 is a fine sight. The nebula-glow is not difficult–very pale, amorphous, but immediately recognizable–and the cluster within it resolves into a moderately-large, moderately-compressed field of delicate star-sparks, unquestionably a cluster and not just an association star-field. The cluster gives the impression of triangularity: an angle of resolved stars is at the north corner, a line of approximately 10th mag stars at the SE corner, and the cluster lucida at the SW corner. The lucida is the eclipsing binary DH Cep, which has a period of 2.111 days and a range between mags 8.6 and 9.0, and consists of two hot O6 stars. The cluster and nebula glow is set off by two long parallel lines of ENE-WSW obscuration in the Milky Way haze just to their north.

The NGC 7380 complex is extremely distant, 11,400 light-years away. It is so far, in fact, that it is in the Perseus Spiral Arm, the next spiral feature out toward the Galaxy's rim from our own Orion-Cygnus Arm (see Figure 3.1 in Chapter 3.) At such great distances the cluster's apparent size of 20' corresponds to a true diameter of 66 l-y, about the same as each component of the famous Perseus Double Cluster. The cluster's integrated absolute mag of –7.5 is deceptively low because most of the energy output of O-type stars like the components of DH Cep is not in visual wavelengths, but in the ultraviolet radiation that fluoresces the emission nebula around the cluster.

NGC 7380 is the main stellar concentration of the huge Cepheus OB1 association, one of the major tracers of the Perseus Spiral Arm. This association spans an 11°×4° area just SE of the galactic equator–a true size of an enormous 2200×800 l-y–and includes the small open cluster Be 94 (discussed below), located 4° SW of NGC 7380 and boasting an O6 star of its own. Cep OB1 is remarkable both for the number, and for the variety in spectral type, of the highly-luminous supergiants it contains. It has three of the extremely rare yellow Ia supergiants, all variables: the mag 5.08 V509 Cassiopeiae, a G0-G5 Ia-0 hyper-supergiant with a visual absolute mag of –9.4 (a luminosity of 440,000 Suns); the mag 6.7 RW Cep, a K0 Ia-0 hyper-supergiant located just NE of Be 94 and of the same visual luminosity as V509 Cas; and the mag 7.3 W Cep located 1° due east of Delta Ceph, an eclipsing binary with a K0 Iaep primary. RW Cep is the most luminous yellow supergiant known in our Galaxy: it has the same *visual* absolute mag as V509 Cas, but is cooler than V509 Cas and therefore radiates most of its energy as *infrared* rather than as visible radiation. Thus the bolometric absolute mag of V509 Cas is only a couple tenths of a mag greater than its visual absolute mag; but the bolometric absolute mag of RW Cep is –10.5, a full mag greater than its visual absolute mag and a total energy output of 1.3 *million* Suns!

Cep OB1 also has four of the rare Wolf-Rayet stars, a couple red M-type supergiants, two white A-type supergiants, several blue B-type supergiants, and two extremely hot O9 Iab and O9 III stars, both of which have bolometric absolute mags of $-9^1/_2$ (a total energy output of 500,000 Suns). The presence of Wolf-Rayets and yellow hyper-supergiants in Cep OB1 might not be a coincidence, for both seem to be stages in the very rapid evolutionary expansion of the most massive stars, and Cep OB1 is obviously well-populated with exceptionally massive stars.

NGC 7419 (Plate XXIV, Photo 2.5): Though its published diameter is a mere 2', the open cluster NGC 7419 is not difficult to see in 15×100 supergiant binoculars, appearing as a spray of six or seven magnitude 10 and 11 stars with some partially-resolved members suggesting richness and helping call attention to the group. The problem is *finding* the cluster, for it is in a very bright-star-poor region about $3^1/_2°$ ENE of Delta (δ) Cephei and the same distance west and a little south of M52 in Cassiopeia. Start at the $^1/_2°$ wide W-E pair of 5th mag 1 and 2 Cassiopeiae, the brightest stars in the region between Delta Ceph and M52. 1 Cas is at the SE end of a conspicuous, almost 2°-long, ESE-WNW line of five stars. (The first star WNW of 1 Cas is a wide double.) NGC 7419 is about $^3/_4$° north of the northwestern-most star in the line and just SW of a 9th mag field star. It is worth the bother of finding because it not only is a young open cluster, it is 19,600 light-years from us and therefore probably a tracer of the outermost spiral feature in our quadrant of the Galaxy, the +II Arm.

NGC 7510 (Plate XXIV, Photo 2.5): Precisely on the galactic equator 2° WSW of M52 in Cassiopeia is the short (7' long) and faint (magnitude 7.9) open cluster NGC 7510. It is a fine sight in telescopes at about 100×, resolving into 16 or more stars packed into a 2' wide ENE-WSW rectangle, the group's two brightest members on the rectangle's opposite ends. The cluster is too compact to be recognized in standard binoculars, but can be partially resolved in giant glasses. In 15×100s it appears as a very thin ENE-WSW dash, the mag 9.7 lucida conspicuous on the NE end and giving the group the look of an exclamation mark. One or two fainter members can be momentarily discerned in the haze of the dash. The extra power of 25×100 instruments resolves NGC 7510 into a little line of four or five stars with a couple more alongside. To the NE of the cluster a NE-SW pair of approximately 8th mag field stars points directly at it. And to its south the star field is fairly rich, with some attractive dark channels of obscuring matter silhouetted upon the background Milky Way glow.

NGC 7510 is small in part simply because it is so far away, 10,200 light-years. This means that it is in the Perseus Spiral Arm with NGC 7380 and the Cepheus OB1 association. And it must be a true tracer of the Perseus Arm because it is a very young cluster: its lucida is a luminous blue B1.5 III giant with an absolute mag around -6.4, and its main sequence goes up to a very hot O6. The group's integrated absolute mag, -8.2 (a luminosity of nearly 160,000 Suns), is about that of each component of the Perseus Double Cluster. NGC 7510 is dimmed a very healthy 3.6 magnitudes by the outlying dust clouds of the 2,700 l-y distant Cep OB3 association, which is centered just north of the cluster (and described in Section 4 of Chapter 3).

NGC 7622 (Plate XXV, Photo 3.17): NGC 7622 is a rather large (11' in diameter) but hard to see and hard to find open cluster in the bright-star-poor spaces of extreme eastern Cepheus. It is 4° due east of Omicron (ο) Cephei, and just NE of the only 5th magnitude star between Beta (β) Cassiopeiae on the south and Gamma (γ) Cep on the north. Though it looks rather populous on photographs, its stars are only mag 11–13 objects and the cluster therefore requires moderately large telescopes for good resolution. And its stars are rather spread out, so it does not offer the binocular observer a well-consolidated moderate-surface-brightness glow. Thus even in 15×100 supergiant glasses, NGC 7622 is barely visible with averted vision as a modest-sized field concentration of mag 11 stars.

NGC 7622 is not particularly distant, about 2,700 light-years away. Its stars are faint largely because they are dimmed around 3.2 magnitudes by the dust involved with the Cepheus OB4 association, a star group spread over an area a couple degrees across just east of the cluster. The brightest member of Cep OB4 is a mag 8.3 G3 Iab yellow supergiant, a star probably several million years old. But the association also includes a very hot, very young O7 member (dimmed 4.3 mags by dust). The Cep OB4 emission nebula NGC 7822, centered less than $1^1/_2$° NE of NGC 7622 and therefore in the same low-power field, is, like the cluster, visible in supergiant binoculars. The two are a challenging, but visually contrasting, object-pair. (NGC 7822 is described in more detail in Chapter 3.) NGC 7622 is, however, not a Cep OB4 group: its very appearance–a swarm of evenly-distributed, comparably-bright stars–is that of an old cluster. It also must be somewhat beyond the association: if it was in *front* of Cep OB4, it would not be dimmed as much as it is; and if it was *in* the association, its stars would be embedded in reflection nebulae from the dense Cep OB4 dust.

Be 94: Though small (only 2.3' in size) and faint (mag 8.7), the open cluster Berkeley 94, located 1.6° SE of Epsilon (ε) Cephei, is worth finding with larger binoculars and small RFTs because it is in the 11,400 light-year distant Perseus Spiral Arm association Cepheus OB1 with the large bright cluster NGC 7380. Be 94 requires at least moderate powers in telescopes to be seen at its (very modest) best, but can be glimpsed with 15×100 supergiant binoculars as a knot of two or three approximately mag 11 stars with a mag 9½ star a couple minutes of arc to their north. The latter is the cluster's extremely hot O6 lucida. The mag 6.7 RW Cephei, a hyper-luminous K0 Ia-0 supergiant and intrinsically the brightest Cep OB1 star, is just NE of Be 94.

II.2 Lacerta

The little constellation of Lacerta occupies a 21°-tall rectangle of the autumn Milky Way between the Delta (δ) + Epsilon (ε) + Zeta (ζ) Cephei triangle on the north and Eta (η) Pegasi on the south. The galactic equator roughly follows the Cepheus/Lacerta border. Southern Lacerta has a splendid 7× binocular field of some 20 magnitude 4½-7 stars. This is the Lacerta OB1 association, centered only about 1600 l-y away. Its full extent is 15°~450 l-y. The dozen brightest Lac OB1 members are listed in the accompanying table. In addition, the 4th mag Nu (ν) Andromedae, though about 30° east of the association center, is thought to be a "runaway" from Lac OB1 ejected some 9 million years ago. The variables 12 and 16 Lacertae are Beta (β) Canis Majoris stars. Because Beta Canis variables are moderately-massive objects just beginning their evolutionary expansion off the stable main sequence, Lac OB1 must be a somewhat evolved OB group, perhaps 20 million years old (though its O9 member would be younger than this. However, association stars do not all form at once.) The bright star population of Lac OB1 resembles that of the Scorpio-

Brightest Members of Lacerta OB1

Star	m_V	Spectrum	M_V
11 Lac	4.50	B2 IV	-4.3
10 Lac	4.87	O9 V	-3.8
12 Lac	5.0–5.1	B2 IV	-3.8
2 And	5.22	B1 III	-3.5
16 Lac	5.3–5.4	B2 IV	-3.5
6 And	5.58	B2 IV	-3.1
8 Lac A	5.75	B1 Ve	-2.9
8 Lac B	6.45	B2 V	-2.2
S 2973 A	5.91	B3 IV	-2.3
Roe 47 A	6.14	B2 IV–V	-2.5
9 And	6.17	B2 IV–V	-2.4

(Source: Garmany and Stencel, *Astron. Astrophs. Suppl.* 94, 223.)

Centaurus Association, particularly in the presence of early-B giants and subgiants and of Beta Canis variables. However, Sco-Cen lacks O-type stars and Lac OB1 has no Antares-type red supergiant.

Northern Lacerta is crossed by a Milky Way stream that begins in NE Cygnus and extends all the way to Kappa (κ) Cassiopeiae. This stream is its brightest in Cygnus and fades gradually farther NE. That is because toward Cygnus we look *ahead* down the length of our star- (and cluster-) rich Orion-Cygnus Spiral Arm, but the farther NE we scan along the Milky Way the more our view is *across* our Arm out toward the exterior of the Galaxy. However, interstellar matter is very thin toward northern Lacerta and adjoining SW Cepheus, so we can see a long way across the Galaxy in this direction and the Milky Way is bright and well-populated with open clusters. Some of the clusters of SW Cep lay as far away as the Perseus Spiral Arm, the next spiral feature out toward the Galaxy's rim from our Ori-Cyg Arm. Lacerta has two moderately-distant open clusters that are easy to see in small binoculars and at their best aesthetically in giant binoculars, NGC 7209 and NGC 7243.

Lacerta Open Clusters

Cluster	RA (2000.0)	Dec	m_V	Size	m_V*	*Spectrum	Distance	A_V	M_V	True Size	Age
N 7209	22^h05^m	+46°30'	7.7	25'	9.0		2900 l-y	0.5	-2.5	21 l-y	410 mil
N 7243	22 15	+49 53	6.4	21	8.5	B5 III	2500	0.8	-3.8	16	250

NGC 7209: The very large (25′ in diameter) and moderately bright (magnitude 7.7) NGC 7209 is located about 5½° SW of Alpha (α) Lacerta. It perhaps is easiest found by scanning 6° ESE from M39 in Cygnus because it is readily seen in 7×50 and 10×50 binoculars as a hazy patch mottled with the partial resolution of its brightest stars, 9th and 10th mag objects. It becomes very well resolved with 100mm aperture, particularly at powers over 20×. At least two dozen stars can be counted in a thick, crude "C" open to the west. They are of comparable brightness (9th, 10th, 11th mag) and rather evenly distributed, which gives the cluster the illusion of richness. However, there is no background glow from unresolved cluster members because in reality NGC 7209 is not particularly populous (and looks it in telescopes). It is estimated to be 2900 light-years distant, so its integrated absolute mag, corrected a half mag for interstellar absorption, is a paltry −2.5, a luminosity of just 830 Suns. Its true size is a very respectable 21 l-y; but that merely implies that the cluster is mostly empty space. It is estimated to be around 410 million years old, but probably will soon fall apart. In age and in luminosity NGC 7209 is similar to the much-nearer M39: but its diameter is three times–hence its volume is *twenty-seven* times–greater than that of the already-sparse M39.

NGC 7243: Easily spotted with even the smallest binoculars 2° west of Alpha (α) Lacertae is NGC 7243, which is almost as large as NGC 7209 but three times (1.3 magnitudes) brighter. The two clusters are only 4° apart, NGC 7209 south and slightly west of NGC 7243, and therefore fit in the same 7× or 10× field. In 10×50 binoculars NGC 7243 is a coarse, loose gathering of 10 or 12 stars, not as visually interesting as the mottled glow of Lacerta's "other" open cluster. The group is virtually completely resolved in 100mm glasses. But even at only 15× it is too scattered, and its stars too unevenly bright and too coarsely distributed, to be impressive aesthetically. It is in two sections: the smaller NE section is triangular, and contains perhaps as many as ten stars (depending upon the aperture of your instrument); and the SW section is trapezoidal, with perhaps two dozen stars visible. The two sections are separated by a narrow, seemingly starless, gap.

NGC 7243 is perhaps a little nearer than NGC 7209, 2500 light-years from us. Its integrated absolute mag is −3.8, so it is intrinsically three times more luminous than the other group. Its mag 8.5 lucida is a B5 III blue giant with an absolute mag of −1.7; and its second brightest member, a mag 8.7 star, has a B7 subgiant spectrum. The presence of such bright blue stars implies NGC 7243 is much younger than NGC 7209. And in fact its main sequence goes up to a mag 9.1 B8 star, suggesting a cluster age of about 110 million years, one-fourth that of NGC 7209. But when NGC 7243 loses its B-type members, in the astronomically near future, it will be even fainter and poorer than its colleague.

II.3 Andromeda

Andromeda is not usually thought of as a Milky Way constellation, especially since its show-piece object is a galaxy, the Andromeda Spiral M31. The galactic equator in fact does not cut through any of the constellation; but some fine star-fields of the outer billows of the Milky Way are to be seen with giant binoculars and RFTs in the extreme NW corner of Andromeda's area (NW of the Iota [ι] + Kappa [κ] + Lambda [λ] + Psi [ψ] asterism. These star fields are described in Sect. 5 of Chap. 3.) Andromeda does have an outstanding open cluster for binoculars, the large, rich-looking NGC 752. Ironically, NGC 752 is located in the extreme SE corner of the constellation, as far from NW Andromeda's best Milky Way fields as it can get.

NGC 752 (Photo 2.6): NGC 752, located 5° south and slightly west of the splendid telescopic double

Andromeda Open Cluster

Cluster	RA (2000.0) Dec		m_V	Size	m_V*	*Spectrum	Distance	A_V	M_V	True Size	Age
N 752	01^h58^m	+37°41′	5.5	50′	8.94	G7 III	1200 l-y	0	−2.3	18 l-y	1.4 bil

Photo 2.6 The evolved open cluster NGC 752 in Andromeda. The two bright stars on the lower (SW) edge of the cluster, the double 56 Andromedae, are not true cluster members, nor even a true physical binary. They are however, a beautiful pair of deep-orange K-type stars

Gamma (γ) Andromedae, is extremely large and bright. Its 50' apparent diameter is almost twice that of the full Moon; and its apparent magnitude of 5.5 implies that it can be seen on dark nights from dark-sky sites with the unaided eye. The cluster contains sixteen stars between mags 8.9 and 10.0 with a total of nearly 80 members down to mag 12.0. Its stars are more-or-less evenly distributed over the group's 50' area. (The 7th mag star on the cluster's east edge is not a true member.) Thus NGC 752 is very easy to resolve even in small binoculars; and virtually every cluster member can be seen with supergiant instruments. However, it is so large, and its stars so evenly distributed, that 25× is just about the maximum power that should be used on it: higher magnifications destroy the appearance of rich "clusterness" that makes NGC 752 such a fine sight even in 7×35 glasses. The aesthetics of the group is enhanced by the double 56 And just outside its SW edge, a pair of pumpkin-orange K-type stars of apparent mags 5.7 and 5.9 separated by 190" and therefore resolvable even at only 7×.

NGC 752's appearance of richness in low-power instruments is very misleading, for the cluster has few members fainter than mag 12.0. In other words, the cluster's 80 or so stars between mags 8.9 and 12.0 are its *total* population, not just a sample limited (as is usually the case) by the light-gathering power of the instrument used to look at it. And these are not very bright stars intrinsically: NGC 752 is estimated to lie about 1200 light-years away, so its lucida, a G7 III yellow giant, has an absolute mag of just +1, a luminosity of only 33 Suns. The cluster's integrated absolute mag is a mere -2.3, equivalent to about 700 Suns. All this–the lack of both bright *and* faint members–as well as NGC 752's rather large angular distance from the galactic equator, are symptoms of its great age: 1.4 billion years. During that long time NGC 752 has lost all its most massive members to supernova explosions and planetary nebula ejections, most of its least massive members have "evaporated" from it (able to escape their parent cluster's ever-weakening gravitational field), and it has been nudged by close encounters with globular clusters and massive giant molecular clouds into a space motion that takes its well off the Galactic plane.

II.3 Cassiopeia

The single most important fact about Cassiopeia is its open cluster richness. Twenty-one open clusters will be described in this section, and of course a good many more can be seen in large telescopes. In addition to Cassiopeia, the best open cluster constellations visible from mid-northern latitudes are Cygnus, Monoceros, and Puppis (with adjoining eastern Canis Major). Toward these constellations we have "windows" between nearby dust clouds through which we can see many thousands of light-years across the spiral disc of our Galaxy. The Cassiopeia Milky Way is a "window" between the dust clouds of central Cepheus to the west (which are between 2000 and 3000 light-years from us) and the dust clouds of Perseus to the east (500 to 1500 l-y away). Through this "window" we see past a sprinkling of nearby clusters (NGC 225, NGC 1027, St 2) and across an inter-arm gap scattered with intermediate-age and

Cassiopeia Open Clusters

Cluster	RA (2000.0)	Dec	m_V	Size	m_V*	*Spectrum	Distance	A_V	M_V	True Size	Age
M52	23h24m	+61°35'	6.9	12'	8.22	gG8	4600 l-y	1.8	−5.8	16 l-y	100 mil
M103	01 33	+60 42	7.0	6	7.15	B5 Ib	9200	1.5	−6.8	16	25
N 129	00 30	+60 14	6.5	21	8.7v	G1 Ib-II	5500	1.6	−6.3	40	
N 133	00 31	+63 22	9.4p	7			9100			19	
N 146	00 33	+63 18	9.6	6	11.6		7800	1.9	−4.7	14	~40
N 225	00 43	+61 47	7.0	12	9.26	B8	2000	0.9	−2.9	7	~200
N 436	01 16	+58 49	8.8	5	11.1		10,700	1.6	−5.4	15	60
N 457	01 19	+58 20	6.4	13	8.65	M0.5 Ib	8200	1.6	−7.2	34	10
N 654	01 44	+61 53	6.5	5	7.29	F5 Ia	9200	2.5	−8.3	13	<20
N 659	01 44	+60 42	7.9	5	10.4		9200	2.0	−6.3	13	22
N 663	01 46	+61 15	7.1	16	8.50	B6 Iab	9200	2.7	−7.9	43	20
N 1027	02 43	+61 33	6.7	20	9.3		3300	1.3	−4.6	19	
N 7788	23 57	+61 24	9.4p	9			7800	0.8		20	~30
N 7789	23 57	+56 44	6.7	25	10.70	K4 III	5900	0.8	−5.4	43	1700
N 7790	23 58	+61 13	8.5	17	10.2v		9500	2	−5.8	44	70
H 21	23 54	+61 46	9.0p	6			9100			16	
K 12	23 53	+61 58	9.0	2	10.4		8100	1.9	−4.9	4.5	10
K 14	00 32	+63 10	8.5	7	11.3		7800	1.5	−4.9	16	~30
Mel 15	02 33	+61 27	6.5	22	7.81	05 IIIf	7800	2.6	−8.0	50	0.7
St 2	02 15	+59 16	4.4	60	8.17	A1 IV	1050	1.0	−4.2	18	170

ancient open clusters (M52, NGC 129, NGC 7789) to an extremely open-cluster-rich stretch of the Perseus Spiral Arm, the next spiral feature out toward the exterior of the Galaxy from our own Orion-Cygnus Arm. (See Fig. 3.1 in Chap. 3.)

The open clusters of Cassiopeia not only come in a variety of distances and structural locations, they also come in a variety of sizes, brightness, and resolvability suitable to virtually any kind of astronomical instrument. NGC 457, NGC 663, NGC 7789, and St 2 are outstanding binocular clusters. All are at or near their best in giant binoculars, with which M52 also begins to come into its own. In wide-field instruments often two, three, or even four clusters are visible at once. M103 and NGCs 654, 659, and 663 all fit in the same giant binocular field. NGC 436 and NGC 457 are less than a degree apart, as are NGC 7788 and NGC 7790. Just north of Kappa (κ) Cassiopeiae is the open cluster triplet NGC 133, NGC 146, and K 14. And in the extreme eastern part of the constellation, centered only 1½° apart, are the contrasting clusters NGC 1027 and Mel 15.

Cassiopeia's open clusters are distributed in large groups. NW of Beta (β) Cas is a 2°-long SE-NW line of a half dozen clusters, including NGC 7790, NGC 7788, H 21, and K 12. Several clusters are around Kappa Cas, including NGC 133, NGC 146, and K 14 to the north and NGCs 129 and 225 to the south. M103 and NGCs 654. 659, and 663 are a distinct set. And the two huge binocular emission nebulae in far eastern Cassiopeia, IC 1805 and IC 1848, have three or four large clusters in or near them, including NGC 1027, Mel 15, and St 2. In part this cluster clumping is from actual physical groupings: NGC 7788 and K 12 NW of Beta Cas are possible members of the Cassiopeia OB5 association; M103 and NGCs 654, 659, and 663 are all members of Cas OB8; and the IC 1805 and IC 1848 emission nebulae are involved with Cas OB6, the central cluster of which is Mel 15.

M52: M52 is a compact (12') but bright (magnitude 6.9) open cluster easily seen with even small binoculars as a hazy "dot" on the northern edge of the Milky Way midway between Kappa (κ) Cassiopeiae and Delta (δ) Cephei. In 10×50 instruments the cluster's mag 8.2 lucida, a yellow G8 giant, is conspicuous on the WSW edge of the group's high-surface-brightness disc. However, because of its populous compactness M52 requires high-power giant binoculars to really begin to resolve. In 15×100 glasses hints of sparkling resolution can be glimpsed in the cluster's tight granular patch, the second brightest M52 star visible on the patch's ENE edge and the third brightest just to the north. In 25×100s more general resolution is achieved, with many faint but distinct stars glittering on a pale background glow. The mag 8.2 lucida on the WSW edge ornaments this concentrated crowd of star-chips like a gemstone. M52 is a fine sight in supergiant binoculars, but so dense that it requires about 100× in moderately-large telescopes to be seen at its spectacular best.

M52 is estimated to be around 4,600 light-years away. It is dimmed 1.8 magnitudes by interstellar dust, much of which must be from outlying clouds of the 2,700 l-y distant Cepheus OB3 association, centered only about 3° to the NW. The cluster's integrated absolute mag, -5.8, corresponds to a luminosity of 20,000 Suns. M52 is located in the inter-arm gap between our Orion-Cygnus Spiral Arm and the Perseus Arm, and therefore must be an intermediate-age or old cluster. However, age estimates for it have ranged between 50 and 150 million years. The truth is probably near the middle. M52 is as concentrated as it looks in the eyepiece, its central star density estimated to be about 50 per cubic parsec.

M103: Only two of Cassiopeia's many open clusters have the honor of being in the comet-hunter Charles Messier's late-18th century catalogue of fuzzy objects that are *not* comets. M52 is worthy of the list; but M103, located 1° ENE of Delta (δ) Cassiopeiae, is inferior to Cassiopeia's NGC 457, NGC 663, and NGC 7789, all of which were overlooked or ignored by Messier and his colleague, Pierre Mechain. M103 is small, only 6' across, and not particularly bright, just mag 7.0. It is dominated by its lucida, a mag 7.15 B5 Ib supergiant triple star, the 9th mag secondary 14" away in PA 142° and the mag $10^1/_2$ third component 28" distant in 145°. The second brightest member of the cluster is a mag 8.54 M0.5 Ib Antares-type supergiant. In 10×50 binoculars the cluster appears only as a knot of two or three stars. With 15×100 supergiant instruments M103 begins to resolve: between the lucida on the NNW and the M-type supergiant on the SSE are a N-S pair of fainter cluster members with a faint haze, to the SW of which is the mag 10.6 star (a B2 main sequence object) that gives the group the triangular shape mentioned in observing guides.

M103 is thought to be 9,200 light-years distant and a member of the extensive Cassiopeia OB8 association, which also includes the clusters NGCs 654, 659, and 663 (all discussed below). Because its brightest members include a Ib red supergiant and its main sequence goes up only to spectral type B2, M103 cannot be extremely young: it is estimated to be already 25 million years old. Cas OB8 as a whole seems to be well-evolved: the brightest stars of NGC 663 are all late-B and M Iab supergiants, none of them Ia supergiants; the lucida of NGC 659 has an absolute mag of less than -4; and in the greater association the brightest star is an F5 Iab supergiant (the mag 6.88 HD 9973) and there are three more M-type supergiants. Consequently Cas OB8 as a whole must also be 20–25 million years old. However, the stars and clusters of associations do not all form simultaneously, but over the course of millions of years, and often in separate subgroups. Thus probably the last-formed Cas OB8 cluster was NGC 654, because its lucida is a Ia supergiant and its main sequence goes all the way up to type B0. And we find one rogue O-type member in Cas OB8, a very hot and luminous O7.5 III giant.

M103 is dimmed a rather modest (modest, given its distance) 1.5 mags by interstellar dust. Its integrated absolute mag is -6.8, a luminosity of almost 50,000 Suns. However, most of the cluster's brilliance is from the lucida, which by itself is as bright as about 40,000 Suns. The M0.5 Ib supergiant contributes most of the rest of the cluster's 10,000-Suns luminosity.

NGC 129: One of the brighter (magnitude 6.5) and larger (21' diameter) open clusters in Cassiopeia is NGC 129. It is easily found right in the center of the square formed by Alpha (α), Beta (β), Gamma (γ), and Kappa (κ) Cassiopeiae, near the intersection of the square's diagonals. A 6th mag field star is just to the cluster's south. NGC 129 can be seen, though not resolved, in 10×50 binoculars. 15×100 supergiant glasses show a small triangle of 9th mag stars in the SE part of a faint, rather compact, area of haze. Another 9th mag star is just NE of the haze. As the cluster's appearance suggests, it is populous with very faint, unresolved stars, but not concentrated in structure.

The lucida of NGC 129, the star at the NW corner of the triangle of 9th mag stars (and thus near the cluster center), is the Cepheid variable DL Cas, which has a range of mags 8.70 to 9.28 in a period of 8.00 days. The Cepheid period-luminosity relation implies a peak absolute mag of –4.2 for this star, and the amount the star is reddened by interstellar dust suggests it is dimmed 1.6 mags. All this converts to a distance of around 5,500 light-years to the star and its cluster. NGC 129 thus has a very respectable integrated absolute mag of –6.4 (a luminosity of nearly 30,000 Suns) and a very respectable true size of 34 l-y. It is positioned out in the inter-arm gap between our Orion-Cygnus Arm and the Perseus Spiral arm, a location consistent for an evolved cluster with a short-period, modest-luminosity Cepheid.

NGC 133, NGC 146, and K 14: Centered about a half-degree NNW of Kappa (κ) Cassiopeiae is a triangle of small (all about 6' in diameter) and faint (mag 8$^1/_2$ to 9$^1/_2$) open clusters, NGC 133, NGC 146, and King 14. K 14 is about $^1/_4$° NW of Kappa, and NGCs 133 and 146 are $^1/_4$° NW and NE of K 14, respectively. These groups are all beyond normal binoculars: especially at low powers the glare of Kappa interferes with seeing them. But from dark sky sites, and when they are near the zenith and thus have the darkest sky background possible, they are an interesting giant binocular set. NGC 133 is a short but distinct line of three mag 9–9$^1/_2$ stars pointing through K 14 SSE at Kappa Cas. NGC 146 appears as a condensation in the general star field, with a 10th mag star on its WSW edge, a slightly fainter star on its ENE edge, and three or four mag 11 stars scattered between them. And K 14 is a condensation in the star field consisting of an 11th mag lucida with three or four fainter stars loosely gathered around it.

An actual physical relation between at least NGC 146 and K 14 is possible. The two clusters are similar in appearance and are estimated to be about the same distance from us, 7,800 light-years, and to be of about the same age, 30 or 40 million years (their main sequences go up to spectral types B2 or B3). Allowing for interstellar absorption, their integrated absolute mags are also virtually identical, around –4.8. If *both* clusters are *exactly* 7,800 l-y away (unlikely even if they are physically related) their true center-to-center separation would be only about 25 l-y. The third cluster of the trio, NGC 133, is much different in appearance from the other two and is in any case estimated to lie over 1,000 l-y beyond them.

NGC 225: About midway between Kappa (κ) and Gamma (γ) Cassiopeiae is a modestly large (12' diameter) field condensation of approximately 18 9th, 10th, and 11th magnitude stars, the open cluster NGC 225. This group can be partially resolved in 10×50 binoculars, though its stars are too faint to make much of an impression. It becomes fully resolved, with no suggestion of background haze, in supergiant glasses. At low power NGC 225 appears spuriously rich because of the similar brightness and even distribution of its stars; but at 50× it is not an interesting sight even in telescopes because it is loose and lacks any faint supporting cast to fill in the gaps between its brighter members. The mag 9.26 lucida, a B8 V star, is on the NE edge of a west-opening arc of four mag 9–9$^1/_2$ cluster members. The main concentration of fainter cluster stars is to the SW of this arc.

NGC 225 is estimated to be only 2,000 light-years distant ("only" with respect to the majority of other Cassiopeia open clusters). Thus its true size is just 7 l-y and its integrated absolute mag (corrected 0.9 mag for dimming by dust) a mere –2.9, a luminosity of 1250 Suns. The B8 V lucida suggests that the cluster is a couple hundred million years old because that is the main sequence lifetime of such objects.

NGC 436: The main reason for looking at NGC 436 is simply that it happens to be just 40' NW, and therefore in the same wide-angle field, with one of

Cassiopeia's front-line open clusters, NGC 457. NGC 436 is small (5'), faint (magnitude 8.8), and has no stars bright than mag 11.1. It requires telescopes to be seen at its (modest) best: even in 15×100 supergiant binoculars it is only a tiny knot of perhaps three 11th mag stars. An 8th mag field star is about 10' to the knot's west; and just to the cluster's SE is an E-W pair of 10th mag stars-which, however, look like the non-members they are. NGC 436 is small and faint in large measure simply because it is so far away, 10,700 light-years. Its integrated absolute mag is a respectable -5.4 (12,000 Suns) and it is not an evolved cluster, for it is estimated to be 42 million years old. It is in, and perhaps can be considered a true tracer of, the Perseus Spiral arm.

NGC 457: In standard 10×50 binoculars NGC 457, located 2° SSW of Delta (δ) Cassiopeiae, is one of the best sights in Cassiopeia, a 13'-long NW-SE bar studded with perhaps 20 magnitude $8^1/_2$–10 stars, the bright mag 5.0 Phi (ϕ) Cas and mag 7.0 HD 7902 two gems at the bar's SE end. Even to the casual glance the cluster is a striking sight. Phi Cas seems to be a true cluster member. It is an F0 Ia yellow supergiant with the extremely high absolute mag of -8.5 (210,000 Suns), and thus one of the most luminous yellow supergiants known. HD 7902, 2' SW of Phi, likewise seems to be a true NGC 457 star: it is a blue B6 Ib supergiant with an absolute mag of -6.5 (33,000 suns). These two stars are a fine color-contrast pair in small telescopes. The brightest star in the "bar" of the cluster is a mag 8.65 M0.5 Ib red supergiant, an Antares-like object with an absolute mag of -5.0. Its color likewise is a fine contrast in small telescopes to the blue-white of HD 7902.

In 15×100 supergiant binoculars NGC 457 remains one of the two or three best Cassiopeia clusters, superior aesthetically even to both of the constellation's Messier groups, M52 and M103. The main line of the cluster "bar," extending NW from the B6 supergiant, is resolved into a half dozen stars. The M supergiant, located a few minutes NW of Phi Cas on the bar's other long side, appears in a shallow arc, concave toward Phi and HD 7902, of another half-dozen stars that crosses the main line and is somewhat longer. A few minutes of arc NW from the main NGC 457 star-bar is a detached pair of 9th mag stars, the SW one an E-W double of mag 10 components. NGC 457 is not as good visually in 25×100s as it is in 15×100s: the extra magnification makes it begin to look rather loose because there is no richness of fainter cluster stars to compensate for the spreading out of its brighter members.

NGC 457 is about 8,200 light-years distant, one of the major open clusters defining the Perseus Spiral Arm. However, like NGC 436 and M103 to its NW and NE, it is dimmed only $1^1/_2$ magnitudes by interstellar dust, a very low value for its distance and testimony to the Galaxy's transparency in this direction. The cluster's integrated absolute mag, excluding Phi Cas and HD 7902, is -7.2, a luminosity of 62,500 Suns. If Phi Cas and HD 7902 are included, the absolute mag of NGC 457 is -8.9, equivalent to 305,000 Suns (though two-thirds of this is from Phi Cas alone). The presence in it of less-luminous Ib supergiants suggests a cluster age of over 10 million years.

NGC 654: Though only 5' in diameter, NGC 654 can be easily spotted in moderate-size binoculars because it is just 40' NNW of the large, conspicuous open cluster NGC 663 (discussed below) and marked by a magnitude 7.3 star, located immediately SE of the cluster core. In 15×100 binoculars the cluster core appears as a small but very distinct, almost conspicuous, haze with a couple 11th mag starspecks glimmering within it. A 9th mag field star is just south of the cluster core and west of the mag 7.3 star.

NGC 654 is 9,200 light-years distant and a member of the sprawling Cassiopeia OB8 association with M103 and NGCs 659 and 663. The mag 7.3 star seems to be a true cluster member because it is an F5 Ia supergiant with an absolute mag near -7.5. Including this star, the cluster's integrated apparent mag is 6.5 and its integrated absolute mag -8.3, the latter corresponding to a luminosity of 174,000 Suns (though half of this is from the F supergiant alone). The main sequence of NGC 654 goes up to spectral type B0. This suggests a cluster age less than the 20–25 million years of the rest of the association, though such internal age differences are the rule rather than the exception for associations. NGC 654 is obscured a rather healthy 2.5 mags by interstellar dust, comparable to the 2.7 mags of obscuration suffered by the nearby NGC 663 but a whole magnitude greater than the obscuration of the other bright Cas OB8 cluster, M103 1½° to the SW. The extra extinction probably is due to outlying dust of the Cassiopeia OB6 complex, centered to the east of Cas OB8 and somewhat nearer to us.

NGC 659: The small (5' diameter) and rather faint (magnitude 7.9) Cassiopeia OB8 open cluster NGC 659 is conveniently located just ½° SSW of the large and bright Cas OB8 group NGC 663 and just a few minutes NE of the 6th mag star 44 Cassiopeiae. It is, like NGC 654, a compact gathering of faint stars, but lacks that group's bright lucida. The high power of giant binoculars is necessary to get the cluster sufficiently distant from the glare of 44 Cas that its little hazy spot can be seen. The integrated absolute mag of NGC 659, −6.3, corresponds to a luminosity of 28,000 Suns.

NGC 663: The three Cassiopeia OB8 association open clusters NGC 654, NGC 659, and NGC 663 are all within an area only 1¼° across and therefore can be seen simultaneously in all binoculars and wide-field telescopes. NGC 663, located 2.7° ENE of Delta (δ) Cassiopeiae, is just a little east of the midpoint of the N-S line joining NGC 654 and NGC 659. The latter two clusters are compact hazy patches difficult or invisible in conventional binoculars. NGC 663, however, is easily spotted in 10×50s as a field condensation, about ¼° across, of a half dozen 9th magnitude stars distributed in coarse lumps without the background haze of faint unresolved cluster members. In 15×100 supergiant glasses NGC 663 has a rich, well-resolved appearance. The cluster's four brightest stars form a wide NE-SW "double-double," the NE pair oriented E-W and the SW pair NW-SE. Most of the fainter members are around and just SE of these brighter star-pairs, and a small cluster subgroup is semi-detached to the west. There still is no background haze. And in fact NGC 663, contrary to its low-power appearance, is not particularly populous, and even at just 25× begins to look more like a field condensation of bright stars than a true cluster. Nevertheless the group is an appealing sight at about 50× in 8-inch telescopes, its stars distributed in loose groups and ragged chains around the six brightest stars, five of which are blue-white and the sixth ruddy-oramge.

Brightest Members of NGC 663

Star	m_V	Spectrum	A_V	M_V
BD+60°339	8.50	B6 Iab	2.6	−6.4
BD+60°333	8.91	B5 Iab	3.0	−6.4
BD+60°331	8.95	B8 Iab	3.1	−6.4
BD+60°351	9.03	B6 Iab	2.6	−5.8
BD+60°336	9.09	B9 Iab	2.7	−5.8
BD+60°335	9.15	M3 Ia-Iab	2.2	−5.4

(From Garmany and Stencel, *Astron. Astrophys. Suppl.* 94, 232 and 237, using the distance from Humphreys, *ApJ Suppl.* 38, 331.)

The six brightest stars of NGC 663 are all Iab supergiants, five of them blue mid-to-late B-types and the sixth a red M-type. The statistics of these interesting stars are given in the accompanying table, the absolute mags computed assuming a distance of 9,200 light-years to the cluster. Notice the heavy obscurations. The cluster-wide average extinction is 2.7 mags. The group's integrated absolute mag of −7.9 corresponds to a true luminosity of some 121,000 Suns. Its size is a very healthy 43 l-y. Like most of the rest of Cas OB8, NGC 663 is at least 20 million years old.

NGC 1027 (Plate XXVI, Photo 2.7): One of the most interesting regions of Cassiopeia is north and NE of the Perseus Double Cluster (NGC 869 + NGC 884, discussed in the section on Perseus). This area contains two large emission nebulae, IC 1805 and IC 1848, which are involved with the Cassiopeia OB6 stellar association. IC 1848 is surprisingly easy in 10×50 binoculars, and IC 1805 can be glimpsed with supergiant glasses; but both are too large and too tenuous for conventional telescopes. They are described in Section 5 of Chapter 3.

In the general area between the two Cas OB6 emission nebulae, 1½° NW of the NW edge of IC 1848 and 1½° due east of the center of IC 1805, is the large (20' diameter), rather bright (magnitude 6.7), but low-surface-brightness open cluster NGC

Photo 2.7 Open clusters NGC 1027 (upper left) and Mel 15 (right center) in Cassiopeia

1027. This group can be seen in 10×50 binoculars as a very pale haze around a 7th mag star, but is difficult to find because this region of Cassiopeia has no bright stars to use as guides. Indeed, the nearest conspicuous object to NGC 1027 is the Double Cluster itself, 5° to the SW. In 15×100 supergiant glasses, however, the cluster is quite noticeable, a fairly crowded gathering of 9th, 10th, and 11th mag stars around the mag 7 central star, their comparable brightness suggesting richness. With 25×100 instruments, some two dozen NGC 1027 stars can be counted in a fairly large area around the 7th mag lucida. They are in straggling knots of three, four, or five stars rather than evenly distributed.

NGC 1027 is not physically involved with the IC 1805 + IC 1848 complex: the cluster is about 3,300 light-years from us whereas Cas OB6 is some 7,800 l-y away. Nor is the mag 7 star the true lucida of the cluster: the brightest actual member of NGC 1027 is a mag 9.3 object. The integrated absolute mag of the cluster, –4.6, corresponds to a luminosity of 6,400 Suns. NGC 1027 has the look of a rather evolved group.

NGC 7788: Beginning 2° NW of Beta (β) Cassiopeiae and extending another $1^1/_2$° NW toward the magnitude 5.4 6 Cas is a peculiarly straight line of five open clusters, Berkeley 58, NGC 7790, NGC 7788, Harvard 21, and King 12. The question of course is whether or not this is a true physical structure or merely a chance alignment. The available data for some of these clusters is rather crude, but the answer seems to be that two of the groups, NGC 7788 and K 12, are probable members (with 6 Cas itself) of the large Perseus Spiral Arm stellar association Cassiopeia OB5, but the other clusters are just in the field and very likely well beyond Cas OB5. Be 58 will not be discussed because it is strictly a large telescope cluster: indeed, its very faintness (mag 9.7, with stars only as bright as mag 11.9) suggests that it is extremely remote.

NGC 7788 is a 9th mag cluster 9' in diameter located just over 1° SE of 6 Cas. It is $^1/_4$° NW of the much larger and brighter NGC 7790, about 40% of the distance from NGC 7790 toward a conspicuous mag 7 field star. NGC 7788 is not an easy cluster to spot even in large binoculars, appearing in 15×100s merely as a ragged field concentration of mag 10 and 11 stars around a lucida of about mag 9$^1/_2$. Telescopes show it as compact and moderately rich.

The membership of NGC 7788 in the 8,200 light-years distant Cas OB5 association is not beyond doubt. However, the cluster's distance has been estimated to be 7,800 l-y and its main sequence goes up to spectral type B1, meaning that it is a rather young group and thus a good candidate for association membership. On the other hand, its 0.8 magnitude of dimming by interstellar dust is low for a cluster in this direction and of the distance of Cas OB5.

Cassiopeia OB5 covers an area approximately 4°=570 l-y across centered somewhat SW of 6 Cas. It contains some unusual stars. The mag 5.42 6 Cas is one of the very rare A-type hyper-supergiants, its spectrum A3 Ia$^+$ and its absolute mag –8.7 (a luminosity of a quarter million Suns). The association has a second star of this type, the mag 6.93 HD 223960, which has an A0 Ia$^+$ spectrum and an absolute mag of –7.4. However, the brightest member of Cas OB5 is in fact the 4th mag variable Rho (ρ) Cas, located 4$^1/_2$° SSE of the center of the association. This interesting object has had magnitude extremes of 4.1 and 6.2, and its color and spectrum vary as well as its luminosity, though not in synch: when Rho Cas is at its hottest, near spectral type F8, its color is excessively red; but when it cools down to spectral type M5, its color remains too blue. All this is the consequence of an expanding gas shell around the star, continuously replenished as Rho Cas sheds mass in its evolution into a red supergiant. The bolometric absolute magnitude of the star, –9$^1/_2$ (a total energy output of some half million Suns), seems to be the peak possible for yellow F, G, and K type Ia supergiants.

NGC 7789: The most beautiful of Cassiopeia's many open clusters is NGC 7789, easily spotted between Rho (ρ) and Sigma (σ) Cassiopeiae. Its 25′ diameter, magnitude 6$^1/_2$, glow is obvious in 7×35 binoculars. In 10×50s it has a grainy texture from the partial resolution of its brightest stars, 11th mag objects. With 15×100s these stars, most of which are on the cluster's west side, are fully resolved, and they ornament a large, modest-surface-brightness circular patch of haze richly sprinkled with uncountable momentarily-resolved star-sparks. The extra magnifying power of 25×100s improves resolution further, a multitude of star-specks being evenly distributed over a large, slightly N-S elongated, area. In any size instrument, the delicate, sparkling beauty of NGC 7789 is best appreciated with a long lingering look to give the eye the chance to acclimatize to the light conditions in the field of view and to become sensitized to the scintillating of the multitude of momentarily visible cluster members.

NGC 7789 is rather distant, 5,900 light-years away, but not as far as the Perseus Spiral Arm. It is in fact a highly-evolved, 2 billion year-old cluster presently in the inter-arm gap between our Orion-Cygnus Arm and the Per Arm. It is large, 43 l-y across, and very bright for a group its age, its integrated absolute mag of –5.4 corresponding to a luminosity of 12,000 Suns.

NGC 7790: The brightest and largest of the five groups in the open cluster chain NW of Beta (β) Cassiopeiae is NGC 7790, 2$^1/_2$° NW of Beta and just west of a 6th magnitude field star. It is a 17′ long oval, but only of mag 8.5. Hence its surface brightness is rather low and it requires larger binoculars to be seen as a small patch of haze with a few of the cluster's brightest members, mag 10–11 stars, resolved. The group's mag 10.2 lucida is on the NW side of the oval. It is a close binary, its components both around mag 11.0 and separated by 2.4″ in PA 264°. The remarkable thing about this system is that both stars are Cepheid variables (designated CE Cassiopeiae a and b), an unusual if not unique situation. In fact NGC 7790 has four Cepheid variables: the other two are CF Cas, which has a range of mags 10.82 to 11.43 in 4.875 days and is 1′ ENE of the CE binary; and the faint CG Cas, which varies a couple tenths mag around 11.3 in 4.37 days and is 2$^1/_2$′ SE of the cluster's core ellipse. The two CE Cas stars are likewise short-period Cepheids (5.14 and 4.48 days).

The Cepheid period-luminosity relation gives a distance to NGC 7790 in excess of 9,500 light-years. Thus the cluster is a very respectable 44 l-y in diameter and has an integrated absolute magnitude (corrected for 2 mags of absorption) of at least –5.8, a luminosity approaching 20,000 Suns. The short-period Cepheids in it imply a cluster age of around

70 million years. NGC 7790 is located within the Perseus Spiral Arm, but is a little too old to be considered a true spiral arm tracer.

H 21 and K 12: Two of the fainter and smaller groups in the open cluster chain NW of Beta (β) Cassiopeiae are Harvard 21 and King 12, both 9th magnitude objects. K 12 is just $^1/_2$° SE of the mag 5.4 6 Cas; and H 21 is $^1/_4$° SE of K 12 midway between a $^1/_4$°-wide pair of 7th mag stars. The physical reality of H 21 has been doubted; and indeed in 15×100 supergiant binoculars all that can be seen is a tiny group of two or three intermittently visible 10th mag stars, too loose to be called a "knot." It is little better in telescopes.

K 12 is a much more promising cluster candidate, appearing in 15×100 glasses as a close NNE-SSW pair of mag 10$^1/_2$ stars, a nice double because tight but cleanly separated, with two or three 11th mag attendants. K 12 is estimated to be about 8,100 light-years away, about the same distance as the Cassiopeia OB5 association, and only 10 million years old, an appropriate age for an OB association cluster. The amount it is dimmed, 1.9 mag, is also consistent with membership in Cas OB5. 6 Cas to its NW, the second brightest Cas OB5 star, is obscured about the same as K 12, 2.1 mags.

Mel 15 (Plate XXVI, Photo 2.7): At the center of the huge (145'×100') but very faint emission nebula IC 1805, located in eastern Cassiopeia 4° NNE of the Perseus Double Cluster, is a $^1/_2$° scattering of nine stars from magnitude 7.8 to mag 10.3, the open cluster Melotte 15. This is not a group that calls attention to itself: even in giant binoculars it appears as a field condensation only by virtue of comparison with the very poor star field around it. In 25×100 glasses it is a coarser, knottier, thinner version of the similar-sized NGC 1027 1$^1/_2$° to its east. The two clusters are an interesting visual contrast pair and conveniently fit in the same field of view.

And the two are an interesting astrophysical contrast pair as well. NGC 1027 is an evolved open cluster, but Mel 15 is very, very young. In fact, the chief interest of Mel 15 are its very young, very hot, very luminous stars, the source of the copious quantities of ultraviolet radiation fluorescing the 330×230 light-year IC 1805. The lucida of Mel 15, the mag 7.81 HD 15558, is an O5 IIIf giant with a bolometric absolute magnitude of −10.6, corresponding to a total energy output of 1.45 million suns. The second brightest cluster member, the mag 8.10 HD 15570, is an even *more* luminous O4 If supergiant: its bolometric absolute mag is estimated to be −10.9, an energy output of 2 million Suns. The cluster also contains one O5, one O6, and two O7 main sequence stars, and a B0.5 Ia supergiant. However, all these stars are dimmed from 2.2 to 3.1 magnitudes by interstellar dust; otherwise this cluster, though thin, would look much more impressive than it does.

Mel 15 is so scattered and large, 50 l-y across, that it perhaps is better thought of as the core of the Cas OB6 association, which includes the 2°×1° emission nebula IC 1848 SE of IC 1805, a good 10×50 binocular object. IC 1848 is fluoresced chiefly by the mag 7.06 HD 17505, an O6.5f star with a bolometric absolute mag equal to that of Mel 15's HD 15558, −10.6. Cas OB6 stars are scattered across a 6$^1/_2$°~900 l-y long area parallel to the galactic equator. In fact, Cas OB6 is along the same stretch of the galactic equator as the Perseus Double Cluster and its giant association, Perseus OB1. The lucida of Cas OB6 outside IC 1805 and IC 1848 is a mag 7.67 O9.5 IIn giant with a very respectable bolometric absolute mag of −10.3. Cas OB6 also includes an O9.7 Ib supergiant and an O5 star with a bolometric absolute mag of −9.5. Thus this is a very, very young, if somewhat thin, association. But the star field here is so poor even in wide-field instruments that, apart from glimpses of the two giant emission nebulae, it does not give away what you are actually looking at.

St 2 (Photo 2.8): One of the best celestial sights in Cassiopeia is the huge (1° in diameter) and populous open cluster Stock 2. This group is often overlooked because of the Perseus Double Cluster, which is only 2° to its SSE. But in 10×50 binoculars St 2 is a stunning swarm of dozens of 8th, 9th, and 10th magnitude stars evenly spread over an area more than twice the diameter of the Moon. In 15×100 glasses at least 80 stars can be counted, their smooth distribution enhancing the cluster's look of richness. And St 2 is indeed so rich that it is at its best with the extra magnification of 25×100 instruments.

St 2 looks so large in part simply because it is relatively nearby, just 1050 light-years away. Thus its true diameter is a respectable, but not impressive, 18 l-y–less than a third the true size of each component of the Double Cluster. Its integrated absolute magnitude, −4.2, corresponds to a luminosity of 4000 Suns. But the cluster's brightest members have absolute mags only between −0.5 and 0, suggesting that St 2 is a rather evolved group. The published age of 170 million years is consistent with St 2's appearance and modestly-luminous lucidae.

II.4 Perseus

In Perseus the Milky Way is at its faintest and narrowest anywhere. This is because most of the constellation is covered by relatively nearby (500 to 1500 light-year distant) clouds of interstellar dust and gas. These clouds, which contain the California Nebula, NGC 1499, one of Perseus' best binocular and RFT objects (described in Sect. 6 of Chap. 3), extend SE over Taurus. However, in front of the Perseus dust is a splendid wide-field open cluster, the Alpha (α) Persei Group.

And the clouds do not quite cover the whole of Perseus. The Cassiopeia Window out to the Perseus Spiral Arm extends into NW Perseus almost as far as Eta (η) Persei. In this extension of the Cas Window we see the famous Perseus Double Cluster, NGC 869 + NGC 884, which has given its name to the Perseus Spiral Arm

But the Double Cluster and the Alpha Per Group are not the only wide-field open clusters in Perseus. Near the SW corner of the constellation (away from the dust clouds) is M34, which is as bright and large as either component of the Double Cluster and, like them, visible to the unaided eye. And around Lambda (λ) Persei on the opposite side of the constellation (and of the Perseus dust clouds) from M34 is a NW-SE, $1\frac{1}{2}$°-wide pair of mag $6\frac{1}{2}$ clusters, NGC 1528 and NGC 1545, with another cluster, NGC 1444, 5° to their west. NGC 1528 is large and just at the limit of naked-eye visibility. But NGC 1444, by contrast, is small, compact, and star-poor–a telescope rather than a wide-field instrument object. However, NGC 1444 is of wide-field interest as one of a trio of compact open clusters all with B0/B1 lucidae, in the huge but sparse stellar association Camelopardalis OB1, the other Cam OB1 clusters being NGC 957 near the Double Cluster and NGC 1502 in SW Camelopardalis.

Per OB3, the Alpha (α) Persei Group (Photo 3.3): Even with the unaided eye, a tiny spray of faint stars can be seen sprinkled SE of Alpha toward Delta (δ) Persei. This group is the Alpha Persei Cluster, and has been designated Melotte 20 and Perseus OB3 (though it has no O-type members). Because the physical reality of this cluster was originally esta-

Perseus Open Clusters

Cluster	RA (2000.0)	Dec	m_V	Size	m_V*	*Spectrum	Distance	A_V	M_V	True Size	Age
Per OB3	03^h28^m	+49°		3.5°	1.79	F5 Ib	540 l-y	0.3		33 l-y	50 mil
M34	02 42	+42 47′	5.2	35′	7.33	B8	1500	0.3	−3.4	15	180
N 869	02 19	+57 09	5.3	30	6.55	B3 Ia	7200	1.6	−8.0	63	6.4
N 884	02 22	+57 07	5.1	30	6.42	A1 Ia	7500	1.9	−8.6	65	14
N 957	02 34	+57 32	7.6	11	8.33	B1 II	3300	2.4	−4.8	11	
N 1245	03 15	+47 15	8.4	10	11.2		7400	0.9	−4.3	21	1000
N 1342	03 32	+37 20	6.7	14	8.44	F0 III	1780	0.9	−2.9	7.3	660
N 1444	03 49	+52 40	6.6	4	6.72	B0.5 III	3300	2	−5.4	3.8	
N 1513	04 11	+49 31	8.4	9	11.2		2600	1.8	−2.9	6.8	
N 1528	04 15	+51 14	6.4	23	8.75	A0 V	950	1.0	−2.0	6.3	270
N 1545	04 21	+50 15	6.2	18	7.13	K5 III	2350	1.2	−4.3	12	
Tr 2	02 37	+55 59	5.9	20	7.4		1900	1.0	−3.9	11	

blished by common proper motion and radial velocity studies, which confirmed the identical true space motion through the Galaxy of these stars, it is also sometimes called the Alpha Persei Moving Group.

The core of the cluster is a $3^1/_2$°-long oval from 29+31 Per on the NW to Psi (ψ) Per on the SE. This is a perfect 7× or 10× binocular star field. The beauty of the view is enhanced by the color contrast of the cream-white magnitude 1.8 Alpha, an F5 Ib supergiant, with the silver-blue of Delta, a mag 3.0 B5 III giant that seems to be a true cluster member, and the orange of mag 4.5 K0 Sigma (σ) Per (not a true Alpha Per Group star). In 25×100 supergiant instruments only the cluster core from 29 to 34 conveniently fits in the field of view; but it is a fine sight, hardly inferior to the Pleiades. Delta, Sigma, and Psi Per comprise another good 25× field, though not as rich as the cluster core.

Some 100 stars from mag 1.8 down to mag 11 have been identified as Alpha Per Group members, including 30 Per some $5^1/_3$° SSW of the cluster core and even mag 2.9 B0.5 IV Epsilon (ε) Per, 10° to the SE. The distance to the group is around 540 light-years. Thus the length of the cluster core is about 33 l-y. Alpha Per itself has an absolute mag of −4.6, a luminosity of 5800 Suns. The group's main sequence goes up to type B3, implying an age of 50 million years. It has over 21 B-type main sequence members from mag 4.66 to mag 7.37.

M34: The open cluster M34 is easily found 5° due east of Gamma (γ) Andromedae and 4° WNW of Algol (Beta [β] Persei). It is very large, over a half degree in diameter, and very bright, with an integrated apparent magnitude of 5.2. Under good sky conditions it is visible to the unaided eye as a dim hazy spot. Standard binoculars show a central core, about 10' across, of seven or eight stars of mags $7^1/_2$ to $8^1/_2$, within a loose but uniform distribution of fainter cluster members.

In supergiant binoculars M34 displays a very intriguing structure. 15×100s show the central core to be a partial ellipse, open to the ESE, of about eight mag 8 to 10 stars, the northernmost being the 20" wide E-W double of mag $9^1/_2$ stars h 1123. Within the ellipse are only two or three very faint stars. The ellipse is enclosed by a large "box" of about a dozen mag $7^1/_2$-10 stars, the cluster's mag 7.3 lucida being at the midpoint of the SE side of the "box." A scattering of fainter cluster members is between the core ellipse and the framing "box." 25× might be near the maximum magnification for the cluster's best aesthetic effect.

M34 is around 1500 light-years away and about average in size (15 l-y) and brightness (absolute mag −3.4 = 1900 Suns) for an intermediate-age (180 million-year-old) open cluster. The cluster core is 4 l-y across–about the distance between the Sun and Alpha (α) Centauri. The lucida is a B8 V star with an absolute mag of −1.3, a luminosity of 275 Suns.

NGC 869 + NGC 884, the Perseus Double Cluster (Photo 2.8): Each component of the Perseus Double Cluster is as large as the Moon (30' in diameter) and has an integrated apparent magnitude of around 5.0. The two clusters are only $^1/_2$° apart center-to-center. Traditionally the western cluster, NGC 869, has been called "h Persei" and the eastern group, NGC 884, "Chi (χ) Persei." These designations go back to Renaissance star charts; for both groups are visible to the unaided eye as a pair of hazy spots within an area of more diffuse Milky Way glow. The ancient Greek astronomers called the Double Cluster the *Harpe*, "Scimitar," of Perseus.

Both NGC 869 and NGC 884 resolve well with the slightest optical aid because they each have at least a dozen mag $6^1/_2$-9 members. And the surrounding Milky Way is likewise rich in bright stars, with over 30 between mags 5.2 and 8.2. Indeed, the two clusters merge imperceptibly into the star field around them. What distinguishes them from the surrounding star field is their slight concentration toward a central star-knot, behind which can be seen a slight haze of unresolved cluster members.

The Double Cluster is at its best aesthetically in large-aperture wide-field instruments like supergiant binoculars which gather the maximum amount of light but allow both clusters, and much of the surrounding star-field, to be seen. 25×100 glasses perhaps give the finest possible view. A curious feature of the field is a conspicuous $1^1/_2$°-long star-arc that begins just north of NGC 869 and curves gently NNW toward the rich, 1°-diameter,

Photo 2.8 The Milky Way around the Perseus Double Cluster (center). The emission nebula IC 1805 in Cassiopeia is in the upper left (NE) corner of the field. The large but loose concentration of stars midway between the Double Cluster and the upper edge of the field is the open cluster St 2

cluster of 8th, 9th, and 10th mag stars, St 2 (described in Cassiopeia).

The question has always been whether NGC 869 + NGC 884 is a true physical double of two clusters at about the same distance from us, or merely the chance alignment of a nearer with a farther group. Though virtually identical in size, brightness, and appearance (structure), the components of the Double Cluster do have significantly different bright-star populations: the dozen brightest members of NGC 884 are all blue early-B giants and supergiants; but eight of the nine brightest stars of NGC 869 are either white late-B/early-A, or red M-type, supergiants. This suggests NGC 869 is a few million years older than NGC 884. However, the uncertainties in the distance estimates for the two clusters are too great to determine which might be nearer. Traditionally it has been thought that NGC 884, especially because it is slightly more dimmed by interstellar dust than NGC 869, might be a bit farther. But the extra extinction could be simply because our line of sight to NGC 884 is through a little more foreground Perseus cloud dust.

Assuming both components of the Double Cluster to be 7,200 light-years away, their true diameters are around 65 l-y and their integrated absolute mags between –8 and –8½ (luminosities around 160,000 Suns). But these are minimum values because both groups blend into the surrounding star-field, which is the major Per Spiral Arm association Perseus OB1. This association covers a huge 8°×6° area, corresponding to true dimensions of at least 1000×750 l-y, and is one of the largest associations in our quadrant of the Galaxy. Its brightest star is the mag 5.17 9 Persei, which has a peculiar A2 Iab shell spectrum and an absolute magnitude of a very high –7.8 (120,000 Suns). Two other Ia A-type supergiants in the association also have absolute mags of –7.8. Per OB1 includes at least a half dozen M-type Iab red supergiants, among them the variables SU Per (mag 7.5v; M3.5 Iab) KK Per (mag 7.7v; M2 Iab), and PR Per (mag 7.8v; M1 Iab). One O5 If$^+$ supergiant is credited to the association; otherwise Per OB1 has no O-type stars. The association's brightest B-type Ia supergiants are 10 Per (mag 6.2; B2 Ia) and 5 Per (mag 6.3; B5 Ia). All this suggests that Per OB1 is a somewhat evolved association, which is confirmed by the lack of emission nebulae anywhere near the Double Cluster: all the gas and dust residual from the giant molecular cloud in which Per OB1 was born has been blown out of the vicinity by stellar winds, radiation pressure, and supernova shock waves.

NGC 957: About 1.6° ENE of the center of the NGC 884 component of the Perseus Double Cluster is NGC 957, a rather small (10') group dominated by a magnitude 8.3 lucida. A 10'-long arc of three mag 9–10 stars curves SW away from the core of the cluster, giving it a conspicuous "comma"-shaped look. The comma is obvious in 10×50 binoculars. In 15×100 glasses a few tiny star-sparks can be resolved in the cluster haze around the lucida. Telescopes reveal that the lucida is in fact on the SE corner of the cluster core, which is elongated E-W.

Telescopes also reveal that the lucida is a double, the mag 9.9 secondary 23" distant in PA 19° (NNE) from the primary. This star is a hot young B1 II giant with an absolute mag of –4.1 (a luminosity of 3600 Suns). Given that the star is dimmed 2.4 mags by interstellar dust, the distance to NGC 957 must be around 3,300 light-years. In fact the cluster is one of three groups in the sprawling Camelopardalis OB1 association, the other two being NGC 1444, described below, and NGC 1502, described in the section on Camelopardalis. All three clusters are small and under-populated, with lucidae that are blue B0/B1 giants in binary systems. NGC 957 is only about 10 l-y long, with a total cluster luminosity just twice that of its B1 giant.

NGC 1245: NGC 1245, located 3° SW of Alpha (α) Persei, is neither large, just 10' across, nor bright, only magnitude 8.4, but can be seen even in 10×50 binoculars as a small hazy patch. Its intrinsic richness becomes evident with 25×100 instruments, in which it appears as a moderately large, roughly oval, area of haze crowded with too many faint or intermittently-resolved stars to be counted. But even when thus partially resolved the cluster is an attractive sight. It looks distant and indeed is–7,400 light-years away. Its true size is a respectable 21 l-y, and its integrated absolute mag, corrected for 0.9 mag of obscuration, is –4.3, a luminosity of 4,400 Suns.

NGC 1245 is of particular interest for its great age, 1 billion years. In telescopes it has the look typical of such evolved groups: a slightly centrally-concentrated swarm of several dozen similarly-bright stars.

NGC 1342: Just over halfway from Zeta (ζ) Persei to the famous variable Algol (Beta [β] Per) is the moderately large (14' long), moderately bright (magnitude 6.7), but coarse and scattered open cluster NGC 1342. Its stars, 17 of which can be counted in 25×100 binoculars, are mostly in clumps along an E-W band. On the west end of the band is a shallow arc, open to the NNE, of four mag 9–10 stars. On the east end is a shallow arc of three more mag 9–10 stars opening NW. North of this end of the cluster-band is a conspicuous NNW-SSE pair of 8th mag stars. A few fainter cluster members can be glimpsed scattered within and around these features. However, the cluster is simply too loose and clumpy to be very attractive.

NGC 1342 is about 1780 light-years away. Thus it is only 7 l-y long. Its mag 8.44 lucida is a yellow F0 III giant. The cluster is estimated to be about the same age as the Hyades, 650-700 million years old. Its integrated absolute mag, –2.9 (corrected 0.9 mag for dust-dimming), is also about the same as that of the Hyades.

NGC 1444: 4.4° NE of Alpha (α) Persei is the triple star Σ446, which consists of a magnitude 6.7 primary with a mag 9.1 secondary 8.6" away in PA 253° and a faint mag 12 third component 12" distant in 39°. This triple system is the main feature of the under-populated open cluster NGC 1444. In 25×100 supergiant binoculars only three or four very faint cluster members can be glimpsed around Σ446, and not many more are added in telescopes. The main interest of NGC 1444 is that, with NGC 957 near the Perseus Double Cluster and NGC 1502 due north in Camelopardalis, it is in the 3,300 light-year distant Camelopardalis OB1 stellar association (described in more detail under NGC 1502 in the next section). The lucidae of all three clusters are luminous B0/B1 giants in binary or multiple star systems, and all three clusters are under-populated. The mag 6.72 lucida of NGC 1444 is a B0.5 III giant with an absolute mag of –5.3 (a luminosity of 11,000 Suns). The cluster as a whole has an integrated absolute mag of just –5.4 (only 0.1 mag greater than that of its lucida) and is less than 4 l-y across.

NGC 1513: Three open clusters lie within 2° of Lambda (λ) Persei: NGC 1513, NGC 1528, and NGC 1545. All three fit in the same field of low-power binoculars. Of the three, the smallest (only 9' in diameter) and faintest (magnitude 8.4) is NGC 1513, 2° SSE of Lambda. Even in supergiant glasses it just begins to resolve into a half-dozen 10th and 11th mag stars embedded in a moderately large area of faint, granular, background glow. Its main interest is that it shares the same binocular field with two other, more visually impressive, clusters.

NGC 1513 is faint in part because it is dimmed 1.8 magnitudes by foreground Perseus dark cloud complex dust–more than twice what would be expected for a cluster of its distance, 2600 light-years. However, intrinsically it is a small and faint group anyway, with a diameter of less than 7 l-y and an integrated absolute mag of only –2.9 (1250 Suns). Its brightest member has an absolute mag only around 0, so NGC 1513 must be a rather old cluster.

NGC 1528: Visually the best of the three NGC open clusters in the area of Lambda (λ) Persei is NGC 1528, centered just $^1/_4$° to the star's NE. This group is larger and brighter than NGC 1513 and more populous than NGC 1545. It is 23' across, has an integrated apparent magnitude of 6.4, and partially resolves even in 10×50 binoculars. With 25× supergiant glasses perhaps three dozen mags 8–10 stars can be seen. Resolution of the cluster is good at this power because the group, though rather rich, is not concentrated toward its center.

NGC 1528 looks large and bright basically because it is so near to us, only 950 light-years away. In reality, however, it is small, only 6$^1/_2$ l-y across, and very faint, its integrated absolute mag of –2.0 corresponding to a luminosity of a mere 275 Suns and the same as that of the large but sparse Coma Berenices Star Cluster (which is about 15 l-y across). The mag 8.15 lucida of NGC 1528 is an A0 main sequence object, implying a cluster age of around 270 million years. The 1.0 magnitude of absorption suffered by the cluster is about three times the average for clusters of its distance and is the consequence of foreground Perseus cloud dust.

NGC 1545: Just over 2° east of Lambda (λ) Persei is NGC 1545, a magnitude 6.2 open cluster 18' in diameter and therefore slightly brighter, and almost as large, as NGC 1528 to its NW. However, NGC 1545 is not such a good sight as NGC 1545 because it is looser and coarser, with fewer stars and those stars of a ragged variety of magnitudes rather than uniformly bright. The lucida of the cluster is a mag 7.1 K5 III orange giant with an 8th mag companion, easily resolvable at 25×, 73" away in PA 330°. On the north edge of the group is a mag 7.9 K2 star with a mag 9.4 secondary 18" distant in PA 346°. The distance of 2,350 light-years published for the cluster no doubt is too high if the K5 giant is a true member: it implies an absolute mag for the star of −3.4, about 3 mags too much for a K-type normal giant. Either the K5 giant is in the foreground of the cluster, or the cluster is more like 1200 l-y away. The latter scenario reduces the cluster's size to 6 l-y and its integrated absolute mag to −2.9–both values comparable to those of its NGC 1513 and NGC 1528 neighbors.

Tr 2: 2° SE of the Perseus Double Cluster, and the same distance due west of Eta (η) Persei, is an approximately 20' long jagged E-W line of nine stars of magnitudes $7^1/_2$ to 9 accompanied by several lesser luminaries. This conspicuous binocular group has been catalogued as the open cluster Trumpler 2. Its published distance of 1,900 light-years puts it well in the foreground of the Double Cluster. It probably is an evolved group presently drifting through the Cassiopeia Window. The true length of Tr 2 is 11 l-y, and its integrated absolute mag of −3.9 corresponds to a luminosity of 3,200 Suns. The cluster's rather high extinction of 1.0 mag is a consequence of its proximity to the western edge of the Perseus dust clouds.

II.5 Camelopardalis

The huge north circumpolar constellation of Camelopardalis has no bright stars nor indeed any conspicuous star-pattern: it is more just a named area of the sky than anything else. It does, however, contain several objects that are good targets for binoculars and wide-field telescopes. Two of these are galaxies, IC 342 and NGC 2403, which will be described in Chap. 4.

Camelopardalis is not usually thought of as a Milky Way constellation, but in fact the galactic equator cuts through its SW corner. The Milky Way glow here is not very bright, nor the star fields exceptionally rich; but there is a remarkable 3°-long chain of some 20 6th to 10th magnitude stars in the Camelopardalis Milky Way (described in Section 5 of the next chapter), with a very interesting open cluster near its SE end.

NGC 1502: The best way to find the rather small (7') but bright (magnitude 5.7) open cluster NGC 1502 is to look for the above-mentioned 3°-long star-chain in SW Camelopardalis, which is *very* conspicuous in a binocular scan WNW of Beta (β) Camelopardalis, a mag 4.2 star due north of Capella. NGC 1502 is NE of the SW end of the star chain. The cluster is 6.7° WNW of Beta Cam and the same distance SW of Alpha (α) Cam.

The mag 6.7 lucida of NGC 1502 can of course be seen in standard binoculars, but the cluster is too tight to be recognized as such at 7× or 10×. However, in supergiant binoculars it proves to be a fine sight, a "double-double" star accompanied by a handful of fainter attendants. The lucida is the binary Σ485, a pair of silver-blue B0.5 III giants of nearly equal brightness (mags 7.0 and 7.3–7.6) only 18" apart but resolvable even at just 15×. (The variable secondary is the Beta Lyrae type eclipsing binary SZ Cam, which has a period of 2.7 days.) The second double of the cluster, north of the lucida, is a pair of 9th mag stars. These two doubles make

Camelopardalis Open Cluster

Cluster	RA (2000.0)	Dec	m_V	Size	m_V*	*Spectrum	Distance	A_V	M_V	True Size
N 1502	04^h08^m	+62°20'	5.7	7'	6.74	B0.5 III	3300 l-y	1.8	−6.1	6.7 l-y

NGC 1502 one of the most attractive low-population open clusters in the sky.

NGC 1502 is thought to be a member with two similar open clusters in Perseus, NGC 957 and NGC 1444 (both discussed above) of a sprawling 3,300 light-year distant association, Camelopardalis OB1. This stellar aggregation, if it exists, consists (in addition to the three open clusters) of one 4th mag, one 5th mag, two 6th mag, and seven 7th mag stars scattered over a $17^1/_2°\times12^1/_2°$ area from north of the IC 1805 + IC 1848 complex of Cassiopeia to NE of the Alpha Persei Cluster–a region measuring 1060×730 l-y. The one stellar concentration of Cam OB1 (outside the clusters) is the 1°-wide N-S pair mag 4.2 HD 21291 and mag 4.6 HD 21389, B9 Ia and A0 Ia supergiants located 10° due north of Alpha Persei. The association also is credited with one other A0 Ia supergiant, one hot O9 subgiant, and a modest-luminosity K3 Ib yellow supergiant; but most of its bright members are early-to-mid B normal giants, stars up to 100 million years old.

Perhaps the best argument for the reality of Cam OB1 is the uniformity in appearance and constitution of its three possible open clusters, NGCs 957, 1444, and 1502. All are small, sparsely-populated groups that feature one or two highly-luminous B0/B1 II or III giants in double or multiple systems. Indeed, these three groups are more like glorified multiple stars than true open clusters. They are reminiscent of the clusters IC 4996 in Cygnus OB1 and NGC 6823 in Vulpecula OB1.

III.1 Auriga

As the Milky Way passes SE from Perseus, where most of it is obscured by nearby dust clouds, into Auriga, it broadens and brightens. Toward the "pentagon" of Auriga–Alpha (α) + Beta (β) + Theta (θ) + Iota (ι) Aurigae with Beta Tauri–transparency through our Galaxy is very good. (The Perseus dust clouds extend down into central Taurus away from the Milky Way.) However, because the Galactic anticenter is on the Auriga/Taurus border (3½° ENE of Beta Tau), in this direction we look directly away from the star-rich Galactic interior toward the nearest stretch of our Galaxy's rim. Consequently the Milky Way in Auriga (and in neighboring eastern Taurus and western Gemini) is not nearly as bright as it is in such Galactic interior directions as Sagittarius and Aquila. Nevertheless, because of the distance we see across the Galaxy here, central Auriga is fairly rich in open clusters. The best of Auriga's open clusters in any size instrument are the famous Messier trio, M36, M37, and M38. M36 and M38 are particularly good in 25× binoculars, and fit together in the same field of view. All three clusters can be viewed simultaneously in 7× extra-wide-angle glasses. And with 10×50 and 15× giant binoculars two NGC groups can be seen in the same field with M38: the small, compact NGC 1907 just ½° to the large cluster's south; and 2½° SSW the moderate-sized gathering of faint stars NGC 1893, which is embedded in a low-surface-brightness emission nebula, IC 410.

M36: Of the trio of Messier open clusters in central Auriga, M36, located just over 2° east and slightly south of Phi (ϕ) Aurigae, is the smallest and least populated but has the brightest stars and therefore is the easiest to resolve. In 10×50 binoculars its seven 9th magnitude lucidae appear in a tight knot with no background haze. With 15× supergiant glasses at least thirty stars can be resolved, but the cluster is not sufficiently magnified for them to be easily counted. To the NE and SE of the central condensation pairs of brighter stars give the group the look of being cupped to the east. A conspicuous NE-SW double of mag 9½–10 stars is on the cluster's NNW periphery, and a chain of a half dozen mag 9½–10½ stars–probably not actually involved with the cluster–extends from its core toward the SW. In telescopes M36 is striking for the beautiful blue-white brilliance of its seven lucidae.

M36 is estimated to be about 4,100 light-years away–which, by coincidence, seems to be about the distance of M37 and perhaps of M38 (though recent estimates put M38 some 1600 l-y beyond M36). The cluster's lucida is a mag 8.86 blue B2 main sequence star, and its other bright members are all also B2 and B3 main sequence objects. This implies that M36 is young, perhaps around 30 million years old–an age sufficiently low that the cluster might be considered a tracer of our Orion-Cygnus Spiral Arm. And in fact in this area of the Milky Way there is an Ori-Cyg Arm association at about the same distance as M36, Auriga OB1, the lucida of which is the mag 4.76 Chi (χ) Aur, a B5 Iab supergiant with an absolute mag of –7.1. Aur OB1 includes about a dozen mag 6, 7, and 8 stars concentrated in a 7½°×3½°~560×260 l-y area along the galactic equator. Among its brighter members are two red M-type supergiants, two hot late-O giants, and a luminous blue B2 Ib supergiant. However, there is no real evidence that M36 is a member of the association.

M36 is only 14 l-y across, which means that it is intrinsically, as well as apparently, the smallest of Auriga's Messier trio. However, its integrated absolute magnitude, –5.5 (a luminosity of 13,200 Suns), is about the same as M37 and M38.

Auriga Open Clusters

Cluster	RA (2000.0)	Dec	m_V	Size	m_V*	*Spectrum	Distance	A_V	M_V	True Size	Age
M36	05^h36^m	+34°08'	6.0	12'	8.86	B2 V	4100 l-y	1.0	–5.5	14 l-y	20 mil
M37	05 52	+32 33	5.6	20	9.21		4100	0.8	–5.7	24	300
M38	05 29	+35 50	6.4	21	7.9	gG0	5700	0.8	–5.6	31	250
N 1893	05 23	+33 24	7.5	12	9.04	O4 V	15,600	1.7	–7.6	54	1
N 1907	05 28	+35 19	8.2	6	10.46		5700	1.3	–4.3	9	400

M37: The brightest, at apparent magnitude 5.6, and the most populous of the three central Auriga Messier open clusters is M37, located about $6^1/_2$° NE of Beta (β) Tauri and 5° SSW of Theta (θ) Aurigae. It is the most difficult of the three clusters to find because it is far from any bright star. (When you are scanning between Beta Tau and Theta Aur for the cluster, keep in mind that M37 is not quite halfway from Theta Aur to Beta Tau and a little SE of the line joining them.) It is also the most difficult of the three to resolve, because its stars are much fainter than those of M36 and M38. In standard binoculars it is easily seen as a large (20') circular patch of haze of fairly high surface brightness; but only its mag 9.2 lucida resolves. With 15×100 supergiant glasses the group becomes a glitteringly granular circular glow, discernably brighter toward its center, whereat sparkles its lucida. Though only partially resolved in such instruments, the cluster is a fine sight.

M37 is estimated to be about 4,100 light-years distant, the same as M36 $3^1/_2$° to its WNW. (If the two groups are both precisely 4,100 l-y from us–very unlikely–their true separation is nearly 250 l-y.) Its main sequence goes up to spectral type B9, which implies a cluster age of 300 million years. M37 is therefore not only the most populous but also the most evolved of the Auriga Messier open clusters. Its integrated absolute mag, –5.7 (a luminosity of 15,800 Suns), is about the same as the other two groups.

M38 (Plate XXIX, Photo 2.9): The best of the three Messier open clusters in Auriga for binoculars and other richest-field instruments is M38. This group is as large as M37, 20' in diameter, but resolves more easily because its stars are brighter. And it is richer and more spread out than the compact M36. Even in 10×50 binoculars a score of M38 stars can be resolved, scattered over a dim, coarsely-textured haze. In 15×100 glasses at least three dozen stars of fairly uniform brightness (magnitudes 9 to 11) are visible. 25× gets them sufficiently separated to be easily counted. The famous "π"-shape of the cluster's stellar distribution is obvious, though it could also be interpreted as an "X" with short SE and SW legs. The mag 7.9 lucida is conspicuous at the end of the NE leg of the "π". The legs and cross-beam of the "π" are thick star streams rather than thin star chains.

Photo 2.9 Open clusters M38 (above) and NGC 1907 in central Auriga

For a long time M38 was thought to be at about the same distance as M36 and M37, 4,100 light-years. Part of the reason was that the extinction of all three clusters due to interstellar dust is about the same, 0.8–1.0 magnitude. However, the most recent estimate places M38 about 5,700 l-y away. At that distance it would be intrinsically the largest of the three clusters, 31 l-y in diameter, but have exactly the same luminosity as the other two. M38 is an intermediate-age open cluster perhaps as much as 250 million years old.

M38 is on the northern edge of a field rich with binocular objects. Just $^1/_2$° due south of it is the small and faint open cluster NGC 1907 (described below). 2.3° to its SE is M36, the two clusters easily fitting together in the same 10× or 15× field of view. A little over $2^1/_2$° SSW of M38 is the large (20') but remote (15,600 l-y distant) open cluster + emission nebula NGC 1893 + IC 410, a surprisingly easy 10×50 binocular object. Just NW of NGC 1893 + IC 410, and about 2° SW of M38, is an interesting double-arc of bright stars, 16-17-18-19-AQ Aurigae, that at low powers looks like an open cluster. And finally, 2° WSW of M38 is the 6th mag star AE Aur, a "runaway" from the Orion Association, embedded in IC 405, the Flaming Star

Nebula, a supergiant binocular and RFT object. Few areas of even the summer Milky Way are so dense with interesting objects as central Auriga. And this is an excellent field for experiencing a sense of "Galactic depth perspective": AE Aur + IC 405 are about 1500 l-y from us, M36 4,100 l-y distant, M38 + NGC 1907 5,700 l-y away, and NGC 1893 + IC 410 about 15,600 l-y out toward the rim of the Galaxy, which must be only some 5,000 l-y farther.

NGC 1893 (Plate XXIX, Photo 2.10): Even with only 10×50 binoculars a very pale 20'-diameter glow overscattered with a handful of 9th and 10th magnitude stars can be glimpsed $1^1/_2$° SW of the 4th mag Phi (ϕ) Aurigae and 1° SE of 19 Aur. This glow is the emission nebula IC 410; and the scattering of stars, concentrated toward the glow's center, is the open cluster NGC 1893, buried within the nebula. The brightest cluster member is a mag 9.1 star with a super-hot O4 V spectrum. The second and third brightest cluster members are both mag 9.4 stars that have O7 V spectra. The presence of an O4 main sequence star implies that NGC 1893 is less than one million years old.

The NGC 1893 + IC 410 complex is very distant, 15,600 light-years away. It seems to lie on the remote out-trailing arc of the Perseus Spiral Arm. The cluster's 12' apparent diameter corresponds to an impressive true size of 54 l-y, not much less than the true diameter of each component of the Perseus Double Cluster. The nebula's actual size is at least 150 l-y. The cluster's integrated absolute mag, −7.6 (a luminosity of 100,000 Suns), is misleadingly low because mid-O stars like the lucidae of NGC 1893 radiate only a few per cent of their total energy output as visible light. The 1.7 magnitudes of absorption suffered by the cluster is *extremely* low for a group three-quarters of the way out to the rim of the Galaxy.

Photo 2.10 The distant open cluster NGC 1893 and its surrounding emission nebula IC 410

NGC 1893 is the SW subgroup of a major Perseus Spiral Arm association, Auriga OB2. The larger, more scattered, NE subgroup of the association is centered about 2° (~540 l-y) NE of the center of the cluster, just east of Phi Aur. The brightest member of the NE subgroup of Aur OB2 is a mag 8.55 O7 V star with an absolute mag of −6.5. However, this section of the association must be the older half because it contains a couple B0 subgiants, stars which have already begun their evolutionary expansion off the main sequence.

NGC 1907 (Plate XXIX, Photo 2.9): Just $^1/_2$° south of M38 is the small (6') and faint (magnitude 8.2) open cluster NGC 1907. It is visible in 10×50 binoculars as a tiny hazy dot, but so compact that even in 15×100 instruments it remains only a small, moderately-high surface brightness, spot. Just to its south is a close WNW-ESE pair of mag $9^1/_2$ non-members: the cluster's true lucida is just mag 10.5.

NGC 1907 might be near M38 in reality as well as in appearance, for both clusters are estimated to lie 5,700 light-years away. Assuming they are both precisely the same distance from us, their actual separation would be only 60 l-y. However, they are in any case *not* physically related: NGC 1907 is about 400 million years old, 150 million years older than M38. The true diameter of NGC 1907, 9 l-y, and its true luminosity, 4400 Suns, are both less than one-third the size and luminosity of M38.

III.2 Taurus

Southeast of Auriga the winter Milky Way passes behind the horn-tips of Taurus (Beta [β] + Zeta [ζ]Tauri), the feet of Gemini (Gamma [γ] + Eta [η] + Mu [μ] Geminorum), and the club of Orion (Mu + Nu [ν] + Xi [ξ] + the two Chi's [χ] Orionis). It is not so bright here as in central Auriga, though the major dust clouds in this direction are well SW of the Milky Way proper, in central Taurus and around the Belt and Sword of Orion.

Because much of it is covered with the relatively nearby (500 to about 1500 light-year distant) dust clouds of the Taurus Dark Cloud Complex, the constellation of the Bull is not one of the more open-cluster-rich regions along the Milky Way. However, in front of the dust of the Taurus DCC are the two finest binocular and RFT open clusters in the entire heavens, the Hyades and the Pleiades.

The Hyades, Melotte 25 (Photo 3.3): Few of the star-patterns that comprise all or part of the Classical constellations are actual physical star-groupings. A couple of the exceptions include the Belt and Sword of Orion, the bright stars of which are all members of the Orion Association, and the stars in the Head and Heart of Scorpius, all members of the Scorpio-Centaurus Association. The "V" that forms the face of Taurus the Bull is almost entirely composed of members of the Hyades Star Cluster. The main non-member is the ruddy-orange 1st magnitude Aldebaran (Alpha [α] Tauri) in the right eye of the Bull, a K5 III giant, variable between mags 0.75 and 0.95, that is only 68 light-years away, less than half the distance to the center of the Hyades, which modern Hipparcos parallax measurements have fixed at 150 l-y. Aldebaran's proper motion on the celestial sphere is toward PA 161° (SSE) whereas that of the Hyades is toward PA 105° (ESE).

The Hyades is the perfect 7×35 or 7×50 binocular star-group, its $4^1/_2$°-tall "V" fitting comfortably in the fields of such instruments. Even 10× is too much. Theta-one (θ^1) and Theta-two Tau are a $5^1/_2$'-wide naked-eye double. In binoculars they form a fine color-contrast pair, the mag 3.84 Theta-one being an orange K0 III giant and the mag 3.40 Theta-two a snow-white A7 III star. The cluster has three more orange giants, mag 3.65 K0 III Gamma (γ), mag 3.76 K0 III Delta (δ), and mag 3.53 G9.5 III Epsilon (ε). Contrast the bluish-white of the mag 4.29 A2 IV 68 Tau with the oranges of Delta to its SW and of Epsilon to its NE. Aldebaran is a somewhat cooler K-type giant than the K0 stars of the Hyades proper and therefore has a noticeably deeper golden-orange tone.

The "V" of the Hyades is only the core of the cluster. Most of the naked-eye stars in the $10^1/_2$° area from Kappa (κ) Tau on the north to 90 Tau on the south are Hyades members. Even Iota (ι) Tau 7° to the ENE belongs to the cluster. The true size of the "V" is 12 l-y, and the projected distance from Kappa to 90 Tau is 27 l-y. Though it is a bit bizarre to talk about the integrated apparent magnitude of such an extended object as the Hyades, if the light of all the cluster members was consolidated into a single stellar point, it would have an apparent magnitude of 0.5–a third magnitude brighter than Aldebaran. The integrated absolute magnitude of the cluster is –3.0, a luminosity of 1320 Suns. This is a very modest value for an open cluster, and indicates how sparse the Hyades really is. The group's low luminosity is also in part the result of its age, between 650 and 700 million years. It has lost all the really massive, and therefore luminous, stars with which it began.

A number of bright stars around the sky share the true space motion through our Galaxy of the Hyades, among them Capella (mag 0.08; spectrum G6 IV + G2 III), Delta Cassiopeiae (2.7var; A5 IV), Alpha Canum Venaticorum (2.9var; A0 IIp), Lambda (λ) Ursae Majoris (3.45; A1 IV), Omega (ω)

Taurus Open Clusters

Cluster	RA (2000.0)	Dec	m_V	Size	m_V*	*Spectrum	Distance	A_V	M_V	True Size	Age
Hyades	04^h27^m	+16°	0.5	330'	3.40	A7 III	150 l-y		–3.0	14.4 l-y	660 mil
Pleiades	03 47	+24 07'	1.2	110	2.87	B7 IIIn	410		–4.4	13	70
N 1647	04 46	+19 04	6.4	45	8.61	B8 III	2600	1.2	–4.5	33	100

Andromedae (4.5; F4 IV), and Xi (ξ) Cephei (4.6 and 6.5; A3 and dF7). These stars are members of the Hyades Stream, the residual of the original stellar association with which the Hyades was born. And that association must have been quite impressive, because it has *two* open cluster cores, the other being M44, the Praesepe Cluster in Cancer, which has a similar space motion, bright-star population, and age to the Hyades. (M44 is described in Section IV of this chapter.) Because the orbital period at our distance from the Center of the Galaxy is about 220 million years, the Hyades Stream has survived as a recognizable star group for *three* complete orbits around the Galaxy, its member stars preserving enough of their parent giant molecular cloud's original space motion that they remain a discernable (if scattered) stellar aggregation. However, the Hyades Stream is not a spiral arm tracer: it is much too old. It is simply part of the general thin disc population of our Galaxy.

About 100 stars brighter than mag 6.5 scattered around the celestial sphere belong to the Hyades Stream. Some of them are 200 l-y past the Hyades. Others are some 300 l-y *behind* us as we look toward the cluster. Thus we must be near the center of the Stream. The Stream as a whole seems younger than its Hyades and Praesepe open cluster cores because, unlike them, it contains early-A type stars.

The Pleiades, M45 (Plate XXX, Photo 2.11): The most beautiful binocular and richest-field telescope object in the heavens is the Pleiades Star Cluster. What makes it so stunning is not only its crowded bright-star richness, but the almost overpowering silver-blue sheen of those stars, which frosts the field in ice-blue. 10×–15× binoculars of 50–100mm aperture are the perfect Pleiades instruments: more power pushes Pleione and Atlas out of the field of view; and more aperture "over-exposes" the stars on the retina, reducing their color to a simple blue-tinged white.

The Pleiades is also a splendid naked-eye object, particularly from high-altitude or desert sites where the transparent air permits at least a dozen stars to be seen. The ancient Babylonians called the cluster the "Seven Stars" or the "Seven Gods". In ancient

Photo 2.11 The Pleiades

Greek mythology the Pleiades were the "Seven Sisters," daughters of the Titans Atlas and Pleione. The problem is that, under average sky conditions, only *six* Pleiades can be seen; and when you *can* see more than six, you will see at least *nine* (16, 17, 19, 20, 21 + 22, 23, 25, 27, and 28 Tauri).

So what happened to the Seventh Sister? This was a puzzle to the ancient Greeks themselves. In the *Phaenomena* of 270 B.C., the earliest complete work on astronomy that survives from Classical antiquity, Aratos wrote, "Seven are they in the songs of men, albeit only six are visible to the eyes. Yet not a star, I ween, has perished from the sky unmarked since the earliest memory of man..." (Lines 257–260 in G. P. Mair's Loeb Classical Library translation.) Aratos' *Phaenomena* was a versification of a prose work written a century earlier by the great early Greek astronomer Eudoxos of Cnidus. Aratos' statement implies that neither he nor Eudoxos before him had seen, or heard legends about, novae, "new stars." The problem of the lost Pleiad, like the problem of the Star of Bethlehem, will perhaps never be solved.

In addition to its bright star richness and to its beautiful star colors, the Pleiades has two fine binocular doubles. Right in the middle of the "Dipper Bowl" of 17, 20, 23, and 25 Tauri is Σ536, a double of two 8th magnitude stars 41" apart in PA 125° (WNW-ESE) that can be split even with only 7×. And 3' west of the brightest Pleiad, mag 2.9 Alcyone at the east corner of the Dipper Bowl, is the mag $6^1/_2$ 24 Tau, which has mag $8^1/_2$ companions 1' distant to the NW and NNW that are comfortably resolved at 15× despite Alcyone's glare. Due south of Alcyone a delicate chain of six stars almost $^1/_2$° long extends south and SE away from the cluster.

The Pleiades is estimated to be 410 light-years away, the fourth nearest star cluster to us after the Ursa Major Moving Group (71 l-y), the Hyades (150 l-y), the Coma Berenices Star Cluster (260 l-y). Thus the true size of the Pleiades, from 19 Tau on the west to 27 Tau on the east, is 7 l-y. However, the cluster extends well beyond this core group. Even in binoculars the field for more than a half degree around the Pleiades Dipper is noticeably star-rich. One study identified 197 Pleiades in a 2°×2° area centered on the cluster center and another 135 possible members out to a 9°×9° area. This implies a full cluster diameter of at least 12°~80 l-y. No doubt most open clusters are surrounded by such halos of low-luminosity members, but few are near enough for their halos to be studied in detail.

The brightest Pleiades are all late-B giant, subgiant, and main sequence stars. The Pleiades main sequence extends up to type B8 and down to type K2. Cluster members later than type K2 are *above* the main sequence–over-luminous for their color–because they are still in their initial gravitational contraction toward the main sequence, a process that for low-mass stars requires tens of millions of years. It is difficult to get a good estimate for the age of the Pleiades from its upper main sequence. B8 stars have main sequence lifetimes of 180 million years; but the bright Pleiades have unusually high rotational velocities (in the case of Pleione, 100 times the Sun's) and this probably has hastened their evolution by mixing material within them so that they are not really as old as late-B subgiants typically are. The age of the cluster has traditionally been thought to be around 70 million years, but more recent estimates are pushing it up to about 100 million years.

One of the outstanding photographic features of the Pleiades are the delicate wisps of filamentary nebulae around the four stars in the "Dipper Bowl." This is dust reflecting the light of those stars. On color photos these nebulae can be seen to be *bluer* than the stars themselves: this is because interstellar dust *absorbs* red light, and thus the light it reflects away is bluer than the star-light that hit it. The filamentary nature of the Pleiades reflection nebulae is probably the result of strong magnetic fields around the stars caused by their rapid rotation (like the armature of an electric motor): the dust is aligned in the same way that a magnet aligns iron filings. The brightest of the Pleiades reflection nebulae is the patch extending south from Merope, the star at the south corner of the "Dipper Bowl." The Merope Nebula, NGC 1435, can be glimpsed even in 10×50 binoculars (from dark-sky sites) curving first SSE from Merope 10' toward a 10th mag star and then for another 10' due south. Its full photographic length is 30'.

The dust of the Pleiades reflection nebulae is not in fact actually involved with the cluster. It is material of the Taurus Dark Cloud Complex, on the edge of which the Pleiades happens to be passing. The Pleiades stars which are not embedded in reflection nebulae, like Atlas and Pleione at the end of the Dipper's "handle," must be somewhat nearer to us than the stars of the Dipper Bowl. The dust of the Taurus DCC is quite thick behind both the Pleiades and the Hyades, which gives the background of those clusters as seen in binoculars and RFTs a peculiarly "thick" or "murky" look.

NGC 1647: In the same wide-angle field with the Hyades, about 4° NE of Aldebaran, is the very large (45' diameter) open cluster NGC 1647. With 10×50 binoculars 15 or 20 cluster members can be seen, scattered in coarse clumps over a very pale background glow. The group's integrated apparent magnitude of 6.4 is not very high given the area it covers. The brightest star in its main body is a mag 8.6 B8 III blue-white giant; but the Cepheid variable SZ Tauri, though 2° WSW of the cluster's center, has been considered a possible NGC 1647 star. SZ Tau is a very short period (4.48 day) Cepheid that would not be out of place in an intermediate-age (roughly 100 million year old) cluster with a B8 III lucida. However, chances are that SZ Tau is somewhat nearer than NGC 1647: its median apparent and absolute mags are 6.5 and –3.1, respectively, and its light is dimmed 1.2 mags by dust; so the star must be about 1800 light-years away.

III.3 Orion

The galactic equator approximately follows the border between Orion and Gemini, with the feet of Gemini (Gamma [γ] + Eta [η] + Mu [μ] Geminorum) to its NE and the club of Orion (Mu [μ] + Nu [ν] + Xi [ξ] + Chi-one [χ^1] + Chi-two Orionis) to its SW. However, most of what is of interest in Orion is well away from the galactic equator, in and around the Hunter's Belt and Sword. And what is of special interest in Orion are not open clusters, of which it has surprisingly few, but bright nebulae, of which it has surprisingly many visible even in "mere" binoculars.

The bright nebulae of Orion are (with the exception of NGC 2174 in the far northeastern part of the constellation) the illuminated sections of the massive dust clouds in which have been born the blue giant and supergiant stars of the Orion Association, Orion OB1. The nebulae and stars of Orion OB1 will be described at the end of the next chapter. The largest star cluster in Orion is in fact the Belt of Orion: in binoculars the Belt supergiants of Orion OB1, Delta (δ), Epsilon (ε), and Zeta (ζ) Orionis, can be seen to be surrounded by a rich field of 6th, 7th, 8th, and 9th magnitude stars. The Belt Cluster is catalogued as Collinder 70. In the Sword of Orion the major open cluster is in fact the very rich, but heavily dust-dimmed, aggregation of pre-main-sequence stars around the multiple star Theta (θ) Orionis. In a few hundred thousand years, after the dust clears, the Theta Ori Cluster should be a very impressive sight. At the north end of the Sword is the more conventional-looking open cluster NGC 1981, another good binocular and RFT group because large and loose. NGC 1981 and Cr 70 will be described with the Orion Association in the next chapter.

NGC 2169: However, even in the part of Orion along or near the galactic equator open clusters are surprisingly few and far between. Only one of them has any pretence of being a suitable object for any type of wide-field instrument, NGC 2169, located a little less than 1° south and slightly east of Nu (ν) Orionis. This is a compact but bright clus-

Orion Open Clusters

Cluster	RA (2000.0)	Dec	m_V	Size	m_V*	*Spectrum	Distance	A_V	M_V	True Size	Age
N 1981	05^h35^m	–04°26'	4.6	25'	6.3		1600 l-y		–4.5	12 l-y	
N 2169	06 08	+13 57	5.9	6	6.92	B1V, B2V	3000	0.5	–4.4	6	
Cr 70	05 34	–01 15			1.70	B0 Ia	1600				4–5 mil

ter, only 6′ in diameter but with an integrated apparent magnitude of 5.9. It is not very populous: its brightness comes from a dozen stars between magnitudes 7 and 10. The cluster lucida is the close binary ADS 4728, which consists of a mag 7.4 B1 V primary accompanied by a mag 8.0 B2 V secondary just 2.5″ way in PA 109°. ADS 4728 requires about 100× to be comfortably split; but NGC 2169 as a whole also requires about that power to be seen at its best because of its compactness. Nevertheless the cluster is visible in higher-power and zoom binoculars as a tight little star knot. NGC 2169 is probably around 3,000 light-years distant and therefore lies beyond the far side of the Orion Association complex. The spectral type of its lucidae imply that it is a young cluster and consequently a true tracer of our Orion-Cygnus Spiral Arm.

III.4 Gemini

Most of the star-pattern of Gemini lies out of the Milky Way. But the galactic equator just nips its SW corner and it shares a stretch of the winter Milky Way with Orion. The bulk of the interstellar dust in this direction is SW of the galactic equator through central Taurus and central Orion, so the Gemini Milky Way – what there is of it – is quite transparent and we can see a fairly great distance out toward the rim of our Galaxy. In fact the highly-evolved Gemini open cluster NGC 2158 is probably two-thirds the way out to the Galactic rim. This small and faint group is just SW of the very large and very bright M35, a front-line binocular cluster. Two other open clusters in Gemini are of interest, though neither is a good binocular or even RFT object: the extremely young NGC 2129, and the extremely old NGC 2420.

Photo 2.12 The open clusters M35 and (to its lower left = SW) NGC 2158 in Gemini

M35 (Plate XXXII, Photo 2.12): Two degrees NW of the 3rd magnitude red giant variable Eta (η) Geminorum is the open cluster M35. Its apparent diameter, 28′, is almost equal to that of the Moon; and its integrated apparent magnitude, 5.1, is half a mag higher than that of the brightest of the Auriga Messier trio, mag 5.6 M37. M35 can be easily seen as a tiny patch of Milky Way haze with the unaided eye on dark evenings when Gemini is near the zenith. The cluster has 20 members between mags 7.5 and 10 and, because it is so large and loose, resolves very well even in 10×50 binoculars. It is an outstanding sight in richest-field telescopes. The cluster's lucida, located just NE of its center, is a mag 7.53 B3 V star. Its second brightest member, a mag 8.18 star towards its SE edge, is another B3 V object. M35 also has two blue giants, B6 and B7 III stars of apparent mags 8.83 and 9.20, and several late-G/early-K yellow giants, the brightest a K0 III star with an apparent mag of 8.6.

M35 is 2,200 light-years away and thus considerably nearer than Auriga's Messier open clusters. Its integrated absolute magnitude, corrected 0.5 mag for dust dimming, is –4.7, a luminosity of 6300 Suns but about one mag fainter than M36, M37, and M38. The absolute mag of its B3 V lucida is –2.3, a luminosity of almost 700 Suns. Its mag 8.6 K0 yellow giant has a luminosity around 250 Suns. The published age of 110 million years for M35 is inconsistent with the presence of B3 main sequence stars in the cluster, which imply an age more like one-half of that.

					Gemini Open Clusters							
Cluster	RA (2000.0) Dec		m_V	Size	m_V*	*Spectrum	Distance	A_V	M_V	True Size	Age
M35	06^h09^m	+24°20'	5.1	28'	7.53	B3 V	2200 l-y	0.5	-4.7	18 l-y	110 mil
N 2129	06 01	+23 18	6.7	6	7.38	B3 Ib	5900	2.6	-7.2	10	
N 2158	06 07	+24 06	8.6	5	12.4		13,000	1.4	-6	19	3 bil
N 2420	07 38	+21 34	8.3	10	11.1		7400	0.2	-3.4	20	2.8 bil

NGC 2129: Located $^3/_4$° due west of the 4th magnitude 1 Geminorum (and 2° SW of M35) is the tiny NGC 2129. This cluster basically is just a N-S pair of mag 7.4 and 8.3 stars several minutes of arc apart, though a handful of much fainter companions can be glimpsed in telescopes. The main interest of NGC 2129 is that its two lucidae are highly-luminous B3 Ib supergiants. Though these two stars are very close together on the celestial sphere, they suffer considerably different amounts of dimming due to interstellar dust, 2.3 mags for the brighter star and 2.9 mags for the fainter one. This is a warning that extinction can be *extremely* variable with even slight changes in direction. The absolute mags of the two stars are -6.2 and -5.6, luminosities of 25,000 and 14,500 Suns, respectively. Given that the cluster is about 5,900 light-years away, its integrated apparent mag of 6.7, corrected 2.6 mags for absorption, implies an integrated absolute mag of -7.2, corresponding to a total luminosity of 62,500 Suns. Thus nearly two-thirds of the light of NGC 2129 comes from the two blue supergiants alone. The high amount of absorption suffered by this little cluster is less due to its distance than to outlying Taurus Dark Cloud Complex dust. A scan in binoculars west from M35 over the position of NGC 2129 into Taurus will show how dramatically the Milky Way star background diminishes toward the Taurus DCC.

NGC 2158 (Plate XXXII, Photo 2.12): Even in only 10×50 binoculars a tiny (5' diameter) magnitude 8.6 smudge of haze can be detected $^1/_2$° SW of the center of M35. This is the very rich but very distant open cluster NGC 2158. Its brightest members are only mag 12$^1/_2$–13 stars, so it requires moderate-size telescopes and, because of the cluster's compactness, moderate powers even to begin to resolve. However, NGC 2158 is worth looking for because, first, it is at least 2 billion years old and therefore one of the most evolved open clusters known in our Galaxy, and second, it is about 13,000 light-years distant and consequently, given that the galactic anti-center (the nearest point of our Galaxy's rim) is only 7° to the NW, is about two-thirds the distance out to the edge of the Galaxy. Thus the M35 field is an excellent region for getting "Galactic depth perspective": M35 itself is about 2,200 l-y away; NGC 2129, 2° to the big cluster's SW, is almost three times farther, 5,900 l-y distant; and NGC 2158 is twice as far as NGC 2129 and six times more remote than M35.

Though twice as far as NGC 2129, NGC 2158 is dimmed 1.2 magnitudes *less* by interstellar dust. The reason is that it is further east from the outer fringes of the Taurus Dark Cloud Complex, which lies a few hundred l-y from us in the foreground of both clusters. The integrated absolute mag of NGC 2158, -6 (22,000 Suns), is very high for a cluster of its age and is the measure of its compact populousness. Indeed, the cluster's compact populousness is what has kept it together these billions of years.

NGC 2420: In east central Gemini, about 4$^1/_2$° east and slightly south of the 4th magnitude Delta (δ) Geminorum, is the moderately large (10') but faint (mag 8.3) open cluster NGC 2420. This group is visually unimpressive in binoculars and RFTs, appearing basically just as a small hazy patch. Some mag 11–12 stars can be resolved in supergiant binoculars. However, NGC 2420 has a curious location, because it is 20° off the galactic equator, an exceptionally large distance for an open cluster. Now, M44, the Praesepe Cluster in Cancer east of Gemini, is even farther off the galactic equator than NGC 2420. But M44 is only 525 light-years from us, whereas NGC 2420 is 7,400 l-y away–fourteen times farther. Thus M44 is only a couple hundred, but NGC 2420 more than *two thousand,* light-years off the plane of the Galaxy. The fact that NGC 2420 is so far from the galactic plane means that it must be very old. It is estimated to be around 3 billion years old and thus even more evolved than NGC 2158 near M35.

III.5 Monoceros

The main stream of the winter Milky Way goes NW-SE east of Orion through Monoceros, which lacks even a single star as bright as magnitude 3.5. However, in central Monoceros, and in central Puppis to its SE, the Milky Way is at its brightest anywhere between the Cepheus/Lacerta star clouds to the NW and the Carina star clouds to the SE–an arc almost halfway around the entire Milky Way. This is because toward Monoceros and Puppis clouds of interstellar gas and dust are scattered and clumpy, with "windows" between them through which we can see long distances across the Galaxy's spiral plane. In NW Monoceros there are dust clouds around the Rosette Nebula, NGC 2237, and its central open cluster, NGC 2244, and around the Christmas Tree Cluster, NGC 2264; and in south central Monoceros there is a dust cloud around the Seagull Nebula, IC 2177. But in between, toward central Monoceros, and SE into Puppis, we have clear views thousand of light-years long out toward the rim of our Galaxy. And along these views an abundance of open clusters can be seen.

Monoceros' many open clusters come in a great variety of sizes, structures, distances, and ages. Aesthetically the constellation's best groups for any size instrument are M50 and NGC 2301, both well-populated 6th magnitude clusters with interesting structures. A 7th mag cluster in the constellation that is also well-populated and has an interesting structure is NGC 2353, one of the youngest Monoceros clusters. But the two youngest Monoceros groups, NGC 2244 in the Rosette Nebula and NGC 2264, are so loose that they look more like field condensations in the star-rich Monoceros Milky Way than real open clusters. The nearest Monoceros cluster, NGC 2232, is neither populous nor concentrated, but at 7×-10× has an appealing appearance. Two clusters, NGC 2335 and NGC 2343, are not visually impressive in wide-field instruments but interesting for being in the foreground of the Seagull Nebula complex. And finally, two of Monoceros' clusters, NGC 2324 and NGC 2506, though compact groups of faint stars requiring telescopes to be seen at their best, are of interest for their great distances and high ages.

M50 (Plate XXXVI, Photo 2.13): Though it is in a bright-star-poor region of the Milky Way, the open cluster M50 is so large (16' in diameter), so bright (magnitude 5.9), and resolves so well even in moderate-power binoculars (with two 8th and a half-dozen 9th mag members) that it is easily spotted halfway between Alpha (α) and Beta (β) Monocerotis. 10×50 binoculars resolve six or seven M50 stars on

Monoceros Open Clusters

Cluster	RA (2000.0)	Dec	m_V	Size	m_V*	*Spectrum	Distance	A_V	M_V	True Size	Age
M50	07^h03^m	−08°20'	5.9	16'	7.85	K2.5 III	2900 l-y	0.8	−4.7	14 l-y	80 mil
N 2232	06 27	−04 45	3.9	30	5.03	B2 V	1200	0.2	−4.1	11	22
N 2244	06 32	+04 52	4.8	24	6.72	03.5 V	4900	1.4	−7.5	34	<1
N 2264	06 41	+09 53	3.9	20	4.5–5.0	07 IV-V	3000	0.3	−6.2	17	3.2
N 2301	06 52	+00 28	6.0	15	8.00	G8 IV	2600	0.1	−3.6	11	110
N 2324	07 04	+01 03	8.4	7	10.3		11,700	0.6	−4.9	23	700
N 2335	07 07	−10 05	7.2	12	9.5		3600	1.2	−4.2	13	~130
N 2343	07 08	−10 39	6.7	6	8.4		2600	0.5	−3.3	4.6	>130
N 2353	07 15	−10 18	7.1	20	9.2		4300	1	−4.5	25	7
N 2506	08 00	−10 47	7.6	6	10.8		10,800	0.2	−5.2	20	2,500
Cr 106	06 37	+05 57		45	6.08	08ep	4900	1.0		63	5
Cr 107	06 38	+04 44	5.1	35	6.15	B1 Ib	4 900	1.3	−6.1	60	15

Photo 2.13 The open cluster M50 in southern Monoceros

a pale, amorphous, rather low-surface-brightness background glow. 100 mm glasses can resolve more than two dozen members, though at least 25× is necessary to make them easy to count. The mag 7.8 lucida, an orange K3.5 III giant, is on the southern edge of the group, with another bright cluster member just to its NW. The second brightest M50 star, a mag 8.3 late-B main sequence object, is in the cluster's NE section. Just to that star's NW is a wide NE-SW double with almost equally-bright mag $9^1/_2$ components. To the SE of the main body of the cluster is a shallowly-curved arc, several minutes long and open to the NW, of three mag 9–$9^1/_2$ stars that look like an outlying subgroup. The overall impression of the cluster is of much subgrouping and clumping, with curious star-vacancies between. This is typical for intermediate-age open clusters, and indeed M50 is at least 100 million years old. It is estimated to be 2,900 light-years away, so its integrated absolute mag (corrected for 0.8 mag of dimming by interstellar dust) is –4.7, a luminosity of 6,300 Suns.

NGC 2232: An open cluster better for binoculars than telescopes is NGC 2232, located 2° due north of the splendid telescopic triple Beta (β) Monocerotis. With 10×50 binoculars this group appears as a 10'-wide, $^1/_2$°-long scattering of perhaps eight 6th to 9th magnitude stars extending north from the cluster lucida, the mag 5.0 B2 V 10 Monocerotis. At moderate powers in telescopes a little gathering of mag 9–11 stars can be seen around 10 Mon. The second brightest NGC 2232 star, 9 Mon at the north end of the group, is a mag 6.1 B3 V object. The cluster is estimated to be about 1200 light-years away, so its integrated absolute mag is –4.1 (a luminosity of 3,600 Suns). The spectral types of 9 and 10 Mon imply that NGC 2232 is not much more than 20 million years old. It therefore must be a true tracer of our Orion-Cygnus Spiral Arm.

NGC 2244 (Plate XXXV, Photo 2.14): One of the most surprisingly easy celestial objects for standard binoculars is the Rosette Nebula, NGC 2237, which can be seen from dark sky sites with just 10×50 glasses. The Rosette Nebula will be described in more detail in the next chapter. The Rosette Star Cluster, NGC 2244, on the other hand, is surprisingly *difficult*. It is large, a 15'×5' NW-SE rectangle, a star at each corner and at the midpoints of the long sides. And its stars are bright, with six from mag 6.7 to 8.2 (not counting the mag 5.9 12 Monocerotis at the SE corner of the rectangle, which is not a true cluster member). However, NGC 2244 is not populous in faint members and therefore offers no tell-tale haze of unresolved stars. Nor is it concentrated. In fact the field of 6th to 9th mag stars around the Rosette is so rich that NGC 2244 appears as little more than a concentration in it, the group's existence betrayed more by its distinctive rectangular shape than by any look of "clusterness."

The Rosette Cluster and Nebula can be difficult to locate because there are no bright stars nearby. NGC 2244 is about $2^1/_2$° almost due south of 13 Monocerotis. But 13 Mon is only a mag $4^1/_2$ star. Probably the best way to find the Rosette Complex is through the "back door"–that is, from Gemini. Start at Gamma (γ) Geminorum and first go 4° SSE to Xi (ξ) Gem, then another 3° south and slightly west to the 5th mag S Mon, the lucida of the Christmas Tree Cluster, NGC 2264. From S Mon go 3° SW to 13 Mon and then $2^1/_2$° south on to the Rosette.

The brightest true Rosette Cluster member is the mag 6.72 HD 16150, located at the middle of the

Photo 2.14 The Rosette Nebula NGC 2237 and its central open cluster NGC 2244. The ragged N-S star-chain on the NE (upper left) edge of the nebula is NGC 2252; and the thinner N-S star-chain toward the east-center edge of the field is Cr 104

rectangle's SW long side. This is a rare ultra-hot O3.5 V object with a surface temperature above 50,000°K and a bolometric absolute mag of –9.9, which means that its total energy output each second is equal to that of 600,000 Suns. The eastern star on the NW end of the Rosette Cluster rectangle, the mag 7.25 HD 40223, is also an O3.5 V object that also has a bolometric absolute mag in excess of $-9^1/_2$. It is the ultraviolet radiation from these two stars that is fluorescing the Rosette Nebula.

The Rosette Complex is estimated to be 4,900 light-years away. Thus the 15'×5' rectangle of NGC 2244 has a true size of 21×7 l-y. Outliers bring the cluster's full extent to 24' = 34 l-y. The integrated visual absolute magnitude of the cluster, –7.5, corresponds to the luminosity of 80,000 Suns, but is misleadingly low: after all, the two mid-O cluster members alone have a total energy output of over 1,000,000 Suns.

The Rosette Cluster is the main stellar subgroup of the Monoceros OB2 association, members of which are scattered for three or four degrees east and NE from NGC 2244. Two of the Mon OB2 star-fields have been catalogued as separate open clusters. About $1^2/_3°$ NE of the center of the Rosette Cluster, and the same distance SE of 13 Mon, is the mag 6.1 HD 47129, "Plaskett's Star," the brightest member of the 45'-diameter Mon OB2 star-field Collinder 106. Plaskett's Star is a spectroscopic binary of two O8 giants, each containing some 50 Solar masses, orbiting each other every 14.414 days at a distance of just 80 million kilometers (about half the distance from the Earth to the Sun). Cr 106 does not look much like an open cluster in the eyepiece, nor is even much of an enhancement in the basic Milky Way background star-density.

$1^1/_2°$ east and slightly south of the Rosette Cluster is another mag 6.1 star, the B1 Ib supergiant HD 47240, lucida of the 35'-large Mon OB2 field Collinder 107. Cr 107 looks a little more like a star cluster than Cr 106 because its three brightest members are arranged in a long, thin, NW-SE triangle. At the SE

Photo 2.15 NGC 2264, the "Christmas Tree" Cluster (upper left = NE) and the distant open cluster Tr 5 (upper right) in northern Monoceros. Tr 5 is a well-populated, but rather evolved, group some 8,000 light-years distant. Its stars are so faint that it requires moderate-aperture telescopes to be seen. The tiny bright triangular nebula on the bottom center is NGC 2261, Hubble's Variable Nebula

end of the triangle is the mag 7.14 B0 III HD 47382, a few minutes of arc to the SW of which is a tight little knot of three or four 8th-9th mag stars. The three open clusters of Mon OB2 are an example of *sequential subgroup formation* in a stellar association, because the Cr 107 subgroup is estimated to be 15 million years old, the Cr 106 subgroup 5 million years old, and NGC 2244 itself, because of its mid-O members and the fact that it is still embedded in the residuals of the cloud of gas and dust in which it was born (the Rosette Nebula), just a few hundred thousand years old.

Two $^1/_2$°-long N-S chains of some 20 faint (mag 10–13) stars near the Rosette Cluster have also been given open cluster designations, though they are probably only chance alignments of nearer with farther stars. They can be seen in large telescopes, and show up very well on photos of the Rosette. One chain, NGC 2252, is on the NE edge of the Nebula, and on its north end has a conspicuous little "hook" to the west. The other star-chain, Collinder 104, is 1° due east of the Rosette Cluster and just west of Cr 107.

NGC 2264 (Plate XXXIII, Photo 2.15): 3° south and slightly west of Xi (ξ) Geminorum is the 5th magnitude variable S Monocerotis. S Mon is a very hot O7 IV-V marginal subgiant that ranges irregularly between mags 4.5 and 5.0 (not a very conspicuous light change visually). The star is at the north end of the 30'-long, very loose, but distinctively-shaped open cluster NGC 2264. The group has sometimes been called the "Christmas Tree" Cluster because most of its 20 brightest members (which range from S Mon's mag 4.5 down to mag 9.3) are arranged in a rather striking arrowhead shape, tip to the south (which is "up" in the inverted field of a telescope). S Mon marks the trunk of the Christmas Tree. The cluster's shape is a great help in identifying it because it is so loose, and this region of the

Monoceros Milky Way is so rich in 7th, 8th, and 9th mag field stars, that NGC 2264 otherwise would be lost in the clutter.

The cluster is almost completely resolved even at only 10×; and at that power, with averted vision, a mag 9.0 B7 companion can be glimpsed 74″ away in PA 140° from S Mon. The second brightest NGC 2264 member is the mag 7.1 B2 star at the tip of the arrowhead. Just south of it is the famous (unfortunately strictly photographic) Cone Nebula, which looks like a celestial volcano that has just belched forth the B2 star. Photos show a concentration of very faint stars at this end of the cluster. Another concentration of fainter cluster stars is around S Mon and SW of it toward the mag 8.4 B5 star $8^1/_2$' distant. The brightest part of the NGC 2264 nebulosity (catalogued as Sh 2-273) is around this B5 star: it is a compact area of combined emission and reflection glow and requires moderate powers in telescopes to be seen.

NGC 2264 is estimated to be 3,000 light-years away. Its true length is therefore 27 l-y. The cluster's integrated apparent mag of 3.9, which needs only 0.3 mag correction for dimming by interstellar dust, thus implies an integrated absolute mag of −6.2, a luminosity of 25,000 Suns. However, much of this is from S Mon, which at maximum has an absolute mag of −5.6, a luminosity of 14,500 Suns. The low absorption to the cluster shows that, though there might be plenty of dust on either side and behind NGC 2264, virtually none is in the 3,000 l-y between here and there. This is one of the *extremely* transparent directions toward the Monoceros, Canis Major, and Puppis Milky Way. The open cluster NGC 2301, some 10° SSE of NGC 2264, is about 2,600 l-y away but dimmed only 0.1 mag.

The transitional IV-V luminosity class of S Mon indicates that it is just beginning its evolutionary expansion off the main sequence. Nevertheless as an O7 object it is a very young star. And NGC 2264 is a very young cluster. Indeed, it is estimated to be only 3.2 million years old, an age consistent with the presence of gas and dust near it. NGC 2264 is the major stellar aggregation of an association, Monoceros OB1. Though star-formation is presently occurring in dust clouds involved with Mon OB1 (particularly around the Cone Nebula, and to the SE in NGC 2261, Hubble's Variable Nebula), the association has only one supergiant, the mag 4.48 13 Mon, an A0 Ib star with an absolute mag of about the same as S Mon, −5.6.

NGC 2301: The best open cluster for telescopes in Monoceros is the rich and strangely-structured NGC 2301, located 5° WNW of Delta (δ) Monocerotis and just 2° SSE of 18 Mon. It is, however, also a fine sight in supergiant binoculars and small RFTs. In 10×50s NGC 2301 appears as a 15'-long N-S line of a half dozen magnitude 8 and 9 stars, quite conspicuous in its star-rich Milky Way field though not accompanied by the background haze of unresolved members that would be expected of an intrinsically populous open cluster. In 15×100 supergiant instruments not only does a cluster haze show up, but the group's peculiar structure begins to be evident. The N-S star line can be seen to consist of a shallow, west-opening arc of four evenly-spaced mag $8^1/_2$–9 stars, north of which is the cluster's mag 8.0 lucida and, further north, along approximately the same line, a tight trio of mag 10–11 stars. The lucida is a close but cleanly-split double with a mag $9^1/_2$ star just to its north, and is embedded in a haze that contains a couple resolved star-sparks. The cluster gives the strong impression of extending toward a bright mag 9 star a few minutes due east of the lucida. This impression is correct, because in telescopes two subgroups can be seen west and SW of this 9th mag star, one subgroup between it and the cluster lucida, and the second subgroup a shallow, SE opening, arc of a half-dozen faint stars.

NGC 2301 is like M50 further SE in its star-arcs and star-clumps. This is a feature of many moderately-rich, intermediate-age open clusters, and NGC 2301 is about the same age as M50, around 100 million years. It is also at about the same distance as M50, 2600 light-years. Thus its true size is 11 l-y and its integrated apparent magnitude of 6.0 corresponds to an integrated absolute mag of −3.6, a luminosity of 2500 Suns. The cluster's integrated absolute mag requires a mere 0.1 mag correction for dimming by interstellar dust–a negligible amount, and testimony to the transparency toward the central Monoceros Milky Way. The mag 8.0 lucida of NGC 2301 is an orange G8 IV subgiant with an absolute mag of −1.6. The second brightest cluster star is a mag 8.2 B9ne object with an absolute mag of −1.4. The group also contains B8 and B9 class III giants with apparent mags of 8.2 and 8.4, respectively, and its main sequence goes up to a mag 9.1 B9 star. However, the mag 8.0 lucida might be a foreground non-member because late-G subgiants usually have absolute magnitudes around $+2^1/_2$ instead of around $-1^1/_2$.

NGC 2324: In the same binocular field with NGC 2301, just 3° ENE of that rich cluster and the same distance NW of Delta (δ) Monocerotis, is the compact (7′) and faint (magnitude 8.4) NGC 2324. The brightest members of this group are only mag $10^1/_2$–11 stars, so it requires supergiant binoculars even to begin to resolve, though it is visible in 10×50 glasses as a tiny shred of haze. It looks distant and indeed is–11,700 light-years away, over four times farther than NGC 2301. However, it is dimmed by interstellar dust only 0.6 mag–impressively low, especially considering the cluster's distance. As a general rule, absorption across our Galaxy's spiral disc is 1 magnitude per kiloparsec (3,260 light-years): NGC 2324 is dimmed 0.6 mag but $3^1/_2$ kiloparsecs away! The Milky Way seems to be practically dust-free in this direction. Telescopes reveal that NGC 2324 has the swarm-like appearance typical of evolved open clusters; and in fact it is estimated to be nearly 1 billion years old.

NGC 2335 and NGC 2343 (Plate XXXVI, Photo 2.16): The two open clusters NGC 2335 and NGC 2343, only about $^2/_3$° apart NW-SE, are located in extreme south central Monoceros just north of the Canis Major border. They are both intermediate-age open clusters that lie in the foreground of the Canis Major OB1 association, in which is involved the "S"-shaped Seagull Nebula, IC 2177. The nebula is visible in 10×50 glasses (and is described in the next chapter), but the two clusters are not easy in anything less than giant binoculars: NGC 2343 is not faint (magnitude 6.7), but very small (6′ in diameter); and NGC 2335 is large (12′) but faint (mag 7.2)

Photo 2.16 Open clusters NGC 2335 (lower right) and NGC 2343 (upper left) and the northern half of the Seagull Nebula, IC 2177, in southern Monoceros. North is to the right, and east is up

with faint stars (mag 9.5 and less) and therefore of very low surface brightness. In 15×100 supergiant binoculars NGC 2343 is a very distinct little cluster of perhaps a half dozen mag 8½–10 stars, the mag 8.4 lucida on its SE edge, but so compact that its members are hard to count. In the same type instrument NGC 2335 is no more difficult than NGC 2343, but much different in appearance: larger, but a mere granular haze between a bright mag 7 field star on the NE and a NW-SE pair of 8th mag stars on the SW, with a close N-S pair of 10th mag cluster members resolved in the haze.

But even with large binoculars NGC 2335 and NGC 2343 are not all that easy to find, for they are in a bright-star-poor (but open-cluster-rich) area of the winter Milky Way. They can be approached from the conspicuous M50, which is just 2° NNW of NGC 2335. But another route to the clusters begins at the 4th mag Theta (θ) Canis Majoris, the star at the "nose" of the Great Dog. From Theta Canis go just over 3° ENE to FN CMa, the only 5th mag star between Theta Canis and Alpha (α) Monocerotis. NGC 2335 is 1° due north of FN Canis, and NGC 2343 is at the eastern vertex of a very shallow isoceles triangle with FN Canis and NGC 2335.

Both NGC 2335 and NGC 2343 are catalogued with main sequences up to spectral type B6 or B7, implying cluster ages of around 130 million years. However, they look like groups of considerably different ages: in particular NGC 2335 has the swarm-like appearance characteristic of evolved clusters. It seems to lie a thousand light-years beyond NGC 2343. In fact, at 3,600 l-y NGC 2335 is almost as far as the 4,300 l-y distant Canis Major OB1 association, and perhaps the cluster's rather high extinction, 1.2 mags (as compared to the 0.5 mag of NGC 2343) is from outlying CMa OB1 dust. On photos NGC 2335 looks like it lies on the northern tip of the IC 2177 "S," but that is just a chance alignment. Intrinsically NGC 2335 is over twice as large as NGC 2343, 13 versus 4½ l-y, and over twice as luminous, 4000 versus 1760 Suns.

NGC 2353 (Plate XXXVI): About 2° due east of NGC 2335 and NGC 2343 is NGC 2353, a much easier cluster to find because it is large (20') and because it has a magnitude 6.0 star on its SW edge. This cluster is located just over halfway from Alpha (α) Monocerotis to Theta (θ) Canis Majoris and about 3° SE of M50. 6th mag field stars are just ½° due north and ½° SE of the mag 6.0 lucida of NGC 2353. This trio of 6th mag stars is a conspicuous asterism as you scan SE from M50 or WSW from Alpha Mon.

In 10×50 binoculars NGC 2353 appears as a spray of faint stars scattered ⅓° NE from the mag 6.0 lucida. 15×100s show this spray to be oval in outline and its star-density to increase toward the lucida, just NE of which is a 20"-wide double of 9th mag stars. With averted vision many threshold cluster members become visible, giving an (accurate) sense of richness.

The mag 6.0 lucida of NGC 2353 is a hot blue O9.5 II-III giant. Given that NGC 2353 is about 4,300 light-years away, and allowing for 0.4 mag of absorption by interstellar dust, the star's absolute mag is near –5, a luminosity of almost 9,000 Suns. This star completely dominates the rest of the cluster, which has an integrated apparent mag of just 7.1 and an integrated absolute mag (correcting for 1 mag of absorption) of only –4.5. However, this same scenario–one or two highly-luminous O-type lucidae attended by a swarm of lesser luminaries–can be seen in NGC 6383 in Scorpius, NGC 2264 in northern Monoceros, and NGC 2362 in Canis Major.

The brightest star of NGC 2353 is also the brightest star of the Canis Major OB1 association, which includes the IC 2177 complex 2° west of the cluster. This is not a particularly populous association, and NGC 2353 seems to be its only open cluster. The second brightest CMa OB1 member is the mag 6.21 HD 54662, an O6.5 main sequence star located ½° west of NGC 2353 toward IC 2177. The third brightest association star is a mag 6.47 O7.5 V object 1° due south of FN Canis. CMa OB1 also includes a B1 Ib supergiant and a B0.5 III giant.

NGC 2506: In extreme SE Monoceros about 5° east and slightly south of Alpha (α) Monocerotis is the very small (6') in diameter) and rather faint (mag 7.6) NGC 2506. The cluster's brightest members are only 11th mag stars, so in 10×50 binoculars it appears only as a tiny patch of haze. However, in reality NGC 2506 is populous and compact, so the surface brightness of its binocular glow is moderately high and the cluster is not difficult to spot. It looks as distant as it is, 8,800 light-years. However, it is particularly interesting for its age, about 5 billion years, which makes it one of the oldest open clusters known. Despite its age, it has the very respectable integrated absolute magnitude of –4.8 (a luminosity of 7,200 Suns), which is the result of its richness. NGC 2506 is located 10° off the galactic equator, which at the cluster's distance corresponds to 1540 l-y off the Galactic plane. It is therefore well out in the thick disc of our Galaxy where only evolved stars and clusters would be expected to be found.

III.6 Canis Major

Milky Way grazes the NE corner of Canis Major. It is rather bright here, as it is in the adjoining areas of Monoceros to the north and Puppis to the east, because only isolated dust clouds block our view through the Galaxy in this direction. However, the brightest part of the Canis Major Milky Way, and the region of the constellation with the richest star fields, is in and around the triangle of Delta (δ), Epsilon (ε), and Eta (η) Canis Majoris. Epsilon is a 1st magnitude star, Delta and Eta 2nd mag stars, Omicron-two (o^2) and Sigma (σ) 3rd mag stars, and Omicron-one, Tau (τ), and Omega (ω) 4th mag stars. And they have a half dozen 5th, a dozen 6th, and numerous 7th, 8th, and 9th mag attendants. It is a splendid 7×–10× binocular field.

Most of the bright stars in southern Canis Major are members of the Canis Major Association, centered about 2,500 light-years away. The brightest stars of the association are supergiants of a wide variety of spectral types, and they offer the binocular and RFT observer a wide variety of colors and color contrasts. Plus Episilon CMa, though only about 680 l-y distant and consequently not a true CMa Association star, is a B2 II giant with a splendid silver-blue color like that of the mid-B supergiant association members Eta and Omicron-two.

The Canis Major Association has one open cluster, the very loose Collinder 121, more simply a star-field enhancement around Omicron-one Canis than a true gravitationally-bound cluster. Cr 121 includes, in addition to Omicron-one, a pair of 6th mag stars about $^1/_3$° south of Omicron-one and the 7th mag variable EZ CMa just to its north. The latter is a peculiarly under-luminous Wolf-Rayet star. However, WR stars like EZ Canis and yellow supergiants like Delta Canis are often found together in the same association. The evolved K- and M-type supergiants of CMa, and the fact that it lacks nebulae, suggests that it is an older association, perhaps 15-20 million years old. The absorption to CMa is extremely low (except for Omicron-one), confirming the remarkable transparency toward much of the Canis Major, Monoceros, and Puppis Milky Way. The $8^1/_2$°×$2^1/_2$° apparent size of the CMa Association corresponds to a true extent of 370×100 l-y.

Southern Canis Major has four open clusters besides Cr 121. Two of them, Cr 132 and Cr 140, respectively SW and south of Eta CMa, are at their best in standard 7×-10× binoculars. Two others, NGC 2354 NE of Delta CMa and NGC 2362 around Tau CMa, are best in telescopes but visible in larger binoculars. NGC 2360 in the far NE part of the constellation east of Gamma (γ) CMa is a good giant binocular cluster. But the best open cluster in the constellation is M41.

M41: M41 is easily found 4° due south of Sirius. In fact this cluster is so large (38' in diameter) and so bright (magnitude 4.5) that on dark, clear nights it is visible to the unaided eye as a patch of haze. It

Supergiant Members of the Canis Major Association

Star	m_V	Spectrum	A_V	M_V	Distance (l-y)	Color
δ CMa	1.84	F8 Ia	0.36	–8.1	2700	Yellow
η	2.45	B5 Ia	0.09	–7.0	2460	Silver-blue
o^2	3.02	B3 Ia	0.09	–6.5	2460	Silver-blue
σ	3.43-3.49	M0 Iab	0.14	–6.0	2350	Orange-red
o^1	3.84	K3 Iab	1.05	–6.5	2350	Chrome orange

Canis Major Open Clusters

Cluster	RA (2000.0)	Dec	m_V	Size	m_V*	*Spectrum	Distance	A_V	M_V	True Size	Age
M41	06h47m	−20°44′	4.5	38′	6.90	K3 II	2250 l-y	0.2	−4.9	24 l-y	200 mil
N 2354	07 14	−25 44	6.5	20	9.1		6200	0.3	−5	36	
N 2360	07 18	−15 37	7.2	12	10.4		3400	0.2	−3.1	12	1000
N 2362	07 19	−24 57	4.1	8	4.39v	O9 II	4900	0.5	−7.3	11	1
Cr 121	06 54	−24 38	2.6	50	3.84	K3 Iab	2350	0.1	−6.8	34	
Cr 132	07 14	−31 10	3.6	95	5.3						
Cr 140	07 24	−32 12	3.5	42	5.4		1470		−4.7	18	30

begins to resolve in even the smallest binoculars; and at least a dozen cluster members can be counted in 10×50 glasses. Scores of M41 stars are visible with 25×100s. The cluster is near its best around 30×, at which power the even distribution and comparable brightness of its stars gives a very convincing illusion of richness. The lucida of M41, located near its center, is a mag 6.90 K3 II orange giant. The 6th mag 12 CMa, just to the SE of the cluster core, is not a true M41 star.

M41 is about 2,250 light-years away, so its true size is 24 l-y and its integrated absolute mag, corrected 0.2 mag for absorption, is −4.9, a luminosity of 8,000 Suns. The 0.2 mag of dimming suffered by the cluster is a very small value for its distance, but not unusual in this stretch of the Milky Way. The six brightest M42 stars are all late-G/early-K class II luminous giants with apparent mags from 6.9 to 7.9 and absolute mags between −2.5 and −1.5. The cluster's main sequence goes up to two mag 8.4 B8 stars. This bright-star population implies a cluster age of nearly 200 million years.

NGC 2354: NGC 2354 is conveniently situated exactly halfway between Delta (δ) and Tau (τ) Canis Majoris. The main interest of this group is that it happens to be located in the star- and cluster-rich southern Canis Major Milky Way, for otherwise it hasn't much to offer visually or astronomically. Though reasonably bright (magnitude 6.5), NGC 2354 is so large (20′ in diameter), and its stars so faint (mag 9.1 and less) and scattered that even in 10×50 binoculars it appears merely as a pale patch of mottled haze, a couple star-sparks resolved if observing conditions are good. The cluster's southerly declination, −26°, will prove a problem for observers in Europe and the northern United States. Though 6,200 light-years away, NGC 2354 is dimmed just 0.3 magnitude by interstellar dust. Its true diameter is 36 l-y, and its integrated absolute mag of -5 corresponds to a luminosity of just under 10,000 Suns.

NGC 2360: All of Canis Major's brighter open clusters are in easily-located positions, including NGC 2360, which is 3½° due east of Gamma (γ) Canis Majoris. A magnitude 5½ field star is just ⅓° to the cluster's west. In 10×50 binoculars the group appears as a 12′-long, mag 7.2, E-W oval of faintly mottled haze. Its brightest half dozen members are mag 10½–11½ stars, so it requires giant binoculars to begin to resolve. The cluster lucida is a mag 10.4 early-K orange giant with an absolute mag near 0. NGC 2360 is about 3,400 light-years distant, 12 l-y long, and has a very modest integrated absolute mag of −3.1 (a luminosity of 1450 Suns). It is dimmed only 0.2 mag by interstellar dust. The cluster's main point of interest is its age, 1 billion years.

NGC 2362: At 100× in 6-inch or 8-inch telescopes, NGC 2362 is a magnificent sight, a swarm of perhaps 40 magnitude 7½-11 diamond chips around the brilliant blue-white multi-carat jewel of mag 4.4 Tau (τ) Canis Majoris. Unfortunately, the cluster is so compact, just 8′ in diameter, that even in the higher powers of zoom binoculars only a couple cluster stars can be coaxed out of the overpowering glare of Tau.

But Tau CMa by itself is worth looking at. It is an O9 II giant with an absolute magnitude of −7.0, corresponding to a luminosity of 52,500 Suns. However, the star's *bolometric* absolute mag is −10.0, a total energy output of around 700,000 Suns. This is a very young star, probably only about 1 million years old. And in fact the H-R diagram of NGC 2362 shows that most of its stars have not even completed their initial gravitational contraction onto the stable main sequence.

Tau CMa and NGC 2362 are about 4,900 light-years away. They are dimmed only 0.5 magnitude by interstellar dust; but this is twice as much as the even more distant NGC 2354 located only 1° to the SW. The additional absorption is because Tau Canis and its cluster are just behind the SW fringes of the Canis Major R1 complex, an isolated, and very clumpy, series of dust clouds centered about 3,600 l-y distant (and therefore in the foreground of NGC 2362) covering a roughly 10°×5° NW-SE area along the Canis Major/Puppis border.

A probable outlying NGC 2362 member is the eclipsing binary EW CMa, located only about $^1/_3$° due north of the cluster. EW Canis ranges between mags 4.7 and 5.0 in 4.395 days, and its primary is a very hot O7 Ia supergiant. The absolute visual and bolometric magnitudes of this star are probably not much inferior to those of Tau Canis itself.

Cr 132 and **Cr 140**: In the star-rich Milky Way south and SW of the magnitude 2.4 Eta (η) Canis Majoris are two large fields, each with a handful of brighter stars, that bear Collinder open cluster designations. (Most Cr numbers apply to such groups.) Centered about $2^1/_3$° SW of Eta Canis is a $1^1/_2$°-diameter area with one 5th, four 6th, and three 7th mag stars, Cr 132. This is a good 7× binocular asterism but probably not a true physically-related star group. On the other hand, Cr 140, a $^2/_3$°-diameter field with three 5th, and three 6th–7th mag members 3° due south of Eta Canis, *looks* more like a true open cluster and indeed is. Cr 140 is estimated to be about 1470 light-years distant. Its true size is 18 l-y, and its integrated absolute mag a surprisingly high (considering its under-populated nature) −4.7, a luminosity of some 7,000 Suns. The rather high luminosity of Cr 140 is because it is a youngish cluster, perhaps about 30 million years old, that still has several bright stars.

III.7 Puppis

The central Puppis Milky Way south of the 3rd magnitude Rho (ρ) and Xi (ξ) Puppis and the central Monoceros Milky Way are the two brightest stretches of the Milky Way between the Cepheus/Lacerta star clouds on the NW and the brilliant Eta (η) Carinae Nebula (NGC 3372) region on the SE. This is because these are very transparent directions through the spiral disc of our Galaxy. In fact, absorption out to the rim of our Galaxy toward Omicron (o) Puppis, a distance of some 30,000 light-years, totals just 4 or 5 magnitudes, and several gal-

Puppis Open Clusters

Cluster	RA (2000.0)	Dec	m_V	Size	m_V*	*Spectrum	Distance	A_V	M_V	True Size	Age
M46	07ʰ42ᵐ	−14° 49′	6.1	27′	8.7		3400 l-y	0.13	−4.0	26 l-y	450 mil
M47	07 37	−14 30	4.4	29	5.7	B2 V	1700	0.2	−4.4	15	~30
M93	07 45	−23 52	6.2	22	8.2	gB9	3400	0.2	−4.2	22	
N 2423	07 37	−13 52	6.7	19	9.0		2400	0.7	−3.3	14	
N 2451	07 45	−37 58	2.8	45	3.60	K4 Ib-II	650		−3.9	9	60
N 2477	07 52	−38 33	5.8	27	9.8		4200	0.8	−5.6	33	1000
N 2527	08 05	−28 10	6.5	16	8.59	A0 V	1800	0.3	−2.5	8.4	
N 2533	08 07	−29 54	7.6	10	9.0		9000	1.1	−6.7	26	180
N 2539	08 11	−12 50	6.5	21	9.15	M4 III	4100	0.3	−4.3	25	660
N 2546	08 12	−37 38	6	40	6.44	O9.5 II	6200	0.5	−6	72	8
N 2567	08 19	−30 38	7.4	10	11.1		5200	0.4	−4.2	15	270
N 2571	08 19	−29 44	7.0	13	8.8	K1.5 II	5900	0.6	−4.9	22	
Cr 135	07 17	−36 50	2.1	50	2.70	K3 Ib	900		−5.1	13	30
Mel 71	07 37	−12 04	7.1	9	10.2		7800	0.3	−5.1	20	1000

axies can be seen in large telescopes right on the galactic equator near the NGC 2467 emission nebula.

But even elsewhere in the Puppis Milky Way interstellar dust clouds are scattered and compact, so through most of the constellation we have clear views across thousands of light-years of the Galaxy's spiral plane. Puppis has one very remote, but surprisingly-bright, emission nebula, NGC 2467 1° ESE of Omicron, an easy binocular object that will be described in the next chapter. But the main type of object the constellation offers observers are open clusters–and lots of them. Several of the Puppis open clusters are in fact naked-eye objects: Cr 135 and NGC 2451, which are very large (at least $^3/_4$° in diameter) and have mag 2.7 and 3.6 lucidae, respectively; M47, which has a mag 5.7 lucida and an integrated apparent mag of 4.4; and M46, M93, and NGCs 2477 and 2546, all of which are at least $^1/_3$° in diameter and have integrated apparent mags of about 6.0. Because M47, NGC 2451, and Cr 135 are large groups of a handful of bright members, they do extremely well in 10× glasses. But M46, M93, and NGC 2546 are more populous and crowded clusters of somewhat fainter stars and therefore are better in 20×–25× supergiant binoculars.

Puppis' open clusters are distributed in four loose groupings from NW to SE along the Puppis Milky Way. Though the clusters in these groupings are not physically related, they are sufficiently near each other that often two, three, or even four of them can be seen in the same wide-field view. In extreme NW Puppis, south of Alpha (α) Monocerotis, is the M46 + M47 cluster "pair" with NGC 2423 and Mel 71 just to its north. M93 is $1^1/_2$° NW of Xi Puppis and the bright emission nebula NGC 2467 about the same distance to the star's SSE. Several degrees south and SSE of Rho Puppis is the quartet of mags $6^1/_2$–$7^1/_2$ NGCs 2527, 2533, 2567, and 2571. And in SE Puppis, from WNW to NNE of Zeta (ζ) Puppis, is a west-to-east line of four open clusters, the two large, loose groups of bright stars Cr 135 and NGC 2451, and the two large, rich groups of faint stars NGCs 2477 and 2546.

M46 and **M47:** Though there are no bright stars near them, the open cluster pair M46 + M47 in extreme NW Puppis are easily spotted 5° due south of Alpha (α) Monocerotis. They are both very large clusters, each almost $^1/_2$° in diameter, and very bright, with integrated apparent magnitudes of 6.1 and 4.4. M47 especially is readily visible with the unaided eye (under good sky conditions) because its brightest members are mag 5.7 and 6.2 stars.

Though so near each other and so similar in size, it is evident even in the smallest binoculars that M46 and M47 cannot be a true open cluster "double" like the Perseus Double Cluster, NGC 869 + NGC 884. The Perseus pair are look-alikes, but M46 and M47 have very different appearances: M46 is a swarm of comparably-faint, evenly-distributed, mag 9, 10, and 11 stars; but M47 consists of two mag 6, two mag $6^1/_2$, and two mag 7.0 stars scattered loosely about, and not much else. M46 has a total population of several hundred, several dozen of which are within 3 magnitudes brightness of the mag 8.7 lucida. But M47 has a total population of only a few dozen, and its appearance is dominated by its half dozen brightest members.

However, the difference between M46 and M47 is more than merely a matter of star numbers. The two groups have significantly different bright-star types: the brightest members of M46 are modest-luminosity red giants and early-A main sequence stars; but the lucida of M47 is a hot blue B2 main sequence star, its second brightest member a B-type giant, and its four mag $6^1/_2$–7.0 stars B5-B7 subgiants and main sequence objects. The difference in their bright-star populations is the result of the two clusters' difference in age: M47, with its blue B-type stars, is young, not more than 30 million years old; M46, with its main sequence up only to early A and its evolved red giants, must be nearly a half billion years old.

Though they appear side-by-side, M46 is probably twice as distant as M47, 3,400 light-years as compared to M47's 1700. Thus M46 is in reality considerably larger than M47, 26 l-y versus 15. But they are not much different in true luminosity. In fact M47 might be a little brighter than M46, its integrated absolute mag of –4.4 corresponding to a luminosity of nearly 5,000 Suns and the –4.0 of M46 to 3,300 Suns. What M47 lacks in star numbers it makes up for in the sheer candle-power of its brightest members: the cluster's B2 lucida and B giant have luminosities of 1250 and 770 Suns, respectively.

Just within the northern edge of M46 is the small (1.1' diameter) mag 10.8 planetary nebula NGC 2438, at 100× in 4- and 6-inch telescopes an aethereal little disc of haze seemingly suspended among M46's swarm of star-sparks. Unfortunately the planetary cannot be a true M46 member: its radial velocity is 80 kilometers per second in recession whereas the cluster's is 42 km/sec (also in recession). By coincidence a faint and distant open cluster in the neighboring constellation of Pyxis, NGC 2818, also seemingly contains a planetary nebula. But there too the radial velocity difference, 21 km/sec in recession vs 1 km/sec in approach, precludes the planetary's membership in the cluster.

A probable outlying member of M47 is the eclipsing binary KQ Puppis, located just $^2/_3$° due west of the cluster's mag 5.7 lucida. This system has the modest range of mags 4.9 to 5.2 and consists of a bloated M2 red giant or supergiant orbited by a compact blue B2 star to which it is losing mass from its extended atmosphere.

That the M2 star is the dominant component is obvious even in binoculars, which show the variable to have a striking chrome-orange color that contrasts (in larger glasses) beautifully with the silver-blue of the mag 5.7 lucida of M47. The absolute mag of KQ at maximum, assuming it to be at the same distance as M47 and, like the cluster, dimmed 0.2 magnitude by interstellar dust, is –3.9.

M93: The bright (magnitude 6.2) and large (22' diameter) M93, a marginally-possible naked-eye object, is conveniently located $1^1/_2$° NW of the golden-yellow mag 3.3 G6 Ia supergiant Xi (ξ) Puppis. In standard binoculars the cluster appears as a comma-shaped haze mottled with partial resolution, the mag 8.2 lucida visible in the middle of the comma-crescent. With 25× supergiant instruments the comma begins to resolve and its horns can be seen to be connected by a semi-circle of faint stars enclosing a peculiarly star-vacant area. Thus M93, like many other moderately-populous intermediate-age open clusters (such as M50 and NGC 2301 in neighboring Monoceros) has a strange structure of star-arcs and straggling subgroups enclosing seemingly star-empty spaces. M93 is about the same distance as M46 to the north, 3400 l-y. It is comparable in size (22 l-y) and luminosity (4,000 Suns) to that group, but probably younger because its lucida is a B9 giant with an absolute mag of –2.2 (a luminosity of 625 Suns).

NGC 2423: In the same binocular field with M46 and M47, just $^2/_3$° north and slightly east of the latter, is the large (19' diameter) but scattered cluster of faint stars, NGC 2423. The lucida of this group is a mag 9.0 object; but most of its members are much fainter. In 10×50 binoculars NGC 2423 appears merely as an amorphous area of low-surface-brightness haze behind the lucida and one or two other 9th mag stars. The cluster's main point of interest is simply in sharing the same field with M46 and M47, for otherwise it offers little visually or astronomically. It is about 2,400 light-years distant and intrinsically neither large (14 l-y) nor luminous (absolute mag = –3.3 = 1760 Suns).

NGC 2451: An ideal cluster for 7×–10× binoculars is NGC 2451, a very large, loose group just over 4° WNW of Zeta (ζ) Puppis. It consists of a dozen magnitude 5 to 8, plus another 20 or so mag 9 and 10, stars scattered in a 45' area around the bright mag 3.6 c Puppis. The lucida is a K4 Ib–II near-supergiant with a fine orange color. The star's absolute mag is about –3.1 and it is dimmed around 0.3 mag by interstellar dust, so its distance must be about 650 light-years. However the physical reality of NGC 2451 has been questioned. One suggestion is that it is the superimposition of two poorly-populated clusters, one about twice as far as the other (the nearer group containing the bright star c Puppis). If it was a single group 650 l-y away, its true size would be 9 l-y and its integrated absolute mag –3.9, a luminosity of 3200 Suns (almost half of which is from c Puppis alone).

This region of the Milky Way happens to have several large, loose, aggregations of brighter stars. About 6° WNW of NGC 2451 is Collinder 135, which consists of the mag 2.7 Pi (π) Puppis and four 5th and 6th mag stars. (Cr 135 is discussed in more detail below.) North of Cr 135 in SE Canis Major are the very large ($1^1/_2$° in diameter) Cr 132 and the slightly more compact ($^3/_4$°) Cr 140, field condensations of a half dozen 5th-7th mag stars respectively SW and south of Eta (η) Canis Majoris. (Cr 132 and Cr 140 are described under Canis Major.) All four of these star groups are perfect for 7×–10× binoculars, and all but Cr 132 seem to be true physical clusters.

NGC 2477: The most populous open cluster in Puppis is NGC 2477, easily found 2° NW of Zeta (ζ) Puppis and just NW of the mag $4^1/_2$ b Puppis (not a true cluster member). This group is sufficiently bright, magnitude 5.8, that it can be glimpsed with the unaided eye as a hazy spot if the sky is sufficiently dark and clear (and if you are sufficiently far south: the cluster's –$38^1/_2$° declination means that at 40° N latitude it gets just $12^1/_2$° above the horizon). However, its brightest members are 10th mag objects, so in 10×50 binocuolars NGC 2477 only just begins to resolve, a few star-sparks embedded in its 27'-diameter, moderately-high-surface-brightness glow. Something of the cluster's crowded richness is evident in supergiant instruments, but it is so populous in faint stars that it requires the light-gathering power and magnifications of at least moderate-aperture telescopes to approach resolution. The central 15' of the group is something of a core around which is a less star-dense halo.

The appearance of the NGC 2477 core when resolved in telescopes–a swarm of more-or-less comparably-bright (mag 10–13) stars distributed in clumps and chains around curiously empty spaces–is typical of well-populated intermediate-age and ancient open clusters, and in fact NGC 2477 is estimated to be 1 billion years old. The 4-magnitude difference between the brightness of the cluster as a whole, mag 5.8, and the brightness of its lucida, mag 9.8, is a measure of the intrinsic populousness of NGC 2477: usually even in rich clusters that difference is only $2^1/_2$-3 magnitudes. Because of its richness, NGC 2477 has a very high integrated absolute magnitude for a billion-year-old open cluster, –5.6, a luminosity of 14,000 Suns. It also is rather large, 33 l-y in true diameter.

NGC 2527, NGC 2533, NGC 2567, NGC 2571: The four open clusters NGC 2527, NGC 2533, NGC 2567, and NGC 2571 will be discussed as a group because their main interest is simply that they are all in a relatively small 4°×2° NW-SE area of the star-rich central Puppis Milky Way. They are all reasonably large, 10' or more in diameter, and bright, magnitudes $6^1/_2$–$7^1/_2$, and thus are not difficult for 10×50 binoculars (except NGC 2533, of which only the mag 9.0 lucida can be seen). The problem is that there are no bright stars nearby to help locate these clusters. The best approach to the field is from the 3rd mag Rho (ρ) Puppis. NGC 2527 is exactly 4° due south of Rho. In 10× binoculars it is a fairly obvious field condensation of six or eight mag $8^1/_2$–$9^1/_2$ stars. NGC 2533, located 2° south and slightly east of NGC 2527 and $^1/_2$° due south of one mag 7 field stars and $^1/_2$° due east of another, has a mag 9.0 lucida, but all its other members are mag 11 and fainter. NGCs 2567 and 2571 are $3^1/_2$° SE of NGC 2527. They are 1° apart N-S and readily visible in 10×50 binoculars as moderately-large but low-surface-brightness patches. The mag 8.8 lucida of NGC 2567 is just $^1/_4$° ESE of a 7th mag field star.

This is, as has been said, a very transparent direction through our Galaxy and these four clusters confirm it by their range of distances, 1800 to 9,000 light-years. They also have an assortment of bright-star populations and ages. But none of these clusters is *young*. The nearest of the four, the 1800 l-y distant NGC 2527, has an A0 main sequence lucida, implying an age of 300 million years. It is an intrinsically small and faint group, just $8^1/_2$ l-y in diameter and only as luminous as 700 Suns: if it was much farther, it would be lost in the rich star field. With telescopes a small NNE-SSW gathering of mag 11–12 stars can be seen in the eastern part of NGC 2527. This is a much more distant, though not much richer, cluster that we happen to see through the foreground NGC 2527 star-gathering. Such line-of-sight cluster pairings are not rare in directions toward which we see long distances across the open-cluster-strewn spiral disc of our Galaxy. NGC 2451, discussed above, seems to be another such superposition of a nearer upon a farther cluster.

NGC 2567 and NGC 2571 are comparably distant, 5200 and 5900 l-y respectively (though there is considerable uncertainty in these numbers). NGC 2567 seems to be about half as luminous as

NGC 2571 (4,000 vs 8,000 Suns) and only two-thirds as large (15 vs 22 l-y). It is also the older of the two: its lucida has an absolute mag of just -0.5 and its main sequence goes up only to spectral type B8, whereas the primary of the mag 8.8 lucida of NGC 2571 is a K1.5 II orange giant with an absolute mag near $-2^1/_2$. NGC 2567 is about 270 million years old and NGC 2571 perhaps half that.

Astronomically the most interesting of these four central Puppis open clusters is also the most remote, 9000 l-y away, and the most difficult to resolve, NGC 2533. This is one of only three Galactic open clusters known to contain a carbon star. The NGC 2533 carbon star, located 5' due north of the cluster's mag 9.0 lucida, has a light range of 11.0–11.7 and a peak absolute mag of -2.3 (a luminosity of 700 Suns). Its color index of +4.3 is exceptionally red even for a carbon star.

NGC 2539: Though in extreme NE Puppis with no bright stars nearby, NGC 2539 is not difficult to find. It is exactly 7° due south of the very large and bright open cluster M48 in Hydra and just NW of the magnitude 4.7 19 Puppis. The latter is not a true cluster member but a foreground K0 star (with a splendid yellow-orange color in binoculars and RFTs). NGC 2539 is bright (mag 6.5), large (21' in diameter), and has a mag 9.1 lucida. In 10×50 binoculars its mag $9–9^1/_2$ members can be seen twinkling on a coarsely-textured background glow. The cluster is sufficiently populous with faint stars that it can take moderate powers in telescopes.

The two brightest members of NGC 2539 are a mag 9.15 M4 III red giant and a mag 9.56 K5 III orange giant. Their absolute mags are about -1.7 and -1.3, respectively. Such modest-luminosity late-type giants imply that this is a rather evolved group; and in fact its age is estimated to be the same the Hyades, 660 million years. The cluster's integrated absolute mag of -4.3 corresponds to a luminosity of 4,400 Suns.

NGC 2546: Centered about $2^1/_2$° NNE of Zeta (ζ) Puppis is a field of some 20 9th magnitude stars between a 40'-wide, NNW-SSE pair of mag $6^1/_2$ stars. This is the open cluster NGC 2546 and is visible to the unaided eye (if you live sufficiently far south: the cluster's declination is $-38^1/_2$°) as a hazy Milky Way patch. It resolves very well in 10×50 binoculars and is a splendid richest-field telescope target because of how bright and evenly-distributed its stars are. The cluster lucida, the mag 6.44 star on its SSE edge, is a hot, luminous O9.5 II giant with an absolute mag of about $-5^1/_2$.

NGC 2546 is around 6,200 light-years distant. Thus its true size is some 72 l-y and its integrated absolute mag higher than -6. It is less a gravitationally-bound open cluster than the main concentration of the Puppis OB3 association. This is not a particularly luminous-star-rich association, but it might include the Cepheid variable RS Puppis, located 3° due north of NGC 2546. RS Puppis is a good candidate for membership in a young OB association like Pup OB3 because it is an extremely long-period (41.414 day), high-luminosity (max absolute mag near $-6^1/_2$), Cepheid: short-period Cephs are more like 100 million years old and consequently found only in intermediate-age stellar aggregations. RS Pup ranges from mag 6.51 to 7.62, and between spectral types F8 and K5. The star's absorption of 1.8 mags is very high for this direction through the Milky Way, but is less from interstellar dust between here and there than from a small circumstellar dust cloud around RS itself.

Cr 135: The possible open cluster Collinder 135 consists of the magnitude 2.70 K3 Ib orange supergiant Pi (π) Puppis and four mag 4.6 to 5.9 stars spread from its NE to its NW. NE of Pi is the 4'-wide E-W common proper motion pair Upsilon-one (υ^1) and Upsilon-two Puppis, the former a mag 4.6 (slightly variable) star with a composite B2 V + B3 Vne spectrum, and the latter a B2 IVne subgiant of mag 5.1 (also slightly variable). Due north of Pi is the mag 5.0 HD 56779, a spectroscopic binary with a B2 IV-V primary. And finally almost 1° NW of Pi is the mag 5.9 B3 V main sequence star HD 55718. At 7×–10× this certainly *looks* like a physical star group; but its reality is still being debated. B2 and B3 main sequence and subgiant stars are often found in clusters that contain orange K or red M Ib supergiants. The absolute mag of K Ib supergiants is usually around $-4^1/_2$, which implies a distance of about 900 light-years to Pi Puppis. The absolute mags of the other four stars, if at that distance, would be between $-2^1/_2$ and -1, consistent with their spectral types. But two of them are variable and two are spectroscopic binaries, so it is hard to say *what* their absolute mags should be. Whatever the true status of Cr 135, in supergiant binoculars and RFTs there is a splendid color contrast between the orange of Pi and the silver-blues of the other four stars.

Mel 71: In 10×50 binoculars a 9' diameter, magnitude 7.1 disc of moderately-high-surface-brightness haze can be seen $2^1/_2°$ due north of M47. This is the distant open cluster Melotte 71. The brightest member of Mel 71 is only mag 10.2; consequently telescopes are required for really good resolution of the group. This cluster is about 7,800 light-years away and therefore its integrated absolute mag (corrected for 0.3 mag of dimming by interstellar dust) is −5.1, a luminosity of near 10,000 Suns. Mel 71 is of interest to the wide-field observer not only for its proximity to the impressive M46 + M47 pair, but also because it is an ancient open cluster, with an age estimated to be 1 billion years.

IV Open Clusters of Spring

During the early evenings of spring, the Milky Way is at its least conspicuous for mid-northern hemisphere observers. At that time of the year the stretches of the Milky Way that are not beneath the horizon are hidden by horizon haze. Thus you might be able to see the brightest stars of the autumn Milky Way constellations Cepheus, Cassiopeia, and Perseus low over the northern horizon, but you can see nothing of the glow of the Milky Way itself. And *beneath* your southern horizon by the same amount that Cepheus, Cassiopeia, and Perseus are *above* your northern horizon are the constellations of the never-rising, forever-invisible Far Southern Milky Way, Carina, Crux, and Centaurus.

What all this means is that toward the spring constellations we are looking out of the dust-laden spiral disc of our Galaxy into transparent intergalactic space. Here we would expect to see other galaxies, if there are any sufficiently bright to be seen with the instrument we are using. And as it happens there are *lots* of galaxies to be seen in this direction, even if we have nothing larger than 10×50 binoculars. These galaxies will be described in Chapter 4.

But open clusters tend to be concentrated along the Milky Way. It was for that reason they were formerly called *galactic* clusters. Thus the spring sky is not the part of the heavens in which we would expect to find open clusters. However, three of the spring constellations, Cancer, Hydra, and Coma Berenices, do in fact contain good binocular and richest-field telescope open clusters. Three of these clusters–M48 in Hydra, and M44 and M67 in Cancer–are actually not all that far away from the winter Milky Way. But the fourth, the Coma Berenices Star Cluster, is located almost at the north galactic pole and therefore as far from the galactic equator and the Milky Way as it can get. However, this too is a bit misleading because the Coma Star Cluster is only 260 light-years from us and consequently not really all that far off the plane of our Galaxy.

Three of the four open clusters of spring, M44, M48, and the Coma Star Cluster, are at or near their best in standard or giant binoculars, and the fourth, M67, is near its best in richest-field telescopes.

Praesepe, M44: Framed within the quadrilateral of 4th and 5th magnitude Gamma (γ), Delta (δ), Eta (η), and Theta (θ) Cancri, the stars forming the "body" of Cancer the Crab, is a N-S oval of dim haze the ancient Greeks called *Praesepe*, "the Manger" (with Gamma and Delta the "Asses" feeding at it). The Greek astronomical poet Aratos wrote in his *Phaenomena* (270 B.C.) that when the stars of Praesepe are obscured, it is a sign that rain is coming. Before him Babylonian astronomers had written that when the cluster becomes "red," strong winds are threatening. Both predictions were based on meteorological fact; for the outlying cirrus clouds that precede weather changes would be sufficient to obscure the hazy cluster itself, but not necessarily the stars around it.

With even the slightest optical aid M44 resolves into a beautiful swarm of stars (hence the cluster's modern name, "Beehive"). It has 24 members from mag 6.3 to mag 8.02, and scores of fainter stars. The central star-condensation of the group is a crude house-shaped asterism, $^2/_3$° E-W, reminiscent of the star-pattern of Cepheus. The stars of the west-pointing "gable" and of the "eaves" of the house-figure are wide, easily-resolved doubles. (The star in the southern "eave" is Epsilon [ε] Cancri.) The cluster is very near its best with 15×100 supergiant binoculars, which have the light-gathering power to make the cluster's brightest members very bright indeed (to the eye as bright as Spica or even Arcturus) but do not magnify it so much that any of it spills out

The Open Clusters of Spring

Cluster	RA (2000.0)	Dec	m_V	Size	m_V*	*Spectrum	Distance	A_V	M_V	True Size	Age
M44	08^h40^m	+19°59′	3.1	95′	6.30	Am	525 l-y		−3.0	15 l-y	660 mil
M48	08 14	−05 48	5.8	54′	8.23	G8 III	1500		−2.7	24	350 mil
M67	08 50	+11 49	6.9	15′	9.69	K3 III	2600	0.1	−2.7	12	6.3 bil
Mel 111	12 24	+26	2.5	3°	4.83	G0III+A3V	260		−2.0	15	500 mil

of the field of view. Though the cluster does not fit in 30× fields, it does improve in telescopes in one respect: the greater light-gathering power brings out the color contrast between M44's three orange giants, G8 III, G9 III, and K0 III stars of apparent mags 6.39, 6.44, and 6.59, and its other bright members, which are snow-white or cream-white A main sequence objects and A9/F0 giants.

M44 is estimated to be about 525 light-years distant. Thus its $1^1/_2$° apparent extent corresponds to a true size of a modest 15 l-y. (The house-shaped asterism is about 6 l-y long.) The cluster's integrated apparent mag of 3.1 implies an integrated absolute mag of just −3.0, a luminosity of only 1320 Suns. Its main sequence goes up to spectral type A5/A6, suggesting a cluster age around 660 million years. M44 is nearly identical to the Hyades in age, luminosity, bright-star types, and even in true space motion through our Galaxy (though at present the two clusters are about 450 l-y apart). The Hyades and Praesepe are the two cluster cores of the *Hyades Stream* of stars, which is described in more detail in the section on the Hyades under Taurus. The two clusters and their star stream are the remnants of an OB association formed in a giant molecular cloud 660 million years ago. Even after three orbits around the Center of the Galaxy (the "Galactic year" of the Sun's neighborhood being about 220 million years) the two clusters and their stream still preserve something of the original space motion of their parent GMC.

M48: Hydra is such a long constellation, spanning 100° of the southern spring sky, that its far SW corner is grazed by the winter Milky Way and its far east end (its tail-tip) dips into the western edge of the summer Milky Way. About 12° SW of the Head of Hydra, right on the Hydra/Monoceros border, is the open cluster M48. This group is easy to see even in the smallest binoculars because it is almost a full degree in diameter, has more than a dozen 8th, 9th, and 10th magnitude members, and its integrated apparent mag is 5.8, above naked-eye visibility. However, it is so far from any really bright star that it can be a little difficult to find the first time you look for it. Start at the Head of Hydra and go 8° SW to the 1°-long, NW-SE line of mag 4.0 C Hydrae flanked by mag $5^1/_2$ 1 and 2 Hya. The cluster is another $3^1/_2$° SW, at the south vertex of an equilateral triangle with C Hya and the mag $4^1/_2$ Zeta (ζ) Monocerotis. But once you see the cluster, there's no mistaking it.

M48 is probably at its best in 15–25×100 supergiant binoculars, with which perhaps four dozen cluster stars can be seen. 25× is preferable because several of the group's brightest members are crowded into a central condensation. Three of them can be seen to be close doubles. A number of the cluster's other bright stars are distributed around the central condensation in something of a large ring.

M48 is about 1500 light-years away, so its true diameter is a respectable 24 l-y but its integrated absolute mag merely −2.7, a luminosity of just 1100 Suns. It has three yellow giants: one of them, a G8 III star, is the mag 8.23 cluster lucida. The group's main sequence goes up to a mag 8.8 A1 star, implying a cluster age of at least 300 million years.

M67: The fairly large (15′) and moderately bright (magnitude 6.9) open cluster M67 is 1.8° due west of Alpha (α) Cancri. In 10×50 binoculars it appears as an oval of mottled haze, elongated NE-SW, with an 8th mag field star just outside the cluster's NE end and the group's mag 9.7 lucida on its SW edge. The resolution of the cluster improves markedly with 15×100 supergiant binoculars, numerous tiny star-sparks being intermittently visible in what are discernably knots of individual stars, not merely clumps of haze. It appears more circular than in standard binoculars because some cluster members can be seen to arc NW of the main oval in a large curve joining the 8th mag field star on the NE to the mag 9.7 cluster lucida on the SW. A scattering of faint star-specks–probably true M67 members–show up north and NE of the 8th mag star. The cluster's center of concentration is in the oval just NE of the lucida. In 15×100s M67 is very compact and very delicate–a superb sight.

M67 is about 2,600 light-years away. Thus its 30° apparent distance off the galactic equator converts to 1300 l-y off the Galactic plane. This is a good deal more than M44 (see above) is off the Galactic plane simply because M67 is five times farther from us than M44. Only old open clusters are found well away from the Galactic plane, and M67, with an age of some 6.3 *billion* years, is one of the oldest. Like most other highly-evolved groups, M67 is neither large nor luminous, though its 12 l-y size and integrated absolute mag of −2.7 are low even for such an ancient cluster.

Coma Berenices Star Cluster, Mel 111: The Coma Berenices Star Cluster, officially catalogued as Melotte 111, is one of the handful of open clusters which can be said to in effect resolve with the unaided eye (excluding intrinsically faint dwarf cluster members). This select group includes of course only the very nearest open clusters: the Ursa Major Moving Group (centered 71 light-years away), the Hyades (150 l-y), Coma (260 l-y), and the Pleiades (410 l-y). The Theta (θ) Carinae and Omicron (o) Velorum clusters in the far southern Milky Way (IC 2602 and IC 1391, both about 490 l-y distant) and the Alpha (α) Persei Cluster (570 l-y) are partially resolved with the unaided eye.

The Coma Star Cluster has eight 5th magnitude, four 6th mag, and six 7th mag stars. Its total population is about 60 down to mag $10^1/_2$, with very few fainter members. The core of the cluster is the 3° from 14 to 22 Comae; but even this part of the group is so loose that it is a better sight in 7× than in 10× binoculars. Some of the bright stars in the area, including Gamma (γ), 7, and 10 Comae, are not true cluster members. But several distant stars are members, among them 31 Comae 6° ENE of the cluster center. Coma has a good 7× double star, the mag 5.3 17 Comae with the mag 6.6 BD+26°2353 145"~$2^1/_2$' to its WSW. In high-power supergiant binoculars the mag 4.8 12 Comae can be seen to have an 8th mag F8 companion 67" distant in PA 167° (almost due south).

If the light from all members of the Coma Star Cluster was gathered into a single stellar point, it would have an apparent magnitude of 2.5. However, at the 260 l-y of the cluster this means that the group's integrated absolute mag is a paltry −2.0, a luminosity of just 525 Suns. Moreover, the cluster has an extremely low mass density. The 3° from 14 to 22 Comae is a projected separation of about 15 l-y. Assuming a cluster diameter of 16 l-y ~ 5 parsecs, and given a total cluster mass of 100 Suns, the mass density of Coma is 0.14 Solar mass per cubic parsec, just three times the mass density in the stars in the Sun's neighborhood. Thus the internal gravitational field of the Coma Star Cluster is very weak, with the consequence that most of its low-mass stars have simply "evaporated" out of it in the 500 million years since it was born.

Though the Coma Star Cluster is practically at the north galactic pole, it is not really very far off the Galactic plane. Because the Sun is about 40 l-y off the Galactic plane toward the NGP, the true distance of Coma off the plane is 260 + 40 = 300 l-y–a far cry from the 1300 l-y of M67!

Chapter 3
The Milky Way and its Bright Nebulae

The Milky Way is the ultimate wide-field celestial object. It is, in fact, as wide-field as a celestial object can get because it extends for a full 360° around the sky. Therefore the Milky Way itself (as opposed to the individual objects, such as clusters, along it) is best observed with as wide-field an instrument as possible–binoculars, giant and supergiant binoculars, and richest-field telescopes ("RFTs." Binoculars can be thought of as two RFTs mounted in parallel.)

This chapter is devoted to the wide-field Milky Way features–star clouds, dust clouds, star chains, star streams, and bright nebulae–best suited to giant binoculars (and small RFTs). Bright nebulae–emission nebulae, reflection nebulae, and supernova remnants–are described in this chapter instead of being given a chapter of their own because they not only are exclusively in the Milky Way, they in fact mark important spiral features of the Milky Way Galaxy. Moreover, these nebulae are often extremely large and therefore at their best in the same types of instrument you would use to look at Milky Way star clouds, dust clouds, and star streams.

Because the Milky Way traces a great circle around the celestial sphere, it naturally accommodates itself to division into degrees, like any other circle. *Galactic longitude* is measured from 0° to 360° along the center line of the Milky Way, the *galactic equator*: 0° is the direction toward the Galactic Center (located in SW Sagittarius 4° WNW of Gamma [γ] Sgr); 90° is the direction of our orbital motion around the Galactic Center (which happens to be toward Deneb in Cygnus); 180° is the *galactic anticenter* (on the Taurus/Auriga border $3^1/_2$° east of Beta [β] Tau); and 270° (in central Vela) is the direction from which our orbital motion is carrying us. (Strictly speaking, our Sun's individual motion through the Galaxy is directed not exactly at longitude 90°, but somewhat off the Galactic plane, toward a point in the constellation Hercules.) *Galactic latitude* gives a celestial object's apparent height above or below the galactic equator and, like terrestrial latitude, is measured from 0° at the galactic equator to 90° at the galactic poles. The north galactic pole is in Coma Berenices some 4° east of the center of the Coma Star Cluster; and the south galactic pole is about 9° south of Beta (β) Ceti in the star-poor wastes of Sculptor.

The Structure of the Milky Way Galaxy

To best appreciate what you are looking at in the Milky Way, you need to know exactly where it fits within the over-all structure of our Galaxy. The Milky Way is not a uniform band of glow around the sky. It is wider along some stretches, narrower along others. It is very bright in certain places, but fades almost to invisibility in others. Some Milky Way constellations are rich in open clusters or emission nebulae, others are not. And for over a third of its circuit the Milky Way is divided into two more-or-less parallel streams by the Great Rift.

All these features reflect the true global and spiral structure of our Milky Way Galaxy. Now, decoding the true form of our Galaxy from what we see along the Milky Way is no easy trick, for we are located fairly within the thick of things, virtually on the dust-rich Galactic plane, and it is difficult to see the forest for the trees (or rather, for the smoke between the trees, because much of the interstellar dust is simply carbon soot). Nevertheless astronomers have been able to determine that we live within a relatively loose-armed Sbc spiral galaxy with a small central bulge (that probably is barred rather than simply spheroidal). The disc of our Galaxy is nearly 100,000 light-years in diameter, our Sun a little less than 30,000 light-years from the Galactic Center on the inner edge of a rather broken spiral arm. The spiral arms are embedded in a star-smooth disc that is (very roughly) around 3,000 light-years thick. The central bulge is a rather flattened structure approaching 10,000 light-years in equatorial diameter. (Both the bulge and the disc thin gradually and therefore cannot be said to have true outer boundaries. Consequently in technical texts and articles astrophysicists are understandably reluctant to commit themselves to firm bulge and disc dimensions.) The total face-on luminosity of our Galaxy is estimated to be about 14 billion Suns, corresponding to an absolute magnitude of $-20^1/_2$. But the Galaxy's total mass in stars, gas, and dust seems to be several hundred billion times the Sun's.

The different structural features of our Galaxy are constituted of different types of stars and other objects. The bulge is composed generally of evolved stars, the brightest of which are K and M orange and

Photo 3.1 The Milky Way toward the Central Bulge of the Galaxy. The bright star pair in the center of the field is Lambda (λ) + Upsilon (υ) Scorpii. Directly below (south) of them, halfway to the bottom of the field, is Theta (θ) Sco. Due west of Theta Sco toward the right edge of the field is the tight open cluster NGC 6231, with the looser cluster Tr 24 and the IC 4628 emission nebula just to its north. The two large open clusters to the NNE and ENE of Lambda + Upsilon Sco are M6 and M7, respectively

red giants. The bulge can be said to be basically free of gas and dust, though there is a central disc of gas and dust immediately around the bulge's strange, and apparently sometimes explosive, nucleus. The spiral arms, by contrast, are rich in gas and dust and are traced by young, hot, blue giant and supergiant stars and the emission nebulae (such as the Lagoon in Sagittarius and the Rosette in Monoceros) which they fluoresce. The spiral arms are, however, embedded within the disc of the Galaxy, which is composed of somewhat evolved stars–A-type and later main sequence stars, and yellow, orange, and red G, K, and M normal giants. (Our Sun, though on the edge of a spiral arm and only 40 light-years off the plane of the Galaxy, is actually a member of the Galaxy's disc population.) Finally, the bulge and disc of the Galaxy are enveloped by a tenuous spherical halo of highly-evolved stars–RR Lyrae variables, K- and M-type asymptotic branch giants, and metal-poor subdwarfs (stars underluminous for their spectral type)–and of globular clusters, which are the most conspicuous constituents of the Galactic halo. The halo population rapidly decreases in density outward from the Galactic bulge. The disc stars of our Galaxy are on circular orbits in the Galactic plane around the Galactic Center; but the halo stars and globular clusters are on much-elongated comet-type orbits that carry them in toward, close around, and then back out in random directions away from the Galactic Center.

The stars of the spiral arms, disc, bulge, and halo form an age sequence. The supergiant and the hot blue O-type stars in the open clusters and stellar associations that trace the spiral arms are at most 20 or 30 million years old, and many are less than a few million years old. The stars and open clusters of the disc are generally from a hundred million to a few billion years old. Bulge stars are roughly 10 billion years old. (The bulge is a very complicated place and some of its stars, particularly in and near the nucleus, are much younger than this.) And the stars and globular clusters of the halo are very ancient, 12 to 15 billion years old.

The Spiral Arms in the Solar Neighborhood

We know that our Galaxy is a rather loose-armed spiral. But tracing the arcs of those spiral arms is no easy trick even in the immediate Solar neighborhood (by which we mean out to 12 or 15,000 light-years from the Sun). Most of our trouble in mapping the spiral structure of our Galaxy comes from our location just 40 light-years off the gas- and dust-rich plane of our Galaxy. Indeed, dust is so thick toward the Galactic interior that it dims the cluster of super-luminous stars known to lie at our Galaxy's Center by *30 magnitudes*: we can "see" these stars only at dust-penetrating infrared wavelengths. Even most directions toward the rim of our Galaxy are anything but transparent. The famous Perseus Double Cluster, for example, is dimmed almost 2 magnitudes by dust: less than 20% of the light of its stars gets through the Galactic murk to us!

Nevertheless, the visible features and the distribution of objects along the Milky Way do hold hints about our Galaxy's global structure and about the course of the local spiral arms. The brightest and broadest stretch of the Milky Way is toward Sagittarius, and therefore it is no surprise that this happens to be the direction of our Galaxy's central bulge, which, in the absence of interstellar dust, could be seen to cover the whole area between Antares on the west, Nunki (Sigma [σ] Sagittarii) on the east, M16 on the NE, and the Tail of Scorpius to the south. The precise location of the Galactic Center is behind the Sagittarius Rift at RA 17^h 45.6^m and declination −28° 56.2', a rather blank spot about 4° WNW of Gamma (γ) Sagittarii. The Great Sagittarius Star Cloud, spread over a roughly triangular area several degrees across north of Gamma and Delta (δ) Sagittarii, is an actual segment of the Galactic bulge and therefore some 25–30,000 light-years away. However, the Small Sagittarius Star Cloud, M24, a $2°\times{}^3/_4°$ NE-SW rectangle located 2° NNE of Mu (μ) Sagittarii and hardly inferior in brilliance to the Great Sgr Star Cloud, is a stretch of a star-rich interior spiral arm (the interior spiral arms in general are broader and brighter than the exterior spiral arms) and consequently somewhat nearer to us. M24 could be the incurving edge of the Norma Spiral Arm, the outcurving edge of which is at the Norma Star Cloud, located in the far southern Milky Way constellation Norma. The nearer edge of the Small Sgr Star Cloud is some 12,000 light-years

Photo 3.2 The Great Sagittarius Star Cloud. The Lagoon Nebula M 8 is on the top-center edge of the field. The bright star on the left-center edge is Epsilon (ε) Sagittarii. The brightest star in the center of the field is Gamma (γ) Sgr. Toward the lower right corner are the large open clusters M6 (right edge) and M7 (lower edge)

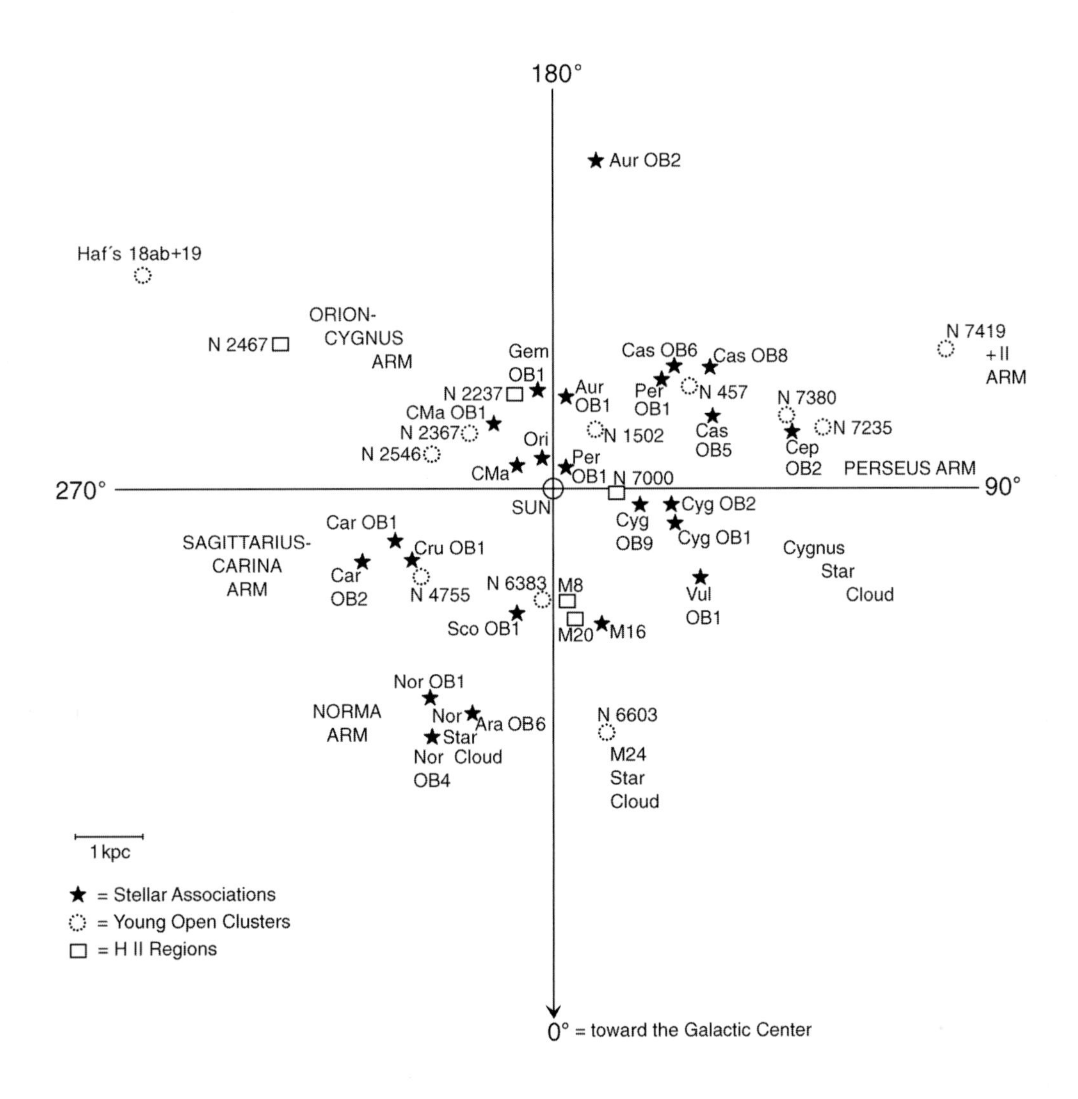

Figure 3.1 The distribution of emission nebulae, OB associations, and young open clusters in the Sun's region of the Galaxy

from us, and the star cloud is probably over 4,000 light-years thick.

Spread from Sagittarius on the NE through the Tail of Scorpius and SW all the way to the extreme southern Milky Way in Crux, SW Centaurus, and NE Carina is a series of famous frequently observed and photographed clusters and nebulae between 5,000 to 8,000 light-years away. They include the Star Queen Nebula M16 (6,500 l-y distant), the Swan Nebula M17 (6,800 l-y), the Trifid Nebula M20 (5,400 l-y), the Lagoon Nebula M8 (5,200 l-y), the open cluster NGC 6231 in the Tail of Scorpius (6,200 l-y), the Jewel Box Cluster NGC 4755 in Crux (8,200 l-y), and the Eta (η) Carinae Nebula NGC 3372 (7,800 l-y). All these objects are tracers of the Sagittarius-Carina Spiral Arm, the next spiral feature in toward the Galactic bulge from our Orion-Cygnus Spiral Arm. The star clouds around the Eta Carinae Nebula are exceptionally bright because in that direction we look down the length of the central ridge of the Sgr-Car Spiral Arm where it arcs out toward the Galactic exterior. Moreover, this part of the Sgr-Car Arm is especially transparent: apparently much of the dust in it was used up in the formation of the numerous associations and open clusters of supergiant stars that lie within, around, and beyond the Eta Car Nebula.

In the foreground of these Sgr-Car Spiral Arm clusters and nebulae are the nearby blue giant and main sequence members of the Scorpio-Centaurus Association, an elongated stellar aggregation centered only about 550 light-years away that includes most of the bright stars in Scorpius, Lupus, Centaurus, and Crux. The Sco-Cen Association marks the very inner edge of our Orion-Cygnus Spiral Arm. Also along the inner edge of the Ori-Cyg Arm is the series of dust clouds comprising the Great Rift which bisects most of the Milky Way from Deneb in Cygnus on the NE to Alpha (α) Centauri on the SW. (In the direction of Sagittarius and Scorpius the Great Rift dust cloud chain arcs off the galactic equator up toward Antares. The rift in the Milky Way through Sagittarius and the Tail of Scorpius is from Sgr-Car Arm dust.) The Coalsack in Crux is a detached Great Rift cloud. It, and the Great Rift as a whole, lies several hundred light-years away just beyond the Sco-Cen Association. We know from photographs of other spiral galaxies that spiral arms are frequently bordered on their inner edges by Great-Rift-type dust lanes.

The incurving arc of the Sgr-Car Spiral Arm seems to be marked by the brilliant Scutum Star Cloud, located nearly 30° NE along the Milky Way from the direction toward the Galactic Center. However, the full width of the incurving arc of the Sgr-Car Arm here is blocked by the Great Rift NW of Scutum in Serpens Cauda. Notice the pronounced asymmetry of the Sgr-Car Arm: the *outcurving* arc of the arm, toward the Eta Car Nebula star clouds, is some 80° from the direction toward the Galactic Center and therefore is, as indeed it should be, considerably farther from the direction toward the Center than its *incurving* arc at the Scutum Star Cloud.

Farther NE along the galactic equator is the Aquila Milky Way, divided into two conspicuous streams by the Great Rift, peculiarly poor in open clusters, and entirely lacking in emission nebulae. The cluster- and nebula-poverty of Aquila is a consequence of the fact that in this direction we look down the length of the interarm gap between the *outer* edge of the Sgr-Car Arm (in Scutum) and the *inner* edge of our Orion-Cygnus Spiral Arm (in Vulpecula and SW Cygnus) and there is nothing to see here but a poor scattering of evolved interarm open clusters in front of the distant wall of our own spiral arm where it curves in toward the Galactic interior. Notice that the NE stretch of the Aquila Milky Way (around Altair), the nearest part of the incurving arc of our spiral arm to us, is, as would be expected, brighter than the presumably much more distant SW Aquila Milky Way.

Strictly speaking, the direction the Solar neighborhood is headed in its orbit around the Center of the Galaxy is almost exactly at Deneb in Cygnus. However, because spiral arms *wind up* in the direction of orbital motion, we find that the brightest part of our Orion-Cygnus Spiral Arm immediately ahead of us is somewhat *within* the direction toward Deneb: it is the Cygnus Star Cloud, a 20°-long oval between Beta (β) Cygni on the SW and Gamma (γ) Cygni on the NE. But ahead of us our spiral arm is very broad and star-rich, so there are also bright star clouds beyond galactic longitude 90° north and NE of Deneb, and in northern Lacerta. Relatively nearby Ori-Cyg Arm associations are in this direction, including Lacerta OB1 (1600 light-years away toward southern Lacerta) and Cepheus OB2 (2,500 l-y distant toward southern Cepheus). And a chain of dust clouds averaging 2,500 l-y from us extends all across central Cepheus.

But farther east along the Milky Way, toward Cassiopeia, the dust thins. Consequently in the direction of Cassiopeia, and as far east as the famous Perseus Double Cluster, we have a more-or-less clear view through our own spiral arm out to the next spiral feature toward the rim of our Galaxy, the Perseus Arm. Hence Cassiopeia and NW Perseus are very rich in open clusters, these clusters lying both in the interarm gap between our Ori-Cyg Arm and the Perseus Arm (including M52 and NGC 7789), and in the Per Arm itself (NGC 7790, NGC 457, M103, NGC 663, the Double Cluster). Moreover, through the Cassiopeia Window we can see several large Perseus Arm emission nebulae (NGC 281, IC 1805, IC 1848). The Perseus Arm objects in the Cassiopeia Window are between 6,000 and 10,000 light-years away. The Cassiopeia Milky Way is decently broad; but it is not especially bright because the star-density toward our Galaxy's rim is much less than the star-density toward our Galaxy's bulge. The exterior spiral arms are both less thick and less star-rich than the interior spiral arms because the interior of the Galaxy contains the bulk of the heavy, dense, cool molecular dust and gas from which stars form.

However, there is at least *some* dust toward our Galaxy's rim, and much of the Perseus Milky Way east and SE of the Double Cluster is obscured by nearby dust clouds. The dust also extends down into northern Taurus. The Alpha (α) Persei Cluster (540 light-years away) and the Pleiades (410 l-y) are just in front of the dust. The Zeta (ζ) Persei Association, which includes Zeta, Xi (ξ), and Omicron (o) Per as well as the California Nebula (NGC 1499), has condensed out of some of this dust and gas just 1.3 million years ago.

Toward Auriga the dust once again thins. But the galactic anticenter, the direction opposite the Galactic Center, is located here (on the Auriga/Taurus border $3^1/_2$° east of Beta [β] Tauri) and consequently there is not a lot of Galaxy in this direction and the Auriga Milky Way, which runs NW-SE through the interior of the Pentagon of Auriga (Alpha [α] - Beta - Theta [θ] - Iota [ι] Aur + Beta Tau) is rather pale and featureless. But because of the thinness of the dust we can see not only the 4,100 light-year distant open clusters M36 and M37, which lie on the far side of our Ori-Cyg Arm, but also the 15,600 l-y distant open cluster + emission nebula NGC 1893 + IC 410, a tracer of a remote arc or spur of the Perseus Arm.

In this stretch of the Milky Way, the foreground objects of our Orion-Cygnus Spiral Arm are actually *beneath* the Milky Way glow proper. These objects include the Zeta Persei Association, the Pleiades, the Perseus/Taurus dust clouds behind the Pleiades, the Orion Association and its involved bright and dark nebulae, and the Canis Major Association (which includes most of the bright stars of southern Canis Major). This arc of young clusters and associations, and their involved dark and bright nebulae, is part of *Gould's Belt*, a ring-like structure in our spiral arm that, on the opposite side of the Milky Way, is traced by the Scorpio-Centaurus Association and the Great Rift chain of dust clouds. Gould's Belt does not lie exactly on the plane of the Galaxy, but is tilted 16°–17° out of it. Consequently it curves below the winter Milky Way through Taurus, Orion, and Canis Major but above the summer Milky Way as far as the stars of the Head and Heart of the Scorpion, all of which are Sco-Cen Association members. In several sections of Gould's Belt star formation is presently occurring (Orion, Perseus/Taurus, and the Rho [ρ] Ophiuchi region). This "ring of fire" is thought to have been ignited some 70 million years ago, perhaps by a supernova explosion, in a "hot spot" between the Pleiades and the Alpha Persei Cluster (which are among the oldest, and therefore earliest-formed, Gould's Belt objects) and to have been expanding outward ever since until it is now nearly 2,500 l-y in diameter.

But over and beyond the Gould's Belt objects from Perseus to Canis Major is the stream of the Milky Way itself, which flows from Auriga on the NW down behind the feet of Gemini and the club of Orion, through Monoceros, over the back of Canis Major, and into Puppis on the SE. The Milky Way glow is at its brightest in the galactic anticenter half of its circuit toward central Auriga, central Monoceros, and central Puppis. These are directions in which the dust of our own spiral arm is very thin, so we can see long distances across the outer disc of our Galaxy. Between these "windows" are relatively nearby Orion-Cygnus Arm complexes of gas and dust with involved open clusters and bright nebulae, including the NGC 2264 "Christmas Tree" cluster in northern Monoceros (3,000 light-years away), the Rosette Nebula NGC 2237 and its central open cluster NGC 2244 in western Monoceros (4,900 l-y), and the IC 2177 emission nebula and its cluster NGC 2353 on the Monoceros/Canis Major border. Through the central Puppis window, far down the outcurving arc of our Ori-Cyg Arm, we see the small but bright emission nebula NGC 2467, 15,000 light-years away. The scattered distribution of these complexes of gas, dust, and young star groups is characteristic of the Galaxy's exterior spiral arms.

Toward the far southern Milky Way constellation Vela we look in the direction of the out-trailing arc of our Ori-Cyg Arm. Vela, like Cygnus and the direction of the incurving arc of our Arm, is rich in

Photo 3.3 Panorama of Gould's Belt from Perseus to Orion, with the Comet Hale-Bopp. Photographed on March 27, 1997 at Nassfeld, Kärnten in the mountains of southern Austria with a tripod-mounted Nikon EM camera and a 50mm lens stopped to f/2.8 using Kodak Ektar 1000 film. This is a mosaic of five 30-second exposures obtained simply by moving the tripod-head for each exposure. This panorama shows how, as seen in the early evenings of spring from mid-northern latitudes, the line of Gould's Belt objects from Perseus to Orion arcs below the Milky Way (which runs along the top of the panorama). The zodiacal light, here visible along the horizon below the Pleiades, is also best seen from northern latitudes during the spring

open clusters, stellar associations, and bright and dark nebulae. The interarm gap between the *inner* edge of the out-trailing arc of our Ori-Cyg Arm and the *outer* edge of the out-trailing arc of the Sagittarius-Carina Spiral Arm is in southern Vela. This gap is not as wide as the Aquila interarm gap on the opposite side of the Milky Way between the in-curving stretches of the Ori-Cyg and Sgr-Car arms, but its Milky Way glow is comparably dim and it is likewise poor in open clusters.

Observing Bright Nebulae

The four types of bright nebulae are: (1) *emission nebulae*, which are (very often large) clouds of interstellar gas glowing from fluorescence; (2) *reflection nebulae*, areas of interstellar dust reflecting the light of nearby stars; (3) *planetary nebulae*, shells or rings of glowing gas recently ejected by a dying red giant, the tiny but hot residual core of which provides the ultraviolet photons ionizing the nebula's gas; and (4) *supernova remnants*, the debris clouds of

Alpha Persei Cluster IC 1805 IC 1848 Double Cluster Cassiopeia Hale-Bopp

catastrophic stellar explosions glowing either by the exotic synchrotron process (in which high-velocity electrons excite the gas) or by the energy released when the supernova shell collides with the surrounding interstellar gas and dust.

The eyepiece appearances of such astrophysically-contrasting objects naturally will be considerably different. Planetary nebulae, because they consist only of the gas from the outer layers of a single star usually not much more massive than the Sun, are intrinsically small; consequently they are almost always visually small as well, though often of relatively high surface brightness. Reflection nebulae are also usually very small in apparent size because the intensity of star-light decreases with the distance from the star *squared* (twice as far, four times less): therefore only from the immediately around even a very bright star will the light reflected by the dust in its vicinity be sufficiently intense to be seen. Moreover, reflection nebulae are often very faint around luminous stars because the radiation pressure and stellar winds from these energetic objects has usually scoured most of the gas and dust out of the region. Emission nebulae, though also of low surface brightness, are usually rather large because the hot stars that power them are giants and supergiants that emit copious quantities of ultraviolet radiation and thus fluoresce the gas in a very large volume of interstellar space. Supernova remnants are rare, briefly-lived objects, and the handful visible in richest-field instruments show a great variety in appearance.

Of the four types of bright nebulae, the one least suited to giant binoculars and other RFTs are reflection nebulae because they are usually faint and always small. A few can be glimpsed in standard 7×50 and 10×50 glasses, and perhaps a score in giant instruments. But the vast majority of reflection nebulae require both large aperture and at least moderate magnification to be seen.

In one sense planetary nebulae are even worse richest-field objects than reflection nebulae: nearly all of them are not only faint – only three are brighter than mag 8.0 – but they are also *extremely* small, less than 1 minute of arc in diameter. Consequently with wide-field instruments even the brightest planetaries usually appear simply "stellar." Nevertheless three or four planetaries show discs even in small binoculars and are quite visually interesting in giant binoculars, including M27 (the Dumbbell Nebula in Vulpecula), M97 (the Owl Nebula in Ursa Major), and NGC 7293 (the Heliacal Nebula in Aquarius). Indeed, NGC 7293 is at its best in binoculars because it is so large and tenuous that it must be viewed with the lowest possible magnification if it is to be seen at all.

But emission nebulae are the ideal giant binocular and RFT object. Because they are usually large, you do not have to magnify them very much to see them: thus you concentrate their tenuous light into a smaller area, making them easier to see. A large emission nebula like IC 1396 in Cepheus is completely invisible at 50× in 8-inch telescopes but easy to see in 10×50 binoculars. And in giant binoculars some of the internal structural detail of several emission nebulae can be glimpsed.

Emission and reflection nebulae are necessarily always found in conjunction with dense clouds of cool, dark interstellar gas and dust, for they are simply the illuminated sections of those clouds. Because the Galaxy's interstellar medium is concentrated along the Galactic plane, virtually all emission and reflection nebulae lie in the Milky Way itself. Supernova remnants are also Milky Way objects because they usually are the debris cloud of the explosion of a young star. Planetary nebulae, however, are the death shrouds of aged stars which have had plenty of time (billions of years) to drift away from the place of their birth in the gas and dust along the Galactic plane and consequently can be seen in practically every direction around the sky, outside as well as within the Milky Way.

I The Star Clouds of Sagittarius and Scutum

The Great Sagittarius Star Cloud, north of Gamma (γ) and Delta (δ) Sagittarii, is the brightest of all the Milky Way star clouds. It is so bright because it is a section of our Galaxy's central bulge which we see over the thick dust of the Sagittarius Rift. Bays of Sagittarius Rift dust can be traced with giant binoculars along the NW side of the star cloud; and a fairly conspicuous triangular extension of the star cloud juts into the Sgr Rift SE of the 4th magnitude Cepheid variable X Sgr. However, the opposite side of the star cloud, to the SE, fades away gradually (as indeed does the periphery of the whole Galactic bulge, if we could see it). A semi-detached lobe of the Great Sgr Star Cloud is centered midway between Epsilon (ε) Sagittarii and the open cluster M6 in Scorpius; and, further SW, the Milky Way glow behind the 80'-diameter open cluster M7 is actually another part of the Galaxy's bulge. We know that these two small star clouds are sections of the bulge because on color photos they have the same yellowish color as the Great Sgr Star Cloud itself. Studies of the bulge have shown that most of its light comes from K-type giants like Arcturus, stars with a yellow-orange or golden-yellow color.

Because of its great distance (roughly 25,000 to 30,000 light-years) and its lack of any really luminous stars (K giants have absolute magnitudes of just −1 to −2), in normal binoculars the Great Sagittarius Star Cloud appears simply as a smooth glow. However, supergiant binoculars bring out from this glow a scintillating partial resolution of multitudes of star-sparks. Notice especially the field 1° NW of Gamma Sgr: this is *Baade's Window*, through which we have a line of sight obscured by only 1.3–2.8 magnitudes of dust to within 1,800 light-years of the Galactic Center itself. (The globular cluster NGC 6528, visible in Baade's Window, is about this distance from the Galactic Center.) Moreover, with giant binoculars a few objects can be glimpsed on the background glow of the star cloud. First is a N-S line of partial obscuration between Gamma Sgr and Y Sgr to Gamma's NW. (Baade's Window is just on the Gamma side of this thin dust lane.) Second, about $2^1/_2$° N and slightly W of Gamma Sgr is the small (5'), but *very* dark, dust cloud B86, a single bright star on its NW edge. Finally, on the east edge of B86 is the compact (6') open cluster NGC 6520, a tight group of six or eight mag 9–10 stars. NGC 6520 is estimated to be about 5,200 light-years away, just the distance of the Sagittarius-Carina Spiral Arm; but it is a somewhat evolved cluster and not a true tracer of that spiral arm. B86 and NGC 6520 are probably not physically involved: the very darkness of B86 suggests it is an isolated dust cloud much nearer to us than the cluster.

But even more spectacular than the Great Sagittarius Star Cloud is the Small Sagittarius Star Cloud, M24, located 2° NNE of Mu (μ) Sagittarii. With even a field glass, the $2° \times {}^3/_4°$ NE-SW bent-rectangle shape of this brilliant star cloud is obvious. On good nights M24 appears in supergiant binoculars *precisely* like it does on the best long-exposure astrophotos. But the binocular appearance of the star cloud is superior to any photo because the vigorous scintillating of the multitude of bright stars sparkling upon its luminous glow gives M24 a living immediacy lacking on static photographs, which simply smear the star images out into small dots. The SW half of M24 has noticeably more stars embedded in it than its NE half.

The density of magnitude 7–10 stars visible on M24 is decidedly greater than that of the surrounding field. This is a consequence of the fact that M24 is a section of one of the luminous-star-rich spiral features of our Galaxy's interior–possibly of the Norma Spiral Arm, the next spiral feature from the Sgr-Car Arm in toward the Galactic bulge. The brightness of the star cloud's glow is because our Galaxy's interior spiral arms are broad and star-dense and because toward M24 we probably look down the incurving arc of the Norma Arm. We view M24 through a "window" in the dust of the Sgr-Car Arm. The Sgr-Car dust lane is about 5,000 to 7,000 light-years away, the star cloud itself centered perhaps twice as far. However, the two small (approximately 12' oval) dark clouds, B92 and B93 (the latter NE of the former) conspicuous along M24's NW long side must be (because of the lack of foreground stars visible on them) much nearer than the Sgr-Car Arm. They could be rogue cloudlets of the Great Rift, which here arcs well to the west away from the central line of the Milky Way. In 100 mm binocs a solitary 12th mag star can be glimpsed in the middle of B92, and a narrow dark channel

Photo 3.4 The Small Sagittarius Star Cloud, M24 (center). To its west (right) and east are the open clusters M23 and M25, respectively. To its north are the small patches of the Swan (M17) and Star-Queen (M16) nebulae. Further north of M16, right on the edge of the Great Rift, is the pale nebula around the open cluster NGC 6604. The Great Sagittarius Star Cloud is on the lower edge of the field and the Scutum Star Cloud in the upper left corner

followed SW from B92. To the west of this channel is a semi-detached star cloud. This cloud is paler than M24, and lacks M24's overscattering of bright stars; but it can be very nicely partially-resolved into a multitude of momentarily-twinkling star-specks.

Near the eastern edge of the "neck" of M24 (where it bends) is the hazy dot of the open cluster NGC 6603. In reality this is a very large, very populous group that would look like M11 in Scutum if it was not 12,000 light-years away and therefore $2^1/_2$ times more distant. 10' SSE of NGC 6603 is the multiple star β639 = HD 168021, the magnitude 6.4 and 7.9 A-C components of which are 18" apart in PA 52° and resolvable at 15×. This star is one of the bright members of the Sagittarius OB4 association, centered about 7,800 light-years away in the Sagittarius-Carina Spiral Arm. The brightest members of the association are hot giant and supergiant late-O/early-B stars (HD 168021A has a B0 Ib spectrum) and it is conceivable that their powerful stellar winds and radiation pressure can be at least partly credited with opening up the window in the Sgr-Car Arm dust through which we see M24.

The next bright star cloud NE along the Milky Way is the Scutum Star Cloud, centered WSW of Lambda (λ) Aquilae. It is roughly 5° across, NW-SE, Beta (β) Scuti on its north corner. Its background glow rivals that of M24, but it lacks the same thick garnish of bright stars. However, in supergiant binoculars the Scutum Star Cloud is visible to its full photographic extent, and, with averted vision, literally *myriads* of tiny sparkling star-chips can be glimpsed twinkling in its glow. The star cloud fades gradually toward the SE away from the galactic equator (because the actual star-density of our Galaxy's disc decreases with distance off the Galactic plane) but is abruptly silhouetted on the NW by the dust of the Great Rift. On the NW corner of the Scutum Star Cloud two NE-SW channels of obscuring matter have cut off a pair of star cloudlets, the larger and rounder cloudlet outside a smaller and narrower one. Both cloudlets are slightly fainter in background glow than the Scu Star Cloud itself, but both (especially the inner one) are more richly overscattered with bright binocular stars.

Several interesting objects can be seen on and around the Scutum Star Cloud. Snaking across the brightest part of the star cloud between the (telescopic) double Σ2350 on the SW and the variable R Scuti on the NE is a chain of four 6th-7th magnitude stars with at least a dozen 8th and 9th magnitude attendants. This chain is easily spotted in giant binoculars; but it does not have the blatant conspicuousness of the kind of star chain that sometimes can be found near young open clusters (such as the chain south of the "dipper" of the Pleiades). To the north of the Scu Star Cloud, and NE into Aquila, is a splendid region of mixed bright and dark Milky Way masses. From this region a dark bay, catalogued B111, bulges down into the Scu Star Cloud. Two long dust lanes can be seen extending ENE from the NE edge of B111 toward Lambda Aquilae. Near the southern edge of B111 is the hazy disc of the rich, compact open cluster M11. In an arm of the Scu Star Cloud that reaches from the ESE into B111 toward (but not all the way to) Beta Scu is a very small granular patch of haze, the distant globular cluster NGC 6704, visible even in supergiant binoculars merely as an enhancement in the background star-cloud glow.

Between the Small Sagittarius Star Cloud and the Scutum Star Cloud are two fainter, but attractive, star clouds (They show up well on Photo 2.4). One is centered due north of M24 and due west of the bright gash of M17, the Swan Nebula. This star cloud has an almost square profile and in supergiant binoculars appears almost as richly star-garnished as M24 itself, but with fainter stars. In 20×60 glasses the square profile of this cloud can be seen, but no stars resolved. The other star cloud in the area is in SW Scutum, with Gamma (γ) Scuti near its SW corner. It has an intriguing triangular shape, two sets of parallel dust lanes, with a streak of pale Milky Way glow separating them, defining its NE and NW sides. Its southern edge is more ambiguous.

I.1 The Bright Nebulae of Sagittarius

The spiral arms of galaxies are so obvious essentially because they are defined by large associations and open clusters of luminous young stars and by the extensive clouds of glowing hydrogen gas, H II

Photo 3.5 The Lagoon and Trifid Nebulae in Sagittarius

Photo 3.6 The Lagoon Nebula, M8 in Sagittarius. North is to the left

Sagittarius Bright Nebulae									
Nebula	Name	Type	RA (2000.0)	Dec	Dimen	m_V*	*Spectrum	Distance	True Size
M8	Lagoon	EN	18^h04^m	−24°23'	45'×30'	5.97	O4V	5200 l-y	75×50 l-y
M16	Eagle	EN	18 19	−13 58	35×28	8.24	O4V	6500	66×53
M17	Swan	EN	18 21	−16 11	40×30	9.3		6800	80×60
M20	Trifid	EN+RN	18 02	−23 20	30×20	6.9	O7V	5400	50×35
N6526		EN	18 05	−23 35	40×30				
N6559		EN+RN	18 10	−24 06	8×5	9.8	B1	6000	14×9
N6726/7		RN	19 02	−36 53	9×7	7.2, 8.8v	A, B2	420	1.1×.85
N6729		EN+RN	19 02	−36 57	1var	9.7, 12	F5pe	420	0.12

regions, around the youngest of these associations and clusters. Toward Sagittarius we look from our location on the Galactic interior edge of the Orion-Cygnus Spiral arm across a relatively empty inter-arm gap at the star clusters, stellar associations, dark dust clouds, and bright emission nebulae of the Sagittarius-Carina Spiral Arm. The major Sgr-Car H II regions toward Sagittarius are, from SW to NE, the Lagoon Nebula M8, the Trifid Nebula M20, the Swan Nebula M17, and, just across the border in extreme SE Serpens Cauda, the Eagle or Star-Queen Nebula M16. In addition to these major emission regions, two other Sgr-Car Arm bright nebulae are visible with large binoculars and richest-field telescopes: the very pale NGC 6526 emission cloud $^2/_3$° due north of M8, and the small patch of combined emission and reflection glow NGC 6559, $1^1/_2$° ENE of M8.

In the tiny constellation of Corona Australis due south of the "Milk Dipper" of Sagittarius is an isolated dense dust cloud containing a couple small patches of emission and reflection glow, NGC 6526/7 and NGC 6729, the former a high-magnification giant binocular object. However, the small dust cloud in which they are embedded, Bernes 158, is not in the Sgr-Car Arm but only 420 light-years away and therefore on the Galactic interior edge of our Ori-Cyg Spiral Arm.

The Lagoon Nebula, M8: One of the few bright nebulae visible to the unaided eye is M8, the Lagoon Nebula. It has an integrated apparent magnitude of 3.6 and thus is four times brighter to the unaided eye than the Orion Nebula, M42, which has an integrated apparent mag around 5.0. The surface brightness of M8 is quite high, so in standard binoculars almost 40'×25' of its 45'×30' photographic extent, and the nebula's basic rectangular shape, can be seen. In the eastern half of the M8 glow 10×50 glasses show a knot of a half dozen members of the NGC 6530 open cluster, which is actually physically involved with the Lagoon. Much of the florescence of M8 is excited by the mag 6.9 lucida of NGC 6530, a hot O6.5 V star.

In the western half of M8 10x50s reveal $^1/_4$° east and slightly south of the mag 5.5 7 Sagittarii (a foreground object), a 3'-wide, NNE-SSW pair of stars, which, like NGC 6530, are actually involved with the nebula. The northern star is an O9 II-III giant with an apparent mag of 7.1 and an absolute mag of −4.8 (8,000 Suns). But the southern star, the mag 6.0 9 Sgr, is an extremely hot O4 V object with a visual absolute mag of an even *more* impressive −10.7 (a total energy output of 1.6 *million* Suns!) Most of the florescence of the Lagoon must be from this single star. Just 3' SW of 9 Sgr is a 9th mag "star." This is in fact the brightest nebulous knot of the Lagoon, the 30"-tall Hourglass Nebula, with the mag $9^1/_2$ O7 star fluorescing it.

The Lagoon Nebula begins to come into its own with 100mm supergiant instruments. In 15×100s the famous 1'-wide, NE-SW rift can be glimpsed with averted vision: it is difficult because of the glare of 9 Sgr to the west and of the NGC 6530 cluster to the east. One could not have "discovered" the Lagoon's rift with 15×100s (at least from mid-northern latitudes: in the southern hemisphere, where the Lagoon can be observed against the black sky near the zenith, the rift probably is not difficult). The Hourglass is well resolved from 9 Sgr. NGC 6530 is exquisite in 15×100s, With which can be seen 15 to

Plate XVII

Plate XVIII

Plate XIX

Plate XX

Plate XXI

Plate XXII

Plate XXIII

Plate XXIV

Plate XXV

Plate XXVI

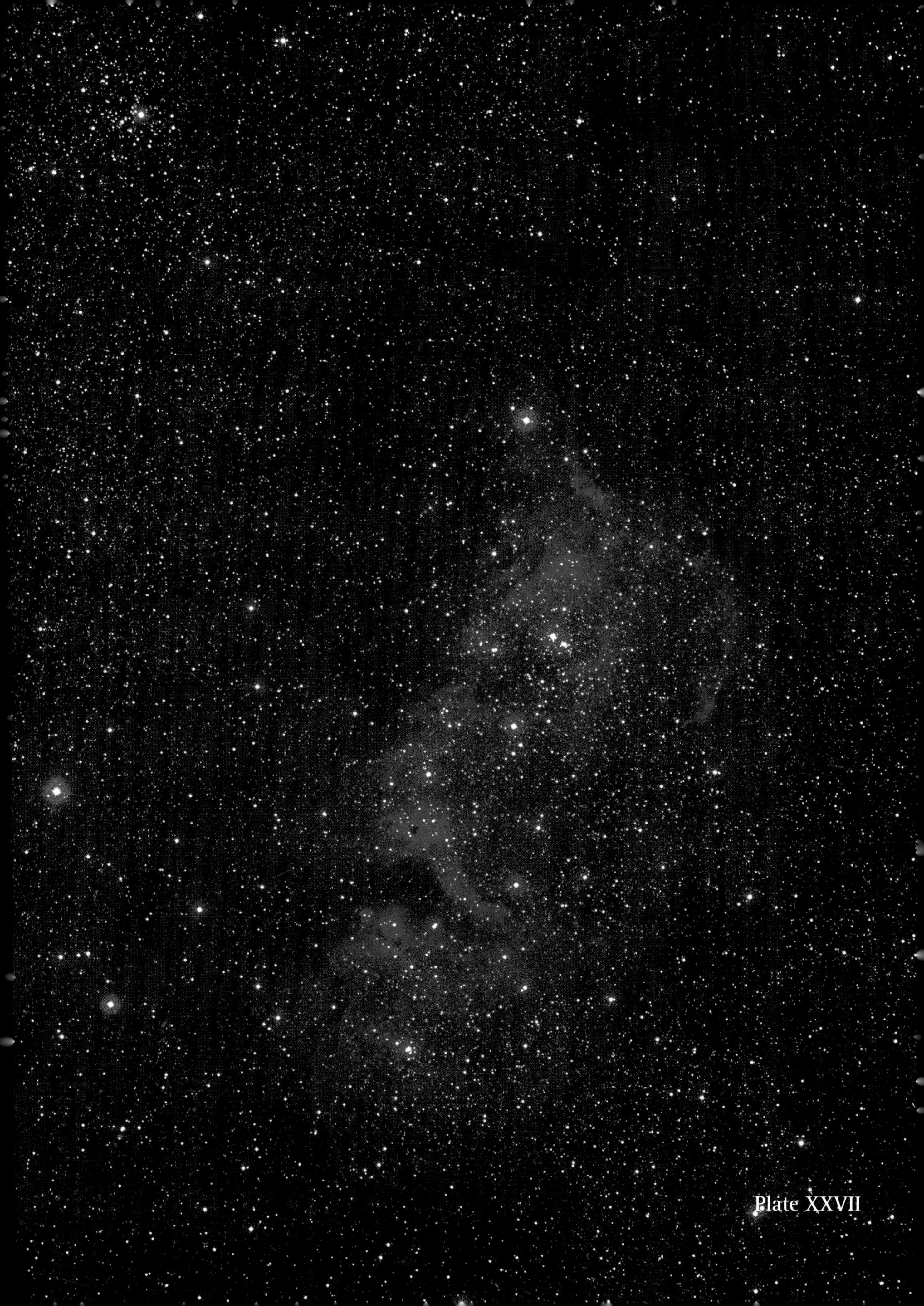

Plate XXVII

Plate XXVIII

Plate XXIX

Plate XXX

Plate XXXI

Plate XXXII

18 tiny stellar pinpricks, most steadily visible but a few twinkling in and out of resolution. The cluster's two brightest members are a N-S double at its center, the mag 6.9 lucida being to the south. The comparable brightness and even distribution of the stars of NGC 6530 makes it a very attractive sight.

In 25×100s the rift of the Lagoon is downright easy and NGC 6530 becomes very large and well-resolved. The Hourglass splits into two bright NE-SW points, the former the mag $9^1/_2$ O7 star, Herschel 36, the southern point the Hourglass Nebula proper. North of the Hourglass/9 Sgr region, a little separated from the main body of the Lagoon, is an outlying area of nebula-glow around a mag $7^1/_2$ star.

The Lagoon Nebula is estimated to be about 5,200 light-years distant. Thus its 45'×30' photographic dimensions translate to a true size of 75×50 l-y. However, faint outlying emission tendrils give M8 a full E-W extent of 80' ~ 130 l-y. The Lagoon is the heart of the Sagittarius OB1 association, which includes the open cluster M21 2° to the north and probably the Trifid Nebula complex $1^1/_2$° to the NNW. The brightest Sgr OB1 star is the mag 3.85 B8 Ia supergiant Mu (μ) Sgr, a small-amplitude eclipsing binary. The second brightest association member is the mag 5.84 HD 164402, a B0 Ia supergiant located $^1/_3$° NNW of M20. Sgr OB1 also has four O7 to O9 giants of apparent mags 7–$7^1/_2$ and absolute mags from –4.8 to –5.9. The two mag $5^1/_2$ stars just NE of Mu Sgr, 15 and 16 Sgr, are also highly-luminous blue giants, 15 Sgr a B0 Ia supergiant and 16 Sgr an O9.5 II-III object. They are, however, not included in Sgr OB1 but considered to be members of a separate Sgr-Car Arm association, Sgr OB7, 5,700 light-years distant.

The Star-Queen Nebula, M16: The traditional name for M16, located in extreme SE Serpens Cauda, is the "Eagle Nebula." However, the late Robert Burnham, Jr. introduced the more evocative title "Star-Queen Nebula," inspired by the complicated dust formation that projects over the center of the nebula from the SE. This dust feature can be seen only in the largest amateur telescopes. But the nebula-glow and its involved open cluster (technically the open cluster is catalogued as M16 and the emission nebula as IC 4703) is $^1/_3$° across, has an integrated apparent magnitude of 6.0, and can be spotted with even the smallest binoculars. Though there are no bright stars nearby, the M16 complex is easily found $2^1/_2$° north and slightly west of the even more conspicuous Swan Nebula M17. It is 4° due north of the northern corner of the Small Sagittarius Star Cloud M24.

In 10×50 binoculars the nebula-glow of M16 is visible over the 20' area from the open cluster (which appears as a knot of a half dozen 8th and 9th mag stars) on the north to the 5'-wide, E-W pair of 8th mag stars on the SSE. The surface brightness is quite low–less than that of M17 to the south. The extent of nebula-glow seen does not appreciably increase with supergiant binoculars and small RFTs, though the star cluster begins to resolve well at just 15×. However, 25×100 supergiant glasses show the broad wedge of obscuring matter that juts over the nebula from the north.

The M16 complex is about 6,500 light-years distant, so the 20' visual size of the nebula corresponds to 40 l-y. The full photographic dimensions of the nebula are 35'×28' = 70×56 l-y. M16 is the core of the Sagittarius-Carina Spiral Arm association Serpens OB1. The M16 open cluster and Ser OB1 are described in more detail in the section on Serpens Cauda in Chapter 2.

The Swan Nebula, M17: In telescopes the M17 emission nebula displays a very interesting shape: it is a bright, 12'-long, E-W bar from the west end of which a short semicircle of nebula-glow, curving back toward the east, extends a few minutes of arc south. Because of its shape M17 has been given a legion of names–"Horseshoe Nebula," "Omega Nebula" (from the Greek capital letter omega: Ω), and "Swan Nebula." The E-W bar, though very narrow, has such high surface brightness that it is visible even in 7×35 binoculars. In 10×50s the bar reveals a distinct northern edge, but fades more ambiguously toward the south; and two faint stars south of the west end of the bar give the impression that the curved feature–the Swan's "neck"–is visible. The glare of these two stars, which lie exactly along the arc of the curved feature, make the actual sighting of the "neck" uncertain even in 15×100 supergiant glasses. But with these instru-

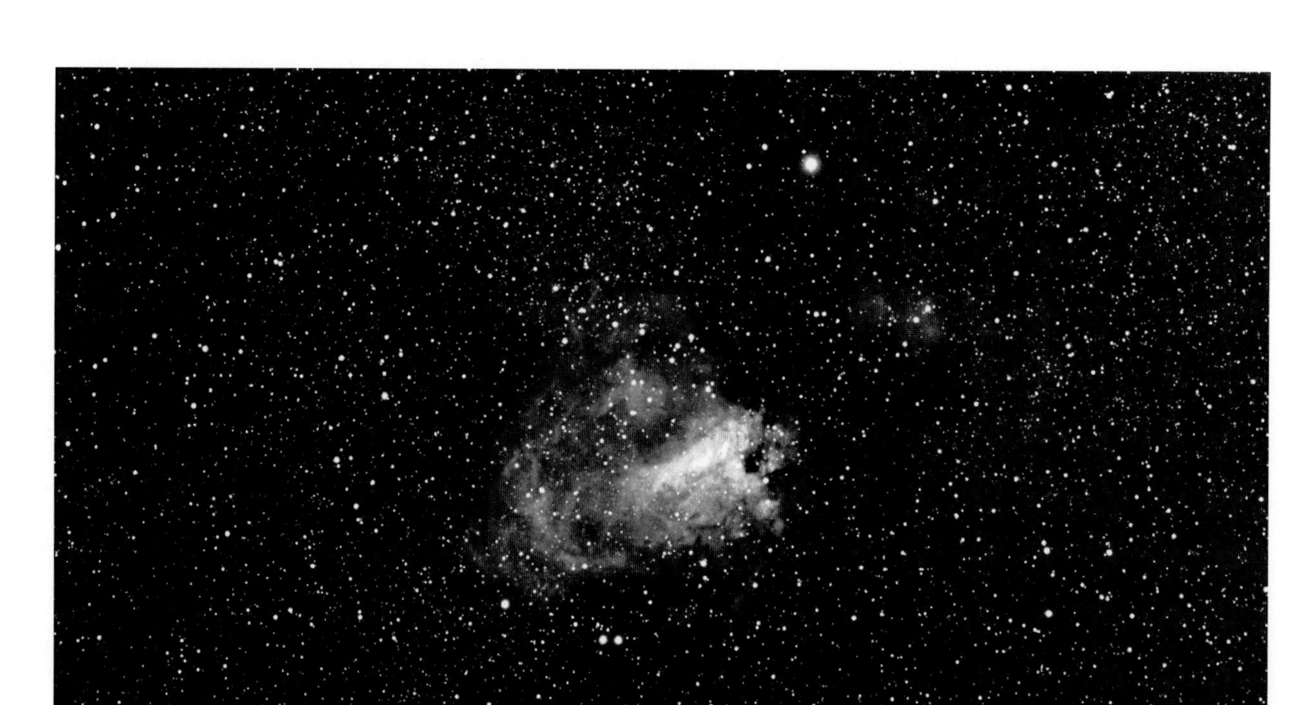

Photo 3.7 M17, the Swan Nebula in Sagittarius

ments the bar can be observed to diminish a little in brightness toward its east end, to be extremely sharply-bordered on the north, and to fade so rapidly toward that south that looks like it must not extend very far in that direction. Just north of the bar is a conspicuous field condensation (though obviously not a true cluster) of mag 10–11 stars.

The sharp northern edge of M17, and the darkness within the "neck" of the Swan, are caused by thick foreground dust of the M17 complex itself. This dust also obscures most of the central star cluster of the complex so that, in contrast to M8, M16, and M20, we do not see (except at dust-penetrating infrared wavelengths) the hot stars whose ultraviolet light is fluorescing the nebula. The hidden core of the M17 cluster is west of the Swan "neck."

M17 is about 6,800 light-years away, so the true size of its 12'-long "bar" is about 24 l-y. Photos reveal that its outlying nebulous tendrils and filaments cover an area of about 40'×30', so its full true size is some 80×60 l-y. The Swan is about the same distance from us as the M16 complex $2^1/_2°$ to the NNW and of the heavily-obscured open cluster NGC 6604 and Ser OB2 association $1^1/_2°$ north of M16. All three complexes are Sagittarius-Carina Spiral Arm tracers and conceivably all part of the same supercloud. At 6,500 l-y the apparent 4° distance from M17 on the south to NGC 6604 on the north corresponds to a projected true distance of 480 l-y.

The Trifid Nebula. M20: The Trifid Nebula can be seen with even the smallest binoculars as a bright little patch $1^1/_2°$ NNE of M8. It is not large, measuring 30'×20', and therefore easier at 10× than at 7×. In 10×50 glasses it is distinctly elongated N-S. The three dark bands of obscuring matter from which the nebula takes its name, collectively catalogued as dark nebula B85, require 6-inch telescopes at about 100× to be seen. They slice the M20 pie into eastern, SW, and NW thirds. On the inner tip of the nebula's eastern segment is its illuminating central star, the multiple HD 164492, the mag 6.9 O7 V primary of which is providing the ultraviolet radiating fluorescing the Trifid. This star has five companions, the easiest to resolve (but requiring telescopes) being the mag 8.8 C component 10.6" distant in PA 212°. This star group is probably a proto-cluster like the Trapezium multiple at the heart of the Orion Nebula M42.

On Plate XII and Photo 3.5 it can be seen that the Trifid actually consists of two distinct parts. The circular southern part, with the three obscuring lanes, appears red on color photos because it is the H II region fluoresced by the O7 primary of HD 164492 and its light is dominated by the red 6562 Å emission line of hydrogen. The smaller northern lobe–which gives M20 its elongation–is a blue reflection nebula around a mag $7^1/_2$ star that is too cool to ionize the hydrogen gas around it. But the dust around this star must be rather dense for the M20 reflection nebula to be visible in 10×50 binoculars: usually reflection nebulae have too low surface brightness to be seen in small-aperture instruments. And the mag $7^1/_2$ star must be rather luminous for its light to penetrate so far into the dense dust around it: most higher-surface-brightness reflection nebulae are small because they are around modest-luminosity stars.

The most recent estimate places M20 about 5,400 light-years away–not much farther than M8. the Trifid is probably in the same complex of giant molecular clouds as the Lagoon and therefore should be considered in the Sgr OB1 association. Its true size is somewhat smaller than M8's, 50×35 l-y. The circular emission glow around HD 164492 is about 35 l-y in diameter. This has been called a "small H II region in an early stage of evolution" and its age estimated to be only 300–400,000 years. Because stars usually radiate equally in all directions, emission regions around hot stars tend to begin spherical. But as they expand (they are hotter than the dark cloud around them), they soon encounter denser pockets of the interstellar medium, and this slows their expansion in that direction and disrupts their spherical symmetry. Hence, as a general rule, the more circular the profile of an H II region, the younger it is. M20 should be thought of as a hot, expanding bubble within a cool dense cloud of molecular gas and dust, some foreground filaments of which are silhouetted upon it. Within the cool dust to the north of the bubble is a bright but cool star, the blue light of which is being scattered by the interstellar dust like the dust in the Earth's atmosphere scatters the blue light of the Sun, giving us our blue sky.

NGC 6526: With 10×50 binoculars a pale Milky Way glow $^1/_2$° across is visible centered about $^2/_3$° north of M8 and the same distance SE of M20. This is catalogued as the emission nebula NGC 6526, though on photographs it can be seen that behind its very pale emission glow is a very rich Milky Way field of faint stars which cumulatively is probably brighter than the nebulosity. In supergiant binoculars NGC 6526 appears as a field of tiny star-chips twinkling in and out of resolution, the stars evenly distributed and comparably bright. In effect NGC 6526 is a window between the dust clouds of the Lagoon and Trifid complexes through which we see past a thin drapery of H II haze deeper into our Galaxy.

NGC 6559: Centered just over 1° ENE of the Lagoon Nebula is a dust cloud within which are embedded several small reflection and emission nebulae, including IC 2174, IC 2175, IC 4685, and NGC 6559. The easiest of these to see is the 8'×5' NGC 6559, visible in 15×100 supergiant binoculars as a small hazy patch around its magnitude 9.8 B1 central star. Like many nebulae around B0 and B1 stars (in contrast to nebulae around hot-O type stars), its glow is a combination of H II emission and of the reflection of starlight off dust. The NGC 6559 complex is within the Sagittarius-Carina Spiral Arm and very probably part of an extensive M8 + M20 + Sgr OB1 supercloud. Assuming it to be 6,000 light-years away, the true size of NGC 6559 itself is 14×9 l-y.

NGC 6726/7: Exactly 1° west of Gamma (γ) Coronae Australis, just south of the Sagittarius/ CrA border, is a NW-SE, 5'-wide, pair of small nebulae, NGC 6726/7 and NGC 6729. They are toward the western edge of a small dust cloud, Bernes 158, that extends between Gamma and Epsilon (ε) CrA. NGC 6729, the SE nebula of the pair, is very small, less than 2' long NW-SE, and thus visible only in telescopes. But NGC 6726/7 measures 9'×7', NE-SW, and is sufficiently large and bright to be a big binocular object (if you are far enough south: the nebula is at –36° declination). It is easy to find because it is just

Photo 3.8 The small nebulae NGC 6726/7 (the brightest patch in the upper center of the field) and NGC 6729 (on the lower right edge of NGC 6726/7) in Corona Australis. North is to the left. Toward the upper right is the globular cluster NGC 6723 in Sagittarius. Epsilon (ε) Coronae Australis is near the upper edge and Gamma (γ) CrA in the lower center of the field. The dark feature extending toward the lower right of the field from the bright nebulae is the dust cloud Bernes 158

over $^1/_2$° SE of the very bright (magnitude 6.1), large (13') globular cluster NGC 6723 in Sagittarius and about 40' ENE of the 4th mag Epsilon (ε) CrA. About $^1/_4$° SW of the nebula is the splendid 25× binocular double Brs 14, which has mag $6^1/_2$ and 7 components separated by 13" in PA 281°.

NGC 6726/7 is a double-lobed reflection nebula. Its SW section, NGC 6726, reflects the light of an A-type star and its NE section, NGC 6727, the light of the erratic nebular variable TY CrA, a B2 star with a range between mags 8.8 and 12.6. NGC 6729, a combination reflection + emission nebula, is illuminated by the nebular variable R CrA. NGC 6729 varies in brightness, size, and internal detail like NGC 2261, Hubble's Variable Nebula, in Monoceros (described in Section 6 of this chapter).

Bernes 158 is only about 420 light-years away. It is an isolated small molecular cloud on the inner edge of our Orion-Cygnus Spiral Arm in which the formation of Solar-mass stars is presently occurring. The length of the cloud is only about 8 l-y, and the size of NGC 6726/7 just 1 l-y. The nebular variables in NGC 6727 and NGC 6729 are stars which have not yet settled down onto the stable main sequence. Another small molecular cloud currently gestating Solar-mass stars is in the south circumpolar constellation Chamaeleon.

II The Aquila Inter-arm Gap and the Great Rift

The most conspicuous feature of the Aquila Milky Way is the Great Rift, which divides its glow into two more-or-less parallel SW-to-NE streams. The Great Rift continues NE through Sagitta and Vulpecula into Cygnus, where it dead-ends at Deneb. It extends SW back from Aquila along the length of Serpens Cauda. However, because it is part of Gould's Belt , the Great Rift is in fact tilted about 16° with respect to the galactic equator and therefore arcs out of the Milky Way from Serpens Cauda across southern Ophiuchus to the Antares region of Scorpius. (The Sagittarius Rift NW of the Great Sagittarius Star Cloud is dust in the Sagittarius-Carina Spiral Arm, not dust associated with the much-nearer Great Rift on the inner edge of our own Orion-Cygnus Arm.) The Great Rift then arcs back toward the Milky Way in Lupus; and in Norma and SE Centaurus it once again divides the Milky Way into two conspicuous streams. Its SW end is at the brilliant mag –0.3 Alpha (α) Centauri (see Plate IV.)

The southeastern stream of the Aquila Milky Way is the brighter of the two: it is the continuation of the Milky Way stream that includes the Great Sagittarius Star Cloud, the Small Sagittarius Star Cloud, and the Scutum Star Cloud. The northwestern stream of the Aquila Milky Way begins SW of Aquila behind the "V" of the Taurus Poniatovii asterism in east central Ophiuchus and continues NE beyond Aquila, becoming brighter through Sagitta and Vulpecula until it climaxes as the Cygnus Star Cloud, the brightest Milky Way feature in the northern heavens.

The finest scan across the Great Rift anywhere is from the Gamma (γ) Aquilae region in the SE stream of the Milky Way to Epsilon (ε) + Zeta (ζ) Aql in the NW stream. The abrupt transition from the brilliance of the Gamma Aql star cloud (a very bright glow embedded with multitudes of momentarily resolved star-grains) to the darkness of the Great Rift really does suggest the interposition of virtually opaque dust, and it is difficult to understand how most astronomers around 1900 could have doubted the existence of interstellar matter. Silhouetted upon the Milky Way glow west of Gamma Aql are two isolated 30'-long dust clouds. The southern one is a simple E-W patch of darkness catalogued as B142. However, the northern one, B143, is more-or-less circular, with two parallel, 15'-long, west-pointing "prongs" extending from it. B142 + B143 are just about at their best in instruments of the 25×100 size.

Another region where the true character of the Great Rift is obvious is in southeastern Serpens Cauda. The field west of the Star Queen Nebula M16 simply *looks* dusty: it has no background star-glitter or Milky Way haze at all, just a poor scattering of field stars. A claustrophobic sight. Just to the north of M16 is an area about $1^1/_2°$ in diameter sprinkled with noticeably more 8th and 9th mag stars than the surrounding Great Rift. This is the 6,500 light-year distant stellar association Serpens OB2. The core cluster of Ser OB2, NGC 6604, is rather obvious on the NW side of the association: it is a NW-SE elongated, granular, partially-resolved group pointing at the association's lucida, the mag 7.5 HD 167971, which lies just outside the SE end of the cluster. NGC 6604 actually *appears* overspread by dust, through which it can glow but feebly. Indeed, the average obscuration of the members of the association is 3.2 magnitudes. The Great Rift between NGC 6604 and Nu (ν) Ophiuchi to the NW is appallingly blank.

Northeast of Aquila the Great Rift remains conspicuous to the unaided eye but in binoculars loses something of its opacity. Around the very large, very loose open cluster Cr 399 in SW Vulpecula, the Rift, though still quite dark, in supergiant glasses is seen to consist of lanes and patches superimposed upon a dim Milky Way background glow. The shapes of these lanes and patches make this an interesting region aesthetically. Cr 399 may or may not be a true physical entity; but its seeming "clusterness" is certainly enhanced by the star-poverty of the Great Rift field around it. The NW stream of the Milky Way here, west of Cr 399, is a pale, unresolved glow cut through by heavy lanes of obscuration intruding off the Great Rift.

Farther NE in Vulpecula the Great Rift's darkness is even more attenuated by the underlying Milky Way glow. A conspicuous stream of magnitude 8-10 stars scattered over a background Milky Way haze begins north of M27, the Dumbbell Nebula, in the vicinity of 13 + 16 Vulpeculae and extends NNW toward Phi (ϕ) Cygni, joining the Cygnus Star Cloud near the small open cluster NGC 6834. Thus the Great Rift in central Vulpecula and extreme southern Cygnus is much compromised, and not as dark and uniform as it is further SW in Sagitta, Aquila, and Serpens Cauda.

The brightness of the Milky Way background glow in the SE stream of the Milky Way slowly fades from the Gamma Aquilae star cloud to the NE through Sagitta and into Vulpecula. But in compensation are increasing numbers of resolved stars. Around Alpha (α) + Beta (β) and Delta (δ) Sagittae the Milky Way background is still quite bright and billowy; and east of Alpha + Beta (SE of Gamma + Delta) is a $3°\times1^1/_2°$ field of fairly faint stars resolved against a relatively bright background glow. But the background glow fades from the vicinity of the globular cluster M71 north into Vulpecula. However, as the background glow fades the field stars proliferate, and the eastern Vulpecula Milky Way is richly strewn with relatively bright binocular stars, an effect that extends north as far as Epsilon (ε) Cygni.

The line joining Alpha, Beta, and Epsilon Sagittae approximately marks the edge of the Great Rift. To the north and west of Alpha + Beta Sag the Milky Way background is broken by heavy lanes and patches–an excellent sight, especially due north of the two stars. Just north of the midpoint of the line joining Alpha and Delta is the southern bay of an irregularly-shaped obscuring cloud that extends south from the Vulpecula Rift. North of Delta Sag, and north and west of Zeta (ζ), the Milky Way glow is mottled by small dark cloudlets outlying from the main obscuring mass. The effect is quite beautiful, because the cloudlets are numerous and rather small.

In Cygnus the Great Rift broadens and darkens from SW to NE along the Swan's body (Beta [β] - Gamma [γ] - Alpha [α] Cygni). However, nowhere in this stretch does it have the appalling opacity of the Serpens Rift NW of M16. The Cygnus Rift consists of darker lanes and channels snaking amongst brighter patches ("brighter" only by comparison), the whole scattered with more foreground stars than the Aquila and Vulpecula Rifts. The Cygnus Rift is very narrow north of M27 in Vulpecula because of the Milky Way star-stream which extends south from the NGC 6834 region of the main Cygnus Star Cloud toward 16 + 13 Vul. Further NE, between Eta (η) Cygni in the Cygnus Star Cloud and NGC 6940 in Vulpecula, the Cygnus Rift is still rather narrow and not particularly dark because of patches of Milky Way glow in it. But still further NE, between Gamma Cyg at the NE end of the Cyg Star Cloud and the 61 Cygni region in the SE Milky Way stream, the Cygnus Rift broadens and darkens, and is sprinkled with noticeably fewer foreground stars. The Rift's terminus, toward Deneb, is a round dark area several degrees across often called the "Northern Coalsack" in imitation of the famous far southern Milky Way dust feature.

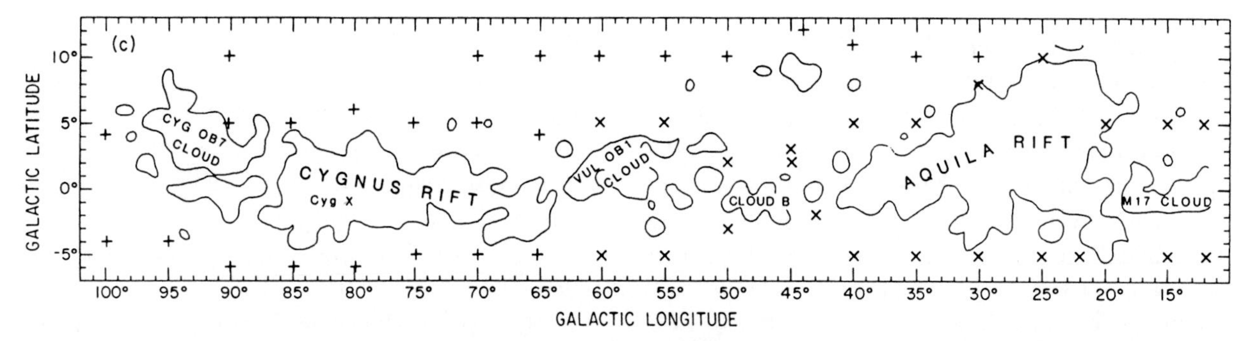

Figure 3.2 The distribution of the giant molecular clouds comprising the Great Rift from Sagittarius (right) to Cygnus. The distance of these clouds increases with galactic longitude: the Aquila Rift clouds are 500 light-years away at longitude 20° (Serpens) and 1000 l-y distant at 40°; the Vul OB1 Cloud (not involved with the stellar association of that designation, merely in the same direction) is 1350 l-y from us; the Cygnus Rift is around 2300 l-y away; and the Cyg OB7 Cloud is at least 2600 l-y distant. (T. M. Dame and P. Thaddeus; *ApJ* 297, 751; Fig. 3.) Courtesy P. Thaddeus and T. M. Dame

III The Star Clouds and Nebulae of Cygnus

The Cygnus Star Cloud, which stretches in a 20°-long oval from Beta (β) Cygni on the SW to Gamma (γ) Cygni on the NE, is the brightest Milky Way feature north of the celestial equator. It is so bright because toward it we are looking thousands of light-years down the incurving arc of our Orion-Cygnus Spiral Arm. But it is not the only brilliant Milky Way feature in Cygnus: the star clouds just NE of Deneb, and around the open cluster M39, are not much inferior in brightness to the Cygnus Star Cloud itself. We shall begin on the SW with the Cygnus Star Cloud in the area around Albireo (Beta [β] Cygni) and sweep up the constellation toward the NE.

The SW half of the Cygnus Star Cloud, between Albireo and Eta (η) Cygni, is a pale glow evenly overspread with an abundance of magnitude 7–10 stars without conspicuous star-chains or field condensations (a rather rare look for a Milky Way star cloud, actually). It also is remarkable for its lack of open clusters, or for any stellar aggregation that even remotely resembles an open cluster. In the SW half of the Cyg Star Cloud we find only NGC 6834, $4\frac{1}{2}$° ENE of Albireo on the Star Cloud's SE edge, a group that appears merely as a small circular intensification of the basic Milky Way background glow. The Milky Way east of Beta and Phi (ϕ) Cygni is impressive for the brightness of its star billows and for their rich, star-sparkling incipient resolution. The brilliance of the Cygnus Star Cloud here, and the myriads of stars that can be resolved or partially-resolved in it, shows that there is very little interstellar material in this direction down our spiral arm. Four especially luminous little Milky Way cloudlets, each vaguely 1° across, are located WSW, south, SSE, and SE of Eta Cyg, the last three in a line that approximately parallels the galactic equator. In supergiant binoculars these cloudlets can be seen to have a very bright background radiance embellished with an exceptional profusion of the twinkling half-resolution of uncountable star-sparks. Even in big glasses these cloudlets may not call attention to themselves on a casual scan through the Eta Cyg region, so carefully look for them with averted vision.

Northeast of Eta Cygni the Cygnus Star Cloud changes character. There is still a profusion of resolved mag 7–10 stars, but the background glow is decidedly paler and it is cut through with heavy lanes of obscuring matter. This is still a star-rich direction down the incurving arc of the Ori-Cyg Spiral Arm; but it is also a direction with more of the cool interstellar dust clouds in which stars form. Many of the mag 7–10 stars in the Eta-Gamma Cyg half of the Cygnus Star Cloud are in fact very young luminous supergiants just condensed from the dust clouds. They are members of the numerous young open clusters and stellar associations strewn ahead of us in a confusing jumble down thc incurving arc of our spiral arm.

Several open clusters are in the field NE of Eta Cygni, though none of them are immediately obvious as open clusters. The most visually intriguing is NGC 6871, which is not a circular-shaped star group at all, but a striking $1\frac{1}{2}$°-long SSW-NNE star chain that begins on the south at a wide SE-NW double, heads north through first one mag $7\frac{1}{2}$ star, then a second mag $7\frac{1}{2}$ star (the unresolved triple OΣ 398), then the multiple β440, and finally to 27 Cygni. Beyond 27 Cyg the NGC 6871 star chain gently arcs first north and then back NE through five mag 6–7 stars to the bright 28 Cyg. This star chain is remarkably reminiscent in appearance to several others elsewhere along the Milky Way, all involved with, or actually in, young open clusters. β440 A-F is an attractive double at 15x; and at this moderate power OΣ 398 to the SSE appears like a β440 component. The primary of β440 is a remarkable star, a spectroscopic binary consisting of a hot O9.5 III giant and one of the rare Wolf-Rayet stars. NGC 6871 is about 5,700 light-years away and estimated to be just 1 million years old.

To the east of β440 is Biurakan 1, a fairly conspicuous E-W oval of seven mag 6 and fainter stars, a mag 8 star within the ellipse's east end. This cluster appears merely as a concentration in the star field, and lacks any background haze suggesting unresolved cluster members.

The open cluster NGC 6883, centered 1° east and slightly north of 27 Cygni, is a condensation in the binocular star field supported by just a hint of sparkling background half-resolution. The cluster's lucida is on its WNW edge, and its most obvious subgroups are located just within its NW and NE edges. NGC 6883 is a vague, ambiguous group at

Photo 3.9 The Milky Way from Eta (η) Cygni (the brightest star near the right center edge) NE to the Crescent Nebula, NGC 6888 (near the upper left corner). North is up, east to the left. The Great Rift is to the lower left. The close optical double star β440 (NNW) + OΣ 398 (SSE) is just east of dead center. It is near the SSW end of the star chain NGC 6871, wich gently curves northeastward toward the star 28 Cygni, the brightest star SSW of NGC 6888. To the SSE of NGC 6888 is 29 Cygni, which is at the northern end of a second star chain described in the text. The conspicuous triangular dark nebula toward the upper right (west and slightly south of NGC 6888) is B145

best; but it is situated within a splendid star field. It seems to be about 4,500 light-years away and therefore slightly nearer than NGC 6871.

The 6th magnitude 29 Cygni marks the northern end of a second peculiar star chain in the Eta Cyg region. From 29 Cyg three stars head south in a shallow arc, open to the west, toward a NE-SW line of four mag 6-8 stars, the most northerly the brightest. This star is, in its turn, at the east end of a nearly straight SE-NW line of five stars, the other four being mag 8-9 objects. The most southerly star in the main 29 Cyg chain is itself at the west end of a long arc, open to the SW, of five mag 8-9 stars. All these lines and arcs are *very* conspicuous in supergiant binoculars; but, except for the NGC 6871 chain, none of them seem to have been catalogued as open clusters.

Both the NGC 6871 star chain and the 29 Cygni chain parallel the galactic equator. And both star chains are superimposed upon channels of bright Milky Way glow. Along the east edge of the NGC 6871 Milky Way channel is a particularly noticeable dark lane. Another conspicuous dust feature in the Eta Cyg area is the flat isoceles triangle, apex pointing south, of B145, located 3° NNE of Eta. B145 is one of the most well-defined small dark nebulae visible with giant binoculars anywhere along the Milky Way. A 7th mag star is at the midpoint of its northern edge.

The contrast between the bright and dark Milky Way masses becomes even more striking around Gamma Cygni (and the multitude of mag 7-10 stars here is no less than farther SW toward Eta Cyg). Indeed, the Gamma Cyg region is an especially beautiful field of dark lanes snaking around various-sized patches of Milky Way glow, the whole liberally garnished with bright field stars. The effect, with careful looking (averted vision at a very dark sky site), is absolutely stunning. Several areas of nebulosity are not difficult to discern in supergiant binocs. They probably all are simply the hot-star-fluoresced portions of the same vast molecular cloud of gas and dust, and three bear the same IC number. The brightest of the group is the conspicuously triangular IC 1318b, centered about 2° NW of Gamma Cyg, almost 1° across and of above-average surface brightness for an emission nebula. In fact the glow of IC 1318b is visibly different in character from that of an unresolved star cloud–"fuzzier" in texture and smoother in radiance. Centered respectively about 1° east and $1\frac{1}{2}°$ ESE of Gamma Cyg, and separated by a 20'-wide NE-SW dark channel, are IC 1318d and IC 1318e. Neither are difficult even in 10×50 binoculars, though both are of inferior surface brightness to IC 1318b. The dark lane that separates them turns sharply west, passing just south of Gamma Cyg. To the lane's south is a broad area of Milky Way glow that must be a mingling of unresolved star-light and of emission from the vast (2°×1°) Sh 2-108 nebulosity that extends SW toward P Cygni. The IC 1318/Sh 2-108 complex is estimated to lie some 4,500 light-years from us.

The Gamma Cygni region has more open clusters than the Eta Cyg region–and more open clusters that actually *look* like open clusters. M29, not quite 2° south and slightly east of Gamma, and IC 4996, almost 3° SSW of Gamma, are described in the section on the Cygnus open clusters in Chapter 2. They are, however, objects better suited to telescopes. But two Milky Way star fields south of Gamma that are at their best in binoculars have been given open cluster designations. Do (Dolidze) 39, $2\frac{1}{2}°$ SSW of Gamma (and just north of IC 4996), is a moderately-large gathering of very faint field stars–recognizable when you know what it is, but otherwise easily overlooked as a mere Milky Way star-condensation. By contrast, Do 41, centered $\frac{1}{2}°$ ESE of Do 39, is a very large gathering–*not* a concentration, though!–of brighter field stars. One would have been inclined to classify it as an open cluster even without knowing that astronomers had already done so.

On the west edge of the Gamma Cygni region, $2\frac{1}{2}°$ WNW of the star, is the 'dual' open cluster IC 1311. The cluster proper is a tiny faint smudge; but it is within, and toward the western rim, of a striking ring of rather bright field stars often mistaken as the cluster. IC 1311 is in a very large area of enhanced Milky Way brightness. Indeed, this area has the glittering partial resolution of distant star-masses otherwise lacking in the Gamma Cyg region. This is a result of the thinning of IC 1318 complex dust this far off the Galactic plane. However, photos show

Photo 3.10 Emission nebulae in the Milky Way around Gamma (γ) Cygni. See the caption to Plate XVII for the identification of the objects in this field

that some of the glow around IC 1311 is from outlying IC 1318 emission.

Further west, NW, and north from the Gamma Cygni region, toward Lyra, the Cygnus Milky Way loses all background glow but becomes *exceedingly* rich in multitudes of bright and faint field stars. Here we look *over* the incurving arc of our Ori-Cyg Spiral Arm, out of the dust-rich spiral plane but at a shallow rising angle through the star-rich thin and thick discs, in which the spiral arms *per se* are embedded. There is very little gas and dust at any distance off the Galactic plane; so, in the absence of nearby dust clouds, any angling view off the galactic equator is *very* clear and therefore star-rich. And that is certainly true in NW Cygnus and adjoining eastern Lyra. The eastern Lyra Milky Way has no dark lanes, but in giant binoculars it can be seen to be richly strewn both with brighter resolved stars and with very faint, intermittently-resolved, star-sparks. (One evening I noticed that a touch of lingering twilight glow overhead made the Lyra Milky Way star-sparks particularly pronounced.) The best Milky Way field in Lyra is around the large, but very low surface brightness open cluster NGC 6791, located just SE of Eta (η) Lyrae. In 25×100 supergiant instruments NGC 6791 can be partially resolved into knots of pale haze, and on good nights even into a fine textured granularity. The cluster gives the peculiar impression of being behind the hundreds of resolved Milky Way stars around it–and it in fact *is*, for it is estimated to be 12,400 light-years away, and most of these stars are undoubtedly nearer. NGC 6791, by its hazy elusiveness, adds a pleasing aesthetic contrast to its glittery star-rich setting. The M56 region in the SE corner of Lyra, though actually closer to the galactic equator, is not nearly as star-rich as the NGC 6791 area: it does, however, have a definite, albeit faint, Milky Way background glow.

The 1st magnitude star Deneb marks the NE end of the Great Rift. Just to Deneb's east, visible to the unaided eye, is a small patch of pale Milky Way glow. This is the famous North America Nebula, NGC 7000, one of the very few emission nebulae that can be seen with the unaided eye. Its surface brightness is so high that its full photographic extent–including even its narrow 'Isthmus of Panama'–is easy with only 10×50 glasses. The nebula measures 180′ × 140′, so it does not fit in 20× fields of view. It appears to have a truly *nebular* glow: it not only lacks a garnishing of over-spread stars, its pale haze is *uniform*, not subtlety granular, as are even unresolved star masses. Contrast NGC 7000 with the Milky Way star clouds just to its north, which are very rich in resolved mag 10–11 stars on a bright background glow. When looking at NGC 7000 in giant binoculars, you can actually *see* its mistiness.

The star-field across the North America Nebula may not be particularly rich, but two open clusters are visible in it. In the east-central part of the nebula, approximately in "northern California," is a large (13′) scattering of 10th and 11th mag stars catalogued as Cr 428–but even in supergiant glasses its faint stars are too broadcast to make much of an impression. More traditionally "cluster-like" in appearance is NGC 6997 in the west central–"West Virginia"–part of NGC 7000, a modest-size (8′) spray of faint stars, not very conspicuous except to averted vision. It is in fact quite loose; but the uniform brightness of its stars gives it the very convincing illusion of richness. More visually impressive than these open clusters are two dust clouds on the north edge of NGC 7000 due north of NGC 6997: B352 and B353 both measure about 20′ across their longer dimension and are $^1/_2$° apart in a N-S direction.

The North America Nebula is silhouetted on its east, south, and west by the dark dust clouds of the complex of interstellar matter of which it is but one of the star-fluoresced sections. A smaller illuminated section of the complex is out in the "Atlantic" west of NGC 7000 around the 5th mag 56 Cygni. This is IC 5067, the "Pelican Nebula," appearing as a large amorphous area of pale haze centered on 56 Cyg (which is in fact not actually involved with the nebula). The Pelican's surface brightness is much less than that of NGC 7000, but it is not ambiguously visible even in only 10×50 binoculars. With 25×100s the two brightest patches of the Pelican, just east and north and just south of 56 Cyg, show up well.

In the middle of the long channel of obscuration off the "Pacific Coast" of NGC 7000 is the beautiful chrome-orange mag 3.9 K5 star Xi (ξ) Cygni. This

Photo 3.11 The North America Nebula, NGC 7000 in Cygnus. The bright star on the left (east) edge of the field is Xi (ξ) Cygni

channel separates the North America Nebula from a fine, rather wide Milky Way star stream that extends between 68 and Nu (ν) Cyg. The background glow of this stream is somewhat fainter than that of the star clouds north of NGC 7000, but it has a very rich garnish of 9th-10th-11th mag stars. From the northern end of the Xi Cyg "bay" a thin, low-contrast dust lane continues about 2° NNE toward, but not quite all the way to, the conspicuous 17'-wide dark cloud B361. This thin lane cuts off a very bright and very star-rich Milky Way cloud on its east, also elongated NNE-SSW, from the main North America Nebula star clouds to its west.

The star clouds NW, north, and particularly NE of NGC 7000 are absolutely stunning–almost the equals of the Small Sagittarius and Scutum star clouds. In fact, given that for mid-northern observers the NGC 7000 star clouds can be viewed near the zenith, where the sky is darkest, they are aesthetically the equals of the Small Sgr and Scu clouds. They are true star *billows*: the multitudes of resolved stars in them is quite simply spectacular. The NGC 7000 star clouds begin to fade beyond 55, 59, and 63 Cygni; and at the NE edge of this area, east of 63 Cyg toward M39, there is a particularly dark, roughly circular, obscuring mass.

To the east of NGC 7000, beyond the 68-Nu Cygni star stream, are several other interesting Milky Way fields. A second splendid star stream, though somewhat narrower than the 68-Nu stream, extends between 68 and Sigma (σ) Cygni. Its background glow is fainter than that of the 68-Nu stream, but it compensates by its garnish of brighter, mag 8–9, stars. (This stellar garnish shows up well on Chart 86 of *Uranometria 2000.0;* but the reality is even more conspicuous than the chart suggests.) Within the triangle formed by Sigma, Tau (τ), and the beautiful binary of chrome-orange stars, 61 Cygni, there is a large field of mag 8–9 stars, many distributed in long chains. This field is very like the star field WNW of the large open cluster Stock 2 in Cassiopeia (to be described in Section 5, below)–or even like a brighter, but somewhat less star-dense, version of St 2 itself. (This field is shown on Chart 69 of *U2000* much as it actually appears on the sky.)

Finally, SE of the Sigma-Tau-61 Cygni area, immediately NW of 71 Cyg, is a peculiar oval of eight or ten stars catalogued as the open cluster NGC 7082. The oval's NE and SW sides are slightly curved rows of three stars each. A double of 9th mag stars are at the figure's east end, and another double of slightly fainter stars on its south edge. The NW end of the oval is open except for a 10th mag star between the NW stars of the two curved rows (nearer the SW row).

Midway between 63 Cygni and the large, loose open cluster M39 5° to its ENE is a circular obscuring mass, an outlying Great Rift dust cloud. This mass is itself enveloped within a larger region of quasi-obscuration, elongated N-S, that has significantly fewer field stars than the Milky Way around it. However, just a little further NE of M39 the dust once again thins, the Milky Way brightens, and the star clouds in the NE quadrant of Cygnus are almost as glorious as those just north of NGC 7000. M39 itself, however, is not one of the more visually impressive Messier open clusters: it is a large but very loose group of approximately two dozen stars, mag 6.8 and fainter, scattered across a 30'-wide triangle–almost too large and loose a group even for 20×. However, between M39 and the two Pi's (π) Cygni is a splendid, very well-defined, little star cloud perhaps 2–2$^1/_2$° in diameter with myriads of brighter and fainter stars resolved on its very bright background glow–a cloud worthy to be in the same constellation with the star clouds north of NGC 7000. A luminous, star-salted channel from this cloud runs north between Pi-1 (π^1) Cyg and the small open cluster NGC 7086 (a sprinkling of faint resolved stars on a pale background haze) to a separate, somewhat paler and less star-rich, but almost equally bright, star cloud occupying the area north of π^1. A lobe of this second star cloud reaches north to a loose group of mag 6–8 stars.

The region NW, west, and SW from NGC 7086 is indeed as blank as star charts suggest, with a murky opacity very like that of the denser stretches of the Great Rift. This area, several degrees across and centered midway between Deneb and Alpha (α) Cephei, has to the unaided eye a circular appearance which has earned it, like the terminus of the Great Rift SW of Deneb, the title "Northern Coalsack." And it is a worthier pretender to the title because in binoculars it can be seen to be much poorer in foreground stars. This is because it is quite near, only 1300 light-years away, about half the distance of the dark clouds of the North America Nebula complex.

III.1 The Bright Nebulae of Cygnus

Because in this direction we look down the dust- and gas-rich forward length of our Orion-Cygnus Spiral arm, Cygnus, and Cepheus to its NE, are rich in bright emission and reflection nebulae. The showpiece object in Cygnus is NGC 7000, the North American Nebula, an unaided-eye object that reveals its famous shape even in 10×50 binoculars. But Cygnus has several other nebulae accessible to moderate-size binoculars: the NGC 6992/5 arc of the Veil Nebula, the Pelican Nebula IC 5067/70, and three patches of the IC 1318 emission complex scattered around Gamma (γ) Cygni at the NE end of the Cygnus Star Cloud. Two of Cygnus' other named nebulae, the Crescent Nebula NGC 6888 and the Cocoon Nebula IC 5146, are visible in supergiant binoculars. However, the Crescent, the NGC 6960 arc of the Veil, and the large but very pale NGC 6820 emission nebula just south of Cygnus in Vulpecula are better viewed in RFTs than even in supergiant binoculars.

Astrophysically, the Cygnus bright nebulae are almost as varied a group as you can get. The North America, Pelican, and IC 1318 nebulae are all standard H II regions fluoresced by hot young stars. The Cocoon is a combination reflection and emission glow around a small cluster that includes a couple modestly-hot, modestly-luminous early-B main sequence stars. IC 5076 is a reflection nebula around a luminous, but not hot, late-B supergiant. The Crescent is the expanding shell of material blown off a Wolf-Rayet star glowing by a combination of florescence stimulated by the W-R star and of heating from its collision with the surrounding interstellar medium. And NGC 6960 and NGC 6992/5 are two separate arcs of a still-expanding supernova shock-wave.

NGC 6820: NGC 6820 is the large (40'×30') but very pale emission glow surrounding the tight open cluster of hot stars NGC 6823, located $3^1/_2$° SE of Alpha (α) Vulpeculae. It is just visible in 15×100 supergiant binoculars, but very difficult, or impossible, in 25×100 glasses, which magnify its tenuous glow into invisibility. The NGC 6820 + 6823 complex, and the association, Vulpecula OB1, of which they are the core, are described in detail in the discussion on NGC 6823 under Vulpecula in Chap. 2. The complex is about 8,200 light-years distant, so the true size of NGC 6820 is a healthy 100×75 l-y. NGC 6820 might be in Vulpecula, but *astrophysically* it is in Cygnus: the NGC 6820 + 6823 complex is a major tracer of the incurving arc of our Orion-Cygnus Spiral arm, the bulk of which lies toward Cygnus.

The Crescent Nebula, NGC 6888: NGC 6888 has an integrated apparent magnitude of 11 and is every

Cygnus Bright Nebulae

Nebula	Name	Type	RA (2000.0)	Dec	Dimen	m_V*	*Spectrum	Distance	True Size
N6820		EN	19^h43^m	+23°17'	40'×30'	9.3,10	O7V,O7V	8200 l-y	100×75 l-y
N6888	Crescent	EN	20 12	+38 21	20×10	7.4	WN6	6000	$35×17^1/_2$
N6960	Veil	SNR	20 46	+30 40	70×6			2000	$41×3^1/_2$
N6992/5	Veil	SNR	20 57	+31 30	72×8			2000	$42×4^1/_2$
N7000	N Amer	EN	21 00	+44 20	180×140	5.96	O6Vf	2000	110×80
I 1318b		EN	20 16	+41 49	50×20			4500	68×24
I 1318d		EN	20 26	+40 30	50×30			4500	68×40
I 1318e		EN	20 28	+40 00	45×30			4500	60×40
I 5067/70	Pelican	EN	20 51	+44 10	80×60			2000	48×36
I 5076		RN	20 56	+47 25	9×7	5.69	B8 Ia	5700	15×12
I 5146	Cocoon	RN+EN	21 53	+47 16	10	9.5	B1	3300	10

bit as difficult as that number implies because it is 20'×10' in size and therefore of extremely low surface brightness. It is located 3° SW of Gamma (γ) Cygni and about the same distance due west of the bright but compact open cluster M29. In 15×100 supergiant binoculars NGC 6888 is evident only as an oval glow extending generally south from a N-S pair of mag 7–7$^1/_2$ stars. It can be difficult to identify even when you do see it because it does not display the crescent shape so conspicuous on photographs and its glow looks just like that of the other Milky Way patches in its vicinity. However, its oval profile is the same as the elliptical outline traced by the photographic Crescent. NGC 6888 is definitely better suited to large-aperture telescopes than richest-field instruments; but use the lowest power possible.

The Crescent Nebula is the expanding bubble of gas blown off by its central star, the mag 7.4 WN6 Wolf-Rayet HD 192163 (the southern of the two bright stars in the northern half of the nebula-glow). The bubble is not exactly centered on the star because the interstellar medium on its north side is denser than toward its south and the bubble's expansion has been slowed in that direction. The glow of the Crescent is brightest around its northern rim simply because its interaction with the interstellar medium is more vigorous there.

The Veil Nebula, NGC 6960 and NGC 6992/5: NGC 6960 and NGC 6992/5 are the two brightest arcs of a group of filamentary emission nebulae collectively called the Veil Nebula. They are about 3$^1/_2$° apart, WSW-ENE, and represent diametrically-opposed segments of the same large structure, the Cygnus Loop, which is the still-expanding shell of an old supernova. The shell is somewhat flattened, its NNW-SSE axis being 2$^3/_4$° long. The brightest part of the Veil Complex outside the NGC 6960 + NGC 6992/5 arcs is a very pale nebulous patch, NGC 6974/9, centered about 1° due west of the northern tip of NGC 6992/5.

Both NGC 6960 and NGC 6992/5 are over 1° long but only a few minutes of arc thick. Nevertheless NGC 6992/5 is not difficult to spot (with averted vision) in 10×50 binoculars as a "fish-hook" shaped condensation in the star-rich Milky Way 3$^1/_2$° SE of Epsilon (ε) Cygni. The shaft of the "fish-hook" points NW directly at Epsilon, and the hook itself, at the southerly end of the nebula, curves west. (The shaft is NGC 6992, the hook NGC 6995.) The NGC 6960 arc of the Veil might also be visible with 10×50s but for the glare of the 4th magnitude 52 Cyg, located on the nebula's west edge just north of its midpoint. (52 Cyg is a foreground star not physically involved with the nebula.) However, it can be seen in 15×100s as a very thin streak extending straight south out of the glare of 52 Cyg. The background Milky Way glow between the NGC 6960 and NGC 6992/5 arcs of the Cygnus Loop is decidedly brighter than it is outside, the contrast especially evident south of 52 Cyg and NW of the north end of NGC 6992. This is because the supernova shock wave has "snowplowed" the dust of the interstellar medium, thus creating a transparent "bubble" in interstellar space. 21-centimeter radio wavelength observations have confirmed that there is less neutral hydrogen gas inside than outside the Cygnus Loop.

The Veil Nebula is centered around 2,000 light-years away. Thus the NGC 6960 and NGC 6992/5 arcs are 40 l-y long and the full size of the Cygnus Loop 120×96 l-y. By astronomical standards the original supernova that generated the Veil Complex was very recent; but age estimates for the Cygnus Loop range from 20,000 down to 5,000 years. Possibly, then, the Veil Supernova occurred right at the dawn of written history, 3000 B.C. However, there is nothing in the mythology, art, or literature of early Egypt or Babylonia which could be interpreted as an allusion to what must have been a truly spectacular celestial event, for the Veil Supernova would have had an apparent magnitude of at least −7, 2$^1/_2$ mags greater than that of the Crab Supernova of 1054 A.D.

The North America Nebula, NGC 7000; the Pelican Nebula, IC 5067/70: Because it is so large, and therefore visually an integral part of the Cygnus Milky Way, the North America/Pelican nebula complex is described in detail in the opening of this section. The complex is estimated to be 2,000 light-years distant on the basis of star counts on the very dense dust cloud LDN 935 that

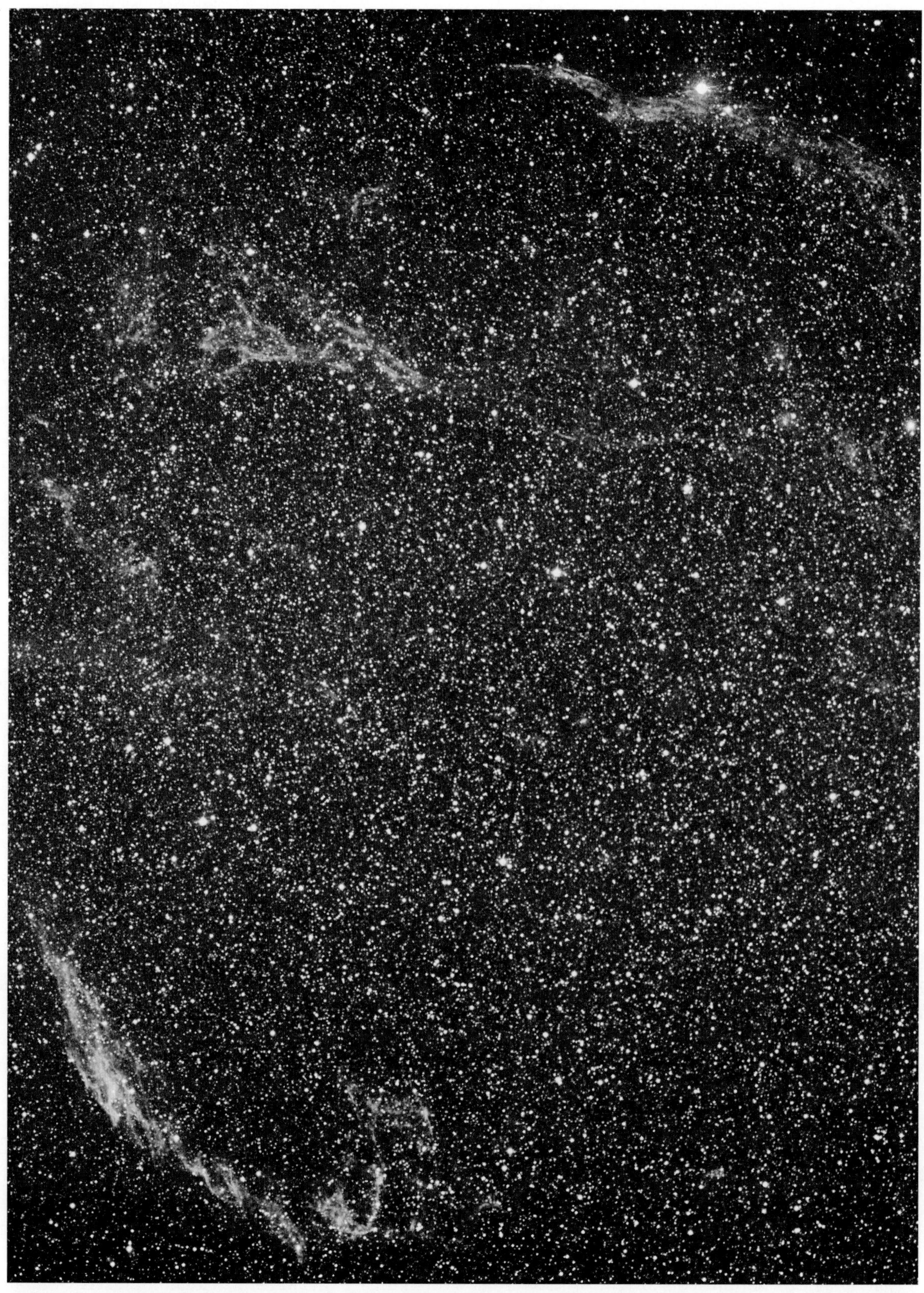

Photo 3.12 The Veil Nebula in Cygnus. North is to the upper left. The NGC 6960 arc of the nebula is to the top right (west), and the thicker NGC 6992/5 arc to the bottom left (east). In the upper left-center is the very faint NGC 6974 patch of the Veil. The background star-field outside the Veil complex, particularly outside the NGC 6960 arc, is noticeably poorer than the star-field seen through the complex because of how the expanding supernova shock-wave has "snowplowed" the dust and gas in front of it, leaving a relatively gas- and dust-free cavity within the bubble of the still-expanding Veil arcs

Photo 3.13 The Pelican Nebula, IC 5067 in Cygnus

separates NGC 7000 and IC 5067/70, and on the degree of reddening suffered by stars that lie just on our side of LDN 935. The North America/Pelican complex has been thought to be involved with a rather scattered association, Cygnus OB7, that would include the 1,600 l-y distant A2 Ia supergiant Deneb (Alpha [α] Cygni), the magnitude 4.83 B2.5 Ia supergiant 55 Cyg, the mag 5.00 O7 IIIfn giant 68 Cyg, and about ten more stars of mags 6.0 to 7.7 strewn over a 11°~400 l-y area–rather large for such a poorly-populated group. One of the stars fluorescing the North America Nebula is the mag 5.96 O6 Vf HD 199579, located $^1/_2$° due north of the open cluster NGC 6997 (a group not involved with the nebula). But other hot stars fluorescing NGC 7000 and IC 5067/70 no doubt lie behind LDN 935. The true size of NGC 7000, 110×80 l-y, is a minimum in that much, or most, of this giant H II region is behind LDN 935 and the other dust clouds to its west and south.

IC 1318: Because the IC 1318 complex of emission nebulae is spread over such a large area ($2^1/_2$°) of the Gamma (γ) Cygni region, it is more of a Milky Way feature than a simple bright nebula. Thus, like NGC 7000, it is described in more detail in the opening of this section. The distance to the IC 1318 complex is uncertain because none of the illuminating stars of the nebulae comprising it have been identified: they must be hidden behind the heavy dust clouds of which the five IC 1318 emission areas are merely the illuminated parts. The complex is probably a little nearer than the 6,000 light-year-distant Cygnus OB1 association with M29 and P Cyg centered to the south, and perhaps farther than the 4,000 l-y distant open cluster NGC 6910 and its association Cygnus OB9–though IC 1318 and Cyg OB9 could be parts of the same supercomplex. The numerous open clusters, OB associations, and bright and dark nebulae in this direction make it a very difficult region to topographically disentangle.

IC 5076: The IC 5076 reflection nebula is involved with the open cluster NGC 6991 and described in the discussion on that cluster in the Cygnus section of Chapter 2.

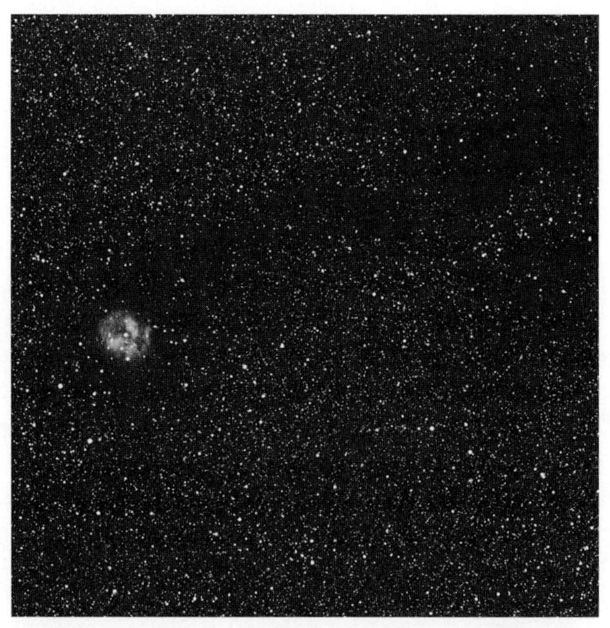

Photo 3.14 The Cocoon Nebula, IC 5168 in Cygnus

The Cocoon Nebula, IC 5146: Centered midway between the large open clusters M39 in Cygnus and NGC 7209 in Lacerta is the 100'-long, 10'-wide WNW-ESE dust lane B158. It is oriented precisely along the line joining the two clusters (which are about $4^1/_2$° apart) and easily seen in 10×50 binoculars because it contrasts so sharply with the Milky Way background glow here. The Cocoon Nebula, a mingled emission + reflection glow about 10' in diameter around a tight N-S pair of magnitude 9.5 stars, is just within the ESE end of B158 about $^1/_4$° NW of a mag 7 field star. IC 5146 is too pale for 10×50 binoculars, but can be glimpsed (with averted vision) in 15x100 super-giant glasses. IC 5146 is usually catalogued as a combination bright nebula/open cluster; but the "cluster" consists only of two mag 9.5 stars with a handful of much fainter companions. B158 + IC 5146 is some 3,300 light-years away, so the true length of B158 is a respectable 100 l-y, but IC 5146 itself is only 10 l-y across. Star-formation is presently still occurring in the dense dust cloud around IC 5146; so eventually, when the dust clears (it is being slowly driven away by the radiation pressure and stellar winds of the two B1 stars), a very nice little open cluster of modest-luminosity stars should be visible here.

IV The Associations of Cepheus and Lacerta

From NE Cygnus the main stream of the Milky Way continues into northern Lacerta. However, a branch of the Milky Way extends straight north from NE Cygnus into Cepheus, dead-ending in the center of the square formed by Alpha (α) - Beta (β) - Iota (ι) - Zeta (ζ) Cep. The Cepheus branch is more conspicuous to the unaided eye than in binoculars because it does not consist, like the Lacerta Stream, of faint resolved stars embedded in a background glow, but of rich fields of mag 6–10 binocular stars: when you look at it with binoculars, you resolve the stars, thus losing most of the haze.

Many of the resolved stars in the Cepheus branch of the Milky Way are members of the Cepheus OB2 association, centered some 2,500 light-years away. The lack of Milky Way background glow in southern Cepheus is largely the fault of Cepheus OB2 dust clouds, which obscure the Milky Way star fields beyond. An illuminated portion of these dust clouds, its hydrogen atoms excited to fluorescence, is the very large (approximately $2^1/_2°$ diameter) emission nebula IC 1396, easily spotted in even the smallest binoculars as an amorphous area of haze extending south of Mu (μ) Cephei, the "Garnet Star," and centered upon the magnitude $5^1/_2$ multiple Σ2816, the hot O6.5 V primary of which is providing the ultraviolet photons fluorescing the nebula. (The mag 8.0 Σ2816D can be resolved from the O6.5 primary with 15× glasses.) IC 1396 is distinctly brighter immediately around Σ2816. However, in supergiant glasses the nebula-glow gives no sense of the circular outline it has in 10×50 binoculars. Though the region around the nebula has a heavy, dark, sooty look because of the dust in which it is embedded, not even 25x100 instruments reveal in it any of the small, seemingly opaque dust features that are so striking on photographs.

The vague concentration of stars within IC 1396 around Σ2816 has been dignified with the open cluster designation Tr 37, though it is hardly more than an enhancement in the star-field scattered across the nebula-glow. Tr 37 consists of three broad star bands of various lengths, the richest and longest extending SSW from Σ2816, the poorest NNE from Σ2816 toward Mu Cep, and the shortest (but intermediate in intensity) reaching from Σ2816 toward the west. Thus the star-group has a "spidery," or crab-like, appearance.

The chrome orange-red Mu Cephei, one of the most brilliantly-colored stars in the sky, and one of the most luminous M-type supergiants known in our Galaxy, is a Cep OB2 member. It is obscured $2^1/_2$ magnitudes by Cep OB2 dust. Indeed, in supergiant glasses a very dark channel of obscuration can be discerned extending NE from near Mu NE toward Nu (ν) Cep (another Cep OB2 star). This channel is elongated to the point of streakiness. Paralleling it immediately to its SE is a thin but bright Milky Way stream richly garnished with mag 9–11 stars. This Milky Way stream extends NE from near Mu to a compact asterism of four mag $6^1/_2$–$7^1/_2$ stars (including 15 Cep) located about one-third the way NW from Lambda (λ) to Nu Cep. On its SE the bright stream is paralleled by a second SW-NE dust streak (which contains the large photographic dark nebulae B169/170 and B174). And farther SE yet is a brighter and broader Milky Way stream that begins at the Delta (δ) - Epsilon (ε) - Zeta (ζ) Cephei triangle and follows the galactic equator NE all the way to Kappa (κ) Cassiopeiae. The whole impression in supergiant binoculars of this region is of long, flat clouds of dark dust paralleling the Galactic plane. And this impression is confirmed by photographs, which show that even the huge dust cloud in which IC 1396 is embedded is not circular but noticeably flattened parallel to the galactic equator.

A second nearby (in Galactic terms) stellar association in this stretch of the Milky Way is Lacerta OB1, the members of which are scattered across southern Lacerta and include 8, 10, 11, 12 and 16 Lacertae and 2, 6, and 9 Andromedae. (Nu [ν] Andromedae, just east of M31, is thought to be a "runaway" from Lacerta OB1 ejected several million years ago. It is a spectroscopic binary with a B5 V primary and an F8 V secondary.) Lac OB1 is only 1,600 light-years distant, and too large and too loose to be a good sight at powers of more than 7×. But the main Milky Way stream goes through the northern half of Lacerta and is a splendid sight in any size binoculars or richest-field telescope. The background glow here is not as bright as the glow of the Cygnus star clouds, but it is exceptionally richly strewn with magnitude 8 and fainter stars.

Indeed, the Milky Way field north of Beta (β) Lacertae is one of the best in the entire northern hemisphere half of the Milky Way. With careful looking, pencil-thin lines of obscuring matter can be seen twining in sharply-curved paths through the background glow. Farther south, around the open cluster NGC 7243, larger obscuring patches are mingled with the Milky Way light. The Milky Way is not exceptionally bright in the NGC 7243 area, so the contrast between the brighter and darker masses in it is not very great and averted vision is required.

The brilliancy and star-richness of the North Lacerta Star Cloud is a consequence of the thinness of obscuring dust in this direction through our spiral arm. Indeed, along the Cepheus-Lacerta-western Cassiopeia stretch of the Milky Way our Ori-Cyg Arm's dust is concentrated NW of the galactic equator in the Cep OB2 association in south-central Cepheus and in a broad band that runs W-to-E through the center of the constellation. That band has produced the 2,700 light-year distant Cepheus OB3 association on the east-central edge of the constellation. Cep OB3 is a 5°-long, $1^1/_2$°-wide, NE-SW stream of mag 7–9 stars centered midway between Iota (ι) Cephei and the open cluster M52 in Cassiopeia. It is a very attractive group that fits perfectly in supergiant binocular fields of view and looks exactly like what it is: a stellar association. No background Milky Way glow can be glimpsed behind the group's star stream–no doubt because of the thick Cepheus Complex dust from which the association has recently condensed. The O6 supergiant Lambda (λ) Cep is a "runaway" from Cep OB3 approaching us with the very high radial velocity of 75 kilometers per second. It probably was ejected from the association, some 4 million years ago, when the massive star around which it was orbiting destroyed itself in a supernova explosion, leaving Lambda with nothing around which to orbit, but still with its high orbital velocity. Between M52 and the center of Cep OB3 is a curious 1°-long N-S arc of mag 6–7 stars–probably not a true physical group, but an attractive binocular asterism nevertheless.

Because of the transparency through our Orion-Cygnus Spiral Arm toward northern Lacerta and extreme SE Cepheus, in this direction we can in fact see out to the next spiral feature toward the Galaxy's rim, the Perseus Spiral Arm. The main tracer of the Perseus Arm in this stretch of the Milky Way is the large open cluster NGC 7380, core of the vast, 11,400 light-year distant, Cepheus OB1 association. This cluster, in a rather isolated spot 2° east and slightly south of Delta Cephei, appears in supergiant binoculars as a moderately-large (20'-diameter), moderately-compressed concentration of star-sparks, unquestionably a cluster and not simply an association star field. Moreover, the emission-glow within which the cluster is embedded is not difficult in binoculars: it is very pale and amorphous, but immediately recognizable. Two long parallel lines of obscuration, oriented WSW-ENE approximately parallel with the galactic equator, are just north of the cluster, setting off the cluster- and nebula-glow.

The main body of the Cep OB1 association, 1° SE of Delta Cephei, appears as a 1°-long, E-W oval of enhanced Milky Way background glow upon which is superimposed a rough zigzag of sixteen mag 7–$9^1/_2$ stars, at its east end the K0 Ia-0 extreme supergiant RW Cep (representative of a rather rare type of star because supergiants evolve *very* rapidly through the F, G, and K stretch of the H-R diagram). Also at the east end of the Cep OB1 oval is the other open cluster core of the association, Berkeley 94, a tiny (3') knot of two or three mag 10 stars, with a mag $9^1/_2$ star just to their north.

Some attractive Milky Way star fields can be seen around NGC 7380. First, just WSW of the cluster are two field condensations of mixed brighter and fainter stars, both about the same size as the cluster (and one directly "behind" the other with respect to NGC 7380). Second, ENE of the cluster is a small but conspicuous concentration of perhaps a half dozen mag 9–11 stars: it looks like it could be classed as a cluster but apparently has not. Finally, south of NGC 7380, just across the border in Lacerta, is a pair of fairly obvious star chains: one is a virtually straight N-S line of four stars, its northern component a double with a companion just to its SW; and the other, immediately to the first chain's south, a shallow E-W "S" of seven faint stars.

IV.1 The Bright Nebulae of Cepheus

Toward the Cepheus Milky Way our view is still basically ahead down the incurving arc of our Orion-Cygnus Spiral Arm, so it is no surprise to find several bright nebulae in the constellation to go with the numerous nebulae in neighboring Cygnus. Cepheus claims one of the easiest reflection nebulae for modest-size binoculars, NGC 7023. This compact but high-surface-brightness object is embedded in a heavy dust cloud at the west end of a chain of dust clouds that extends west to east across central Cepheus at a distance of between 2,000 and 3,000 light-years. This cloud chain contains several more bright nebulae and a couple OB associations. About 4° SE of NGC 7023 is a nest of small reflection nebulae with an involved open cluster, NGC 7129. These are strictly telescope objects. Farther east, near the Cepheus/Cassiopeia border, is a $3^1/_2$°×$1^1/_2$° NE-SW field of over a dozen mag 7.4–9.0 stars centered about midway between Iota (ι) Cephei and the bright open cluster M52 in Cassiopeia: this is the Cepheus OB3 association, which has a photographic emission nebula, Sh 2-155, at its SE end. (Sh 2-155 can be seen on Plate XXIV.) Finally, in the bright-star-poor spaces of far eastern Cepheus, centered about 4° due east of Omicron (o) Cep and 8° due north of Beta Cassiopeia, is the Cepheus OB4 association and its two extremely large but low-surface-brightness emission nebulae NGC 7822 and Cederblad 214, the former a supergiant binocular object. Eastern Cepheus is star-poor in part because of the thickness of the dust of the Cepheus OB4 complex, which dims the light of some members of the association by over 4 magnitudes.

But the showpiece bright nebula of Cepheus is the huge IC 1396, which measures almost 3°×$2^1/_2$° and is not difficult for 10×50 glasses. IC 1396 is the heart of the large and supergiant-star-rich Cepheus OB2, the stellar aggregation which makes south-central Cepheus such a star-rich field for binoculars and RFTs.

Photo 3.15 Reflection nebula NGC 7023 in Cepheus

NGC 7023; About $3^1/_2$° SW of Beta (β) Cephei is one of the brightest reflection nebulae in the sky, NGC 7023. It is not large, only about 10'×8', but has exceptionally high surface brightness for a reflection nebula. Thus, with sufficient magnification to get it out of the glare of its magnitude 6.8 B5e central star, it is not difficult with 50–60 mm diameter optics. NGC 7023 is extremely easy with 25×100 supergiant binoculars, resembling the discs of haze that surround brighter stars on a night with lots of humidity, or nights with thin, high-level clouds. And the sense of the large obscuring mass in which the nebula is embedded is also very strong because the surrounding field is not only peculiarly poor in faint stars, but has a murky, even a sooty, look (an effect lost in telescopes).

Cepheus Bright Nebulae

Nebula	Type	RA (2000.0)	Dec	Dimensions	m_V*	*Spectrum	Distance	True Size
N7023	RN	21^h00^m	+68° 10'	10'×8'	6.8	B5e	2000 l-y	5.8×4.6 l-y
N7822	EN	00 04	+68 37	65×20			2700	51×16
I 1396	EN	21 40	+57 30	170×140	5.62	06.5 V	2500	125×100

Photo 3.16 Emission nebula IC 1396 in Cepheus. The bright star on the upper edge of the field is the "Garnet Star", Mu (μ) Cephei

NGC 7023 is about 2,000 light-years away and therefore a little less than 6×5 l-y in size. The NGC 7023 dust cloud is rather far off the galactic equator, around 14°. Thus it is about 500 l-y off the Galactic plane–definitely distant for a dust cloud, because our Galaxy's dust tends to be strongly concentrated along the Galactic plane. The distance of this dust cloud from the galactic equator contributes to the star-poverty of the field around NGC 7023, but prevents the cloud from being as conspicuous as the "Northern Coalsack" between Deneb and Alpha (α) Cephei, which stands out well because it is silhouetted against a rich Milky Way star field.

NGC 7822: Two large emission nebulae are involved with the heavily dust-dimmed association Cepheus OB4, Cederblad 214 and NGC 7822. Both are extremely low-surface-brightness objects. NGC 7822, however, can be glimpsed with averted vision in 15×100 supergiant binoculars as a large, hazy, amorphous patch extending north, NE, and east from a NE-SW pair of 8th magnitude stars separated by a few minutes of arc. On photos it displays a "~"-shape about 65' long E-W and 20' wide N-S. Ced 214, a 50'-diameter mass of bright and dark nebulae, is reported to be visible in 10-inch telescopes with UHC eyepiece filters. Its eastern half appears to have a higher surface brightness than NGC 7822; but sighting it is a problem because of the 6th mag field star (physically unrelated to it) near its center.

The main problem with NGC 7822 is not so much seeing it (if you have supergiant binoculars or an RFT) as *finding* it, because it is in a very bright-star-poor area of the northern sky. Start at the 4th mag Omicron (o) Cephei (which is about 3½° NE of Iota [ι] Cep). Not quite 3° east and slightly south of Omicron Cep is an isolated 5th mag star–it is not difficult to identify if you know your instrument's field diameter because there is no other 5th mag star anywhere near it. Just NE of this star is the 11'-diameter open cluster NGC 7762–another supergiant binocular object. Finally, just about 2½° NE of the 5th mag star is the NE-SW pair of 8th mag stars on the SW edge of the brightest patch of NGC 7822.

The Cepheus OB4 association is some 2,700 light-years distant, so the true size of NGC 7822 is about 51×16 l-y. Ced 214 is around 40 l-y across. The lucida of the association, the mag 8.33 HD 224273, is an evolved G3 Iab yellow supergiant; but Cep OB4 also has two hot O-type members, a mag 8.72 O9.5 V star and a mag 9.05 O7 giant, the latter dimmed 4.3 mags by dust. NGC 7762 is not involved with the association. It seems to be an intermediate-age open cluster somewhat beyond the Cep OB4 dust.

IC 1396: The famous Garnet Star, Mu (μ) Cephei, is on the northern edge of the 2½°-diameter emission nebula IC 1396. This H II region has a very low surface brightness but is not at all difficult (with averted vision) in 10×50 binoculars. 7×, however, does not sufficiently darken the sky background to provide contrast for the tenuous nebula-glow. In 15×100s IC 1396 is easy, but lacks the sense of circularity it has in 10×50s. With 15×100s it is distinctly brighter in the center, and the surrounding region has a peculiarly sooty look. Under very dark sky conditions 10×50s reveal that this surrounding dust cloud makes bays and broad channels into the IC 1396 haze.

The central star of IC 1396 is the multiple HD 206267 = Σ2816. The magnitude 5.62 primary of this system is a very hot O6 Vf star providing most of the ultraviolet radiation fluorescing the nebula. HD 206267 C and D, mags 7.90 and 8.02 and respectively 12" in PA 121° and 20" in PA 339° from the primary, are both B0 V stars. The mag 8.02 component can be resolved in 15x100 binoculars. The star field around HD 206267 has been given the open cluster designation Trumpler 37. It is, however, more simply the richest stellar concentration of the Cepheus OB2 association than a true gravitationally-bound open cluster. Tr 37, as a Milky Way star field, is discussed in more detail in the opening part of this section.

The members of Cepheus OB2 are strewn in an 8°-long oval from the IC 1396 complex on the SSW to 19 and VV Cep on the NNE. The association includes the compact open cluster NGC 7160, located 1° SSW of VV Cep. NGC 7160 is over 20 million years old whereas the O6.5 central star of Tr 37 in IC 1396 must be only a couple million years old, implying that the NE half of Cep OB2 formed first. In fact NGC 7160 is part of an older Cep OB2 subgroup surrounded by a 9°-diameter ring of dust

Photo 3.17 Emission nebulae NGC 7822 (above = north) and Cederblad 214 in eastern Cepheus. The evolved open cluster NGC 7622 is on the WSW edge of NGC 7822 just NE of a 5th magnitude field star

called the Cepheus Bubble, on the margins of which are IC 1396 and the NGC 7129 open cluster + reflection nebula. The association has a rich mixture of luminous-star types: Mu Cep and the primary of the eclipsing binary VV Cep are red M2 Ia supergiants; the mag 4.20 Nu (ν) Cep is an A2 Ia white supergiant like Deneb in Cygnus; the mag 4.74 9 Cep is a hot B2 Ib supergiant and the mag 5.11 19 Cep an even hotter O9.5 Ib supergiant with a bolometric absolute magnitude (total energy output) of $-8^1/_2$ (200,000 Suns); 14 Cep is a mag 5.54 O9 main sequence star which also has a bolometric absolute mag near $-8^1/_2$; and 13 Cep is a mag B8 Ib supergiant. Cep OB2 also contains two red M-type Ib supergiants, a G8 Iab yellow supergiant, and an O9 Ib-II near-supergiant. The distance to the association is about 2,500 light-years (some estimates place it nearer 2,000 l-y), so its full length must be over 400 l-y and the 170'×140' photographic dimensions of IC 1396 correspond to a true size of 125×100 l-y.

V The Cassiopeia Window

The wedge opened in the dust of our Orion-Cygnus Spiral Arm in the N Lacerta/SE Cepheus Milky Way widens in Cassiopeia, and the Cassiopeia Window extends all the way east to the Perseus Double Cluster (NGC 869 + NGC 884). Because of the transparency in this direction, the Cassiopeia Milky Way is remarkably rich in open clusters. A couple of these (NGC 225, Stock 2) are sufficiently near to us that they could be thought of as *in* our Ori-Cyg Arm, though they are too old to be true spiral-arm tracers. A few of the Cassiopeia clusters (M52, NGC 129, NGC 7789) lie in the interarm gap between our Ori-Cyg Arm and the next spiral arm out in our Galaxy, the Perseus Arm. But most of the open clusters in the Cassiopeia Window are young (25 million years old and less) tracers of the Perseus Spiral Arm that have recently condensed from the relatively thick clouds of interstellar matter that define this arc of the Perseus Arm. (The average obscuration to the Perseus Arm open clusters in Cassiopeia is a rather modest 2-2$^1/_2$ magnitudes; but radio observations suggest that the bulk of the interstellar material along this stretch of the Perseus Arm is *behind* the clusters. This is unusual: the dust is more commonly on the *inner* edge of an arm.)

Our view in this direction is *ahead*–that is, in the direction of our orbit around the Center of the Galaxy–but at an angle *out*. Thus there is a good deal less Galaxy to look through here than even toward adjoining Cepheus and Lacerta. Moreover, because the average density of the interstellar material from which stars form *lessens*, and lessens significantly, beyond the circle of the Solar orbit around the Galactic Center, the spiral arms toward our Galaxy's rim are in general less star-rich than those toward the Galaxy's bulge. Consequently, for both these reasons, though we are looking through a relatively dust-free "window" here, the Cassiopeia Milky Way is not especially broad or bright. Its most conspicuous feature is the earlier-mentioned stream that originates at the Delta (δ) - Epsilon (ε) - Zeta (ζ) Cephei triangle, follows the galactic equator ENE along the Cepheus/Lacerta border, passes just to the south of M52, and dead-ends at Kappa (κ) Cassiopeiae, the star at the northern corner of the Square of Cassiopeia. The stream's background glow is actually quite faint, but it is richly scattered with brighter field stars.

However, one of the best regions for binocular-scanning in the Cassiopeia Milky Way is in fact well *off* the M52/Kappa Cas stream in far SW Cassiopeia and neighboring NW Andromeda. This is a rather obscure, little-known area of the Milky Way because it contains no outstanding objects. The open cluster NGC 7686, NNW of Lambda (λ) Andromedae, is very attractive in moderate-sized telescopes around 50x; but even with supergiant binoculars it appears merely as a slight field condensation of mag 10 stars around the group's overpoweringly-bright 6th mag lucida, which drowns out most of the fainter cluster members. Two other star-condensations in the area look more like real open clusters in binoculars than NGC 7686, one north of Lambda Andromedae and the other south of 18 Andromedae. (They are SE and ESE of NGC 7686, respectively.) Both are rough rings of eight or ten stars, the Lambda ring comprised of mag 9–11, and the 18 And ring of mag 8–10 members. In the Lambda And ring a close double of 10th mag stars can be resolved.

The whole SW Cas/NW And region is surprisingly star-rich: though there is not much Milky Way background glow here, it has an abundance of mag 9-10-11 stars in loose groups and ragged chains. Very good binocular viewing. A star stream, rather wide and very rich in mag 8–9 stars, meanders NNW from just north of 18 Andromedae toward the 4th mag AR Cas (the only reasonably bright star west of Beta [β] Cas). The stream curves first just slightly west, then back east, then west again, then east, and finally crests in one last arc west. Its full length is about 5°. It is *very* conspicuous in supergiant glasses and an outstanding feature in an otherwise overlooked corner of the Milky Way.

The Square of Cassiopeia (Alpha [α] - Beta [β] - Gamma [γ] - Kappa [κ] Cas) does not contain either a very bright Milky Way background glow or any rich star fields. Its SE third boasts the greater concentration of mag 7–9 field stars (which is *not* saying much); but, strange to say, its greatest open cluster concentration–NGCs 103, 129, 136, 189, 225, and St 24–is in its relatively star-poor northern quadrant south and SE of Kappa. The best star fields in central Cassiopeia are in fact outside the Square. A remarkably conspicuous 4°-long, 1°-wide stream of mag 7–10 stars extends south from Beta Cas (the star at the western corner of the Square). It has at least 30 members (not counting the usual faint Milky Way background stars), including (in its northern half) four wide doubles, each with components several minutes of arc apart but similarly bright and consequently appearing like real, physically-related pairs. A second star-stream begins NNE of Beta Cas and extends for $2^1/_2$° to a 6th magnitude star, where it sharply angles to the NE and continues the 3° on to Kappa Cas, ending at the ENE terminus of the long, broad Milky Way glow that originated far to the WSW at the Delta (δ) - Epsilon (ε) - Zeta (ζ) Cephei triangle.

Just NW of Beta Cas begins yet a third star stream, rather broad, with faint stars on a pale Milky Way background glow. This stream includes the curiously-straight, 1°-long, SE-NW line of five open clusters, Be 58, NGCs 7790 and 7788, H21, and K12. The stream itself is fairly straight from Beta as far as the 6th magnitude 6 Cas, at the NW end of the open cluster chain; but there the stream arcs first WNW, then west, and finally SW around the northern periphery of a very distinct dark cloud. This cloud is the ENE extremity of a large oval dark region south of the M52/Kappa Cas star stream. A faint star field can be glimpsed due west of NGC 7790 out in this dark region.

But the finest star field in the Cassiopeia Window is of course around the glorious Perseus Double Cluster, NGC 869 + NGC 884. The Double Cluster is probably at its very best possible in 25×100 supergiant binoculars: the two groups' mag 5, 6, and 7 members are made very bright by the light-gathering power of the 100mm objectives, which also pull in many fainter stars; but 25× does not overly-magnify the field, in which not only both clusters, but the inner part of the surrounding stellar association, comfortably fits. Neither of the two clusters is especially concentrated, and there is no appreciable haze of unresolved cluster stars behind even their central knots: the beauty of the sight is from the bright-star-richness of the field in which the central knots of the clusters are embedded, which is so great that it is impossible to say where the clusters themselves end and the surrounding stellar association begins.

The Perseus Double Cluster is probably a true open cluster "double" in the sense that both groups formed from the same giant molecular cloud of interstellar material. They are, however, not in orbit around one another. Indeed, NGC 884, because its stars are dimmed an average of 1.9 magnitudes by interstellar dust whereas those of NGC 869 are dimmed only 1.6 mags, might be somewhat farther than its companion, 7,500 light-years as compared to 7,200 l-y. The two groups also have quite different bright star populations: NGC 869's brightest stars are all blue early-B giants and supergiants, whereas those of NGC 884 are white early-A and red M-type supergiants. This suggests that NGC 884 is not only farther, but older, than NGC 869: 14 million as opposed to 6.4 million years.

In the star field due north of NGC 869 is one of the most conspicuous star chains in the sky. It is about 1° long, consists of nine mag 6–9 stars, and is oriented SE-NW (arcing gently toward the west at its northern end). Several other stand-out star chains can be seen around the Milky Way. One is just south of the Pleiades "Dipper." Another is not far from the Double Cluster in the west-central part of the vast but dim constellation Camelopardalis and is at least visually associated with the small but very beautiful open cluster NGC 1502, which in 25×100 glasses consists of a close double of 6th mag blue-white stars (Σ485) and a close double of 9th mag stars, with two or three fainter cluster members scattered about. (NGC 1502 can be found about 7° SW of Alpha [α] Camelopardalis and the same distance west of Beta [β] Cam.) The star chain is NW of NGC 1502 and oriented generally NW-SE. Its NW section contains eleven stars, including one of the 5th magnitude at its SE end. A lone 8th mag star is south of the 5th mag star between the two chain-sections. The second section contains seven stars and extends SE to a point just south of NGC 1502.

North of the NW end of the NGC 869 star chain is the huge (1° in diameter) but very rich open cluster Stock 2, which, like the Double Cluster itself, is at its best in supergiant binoculars. At least 80 St 2 stars can be counted, and they are a fine sight because they are of comparable brightness (8th to 10th magnitude) and evenly distributed around the cluster area, giving it its strong sense of richness. St 2 is not exceptionally distant, only 1050 light-years away, so despite its large apparent diameter (twice that of the Moon!) its true size is only 18 l-y. (The true size of each component of the Double Cluster, by comparison, is over 65 l-y.) To the WNW of St 2 is a $1^1/_2°$ field of mag 6–10 stars that, though a little looser and more ragged than St 2, is a fine binocular sight and sufficiently concentrated that it looks like it has as much right to be catalogued as an open cluster as St 2.

The Milky Way south and SW of the Double Cluster, toward Phi (ϕ) Persei, is likewise very rich in resolved stars (with just a hint of background glow). To the east the Double Cluster star fields continue for 3°. But then they are abruptly terminated by the N-S border of the Perseus dust clouds. The edge of the dust clouds is impressively sharp. In fact, with averted vision (and under very good sky conditions) two long, thin N-S streaks of outlying dark matter can be seen paralleling the western edge of the main obscuring mass. They are reminiscent of the long cirrus clouds that border the perimeters of approaching winter low pressure systems.

V.1 The Bright Nebulae of Cassiopeia

The stretch of the Perseus Spiral Arm which we see through the Cassiopeia Window is not only rich in Perseus Arm open clusters but contains several large emission nebulae embedded in the fairly thick dust from which the clusters have recently condensed. Its highest-surface-brightness nebula is the compact NGC 281, $1^1/_2°$ due east of Alpha (α) Cassiopeiae in a spot curiously isolated from Cassiopeia's open clusters. Of entirely different character are the two extremely large, extremely low-surface-brightness IC 1805 and IC 1848 in far eastern Cassiopeia, both of which are involved with the sprawling Cassiopeia OB6 association. IC 1848 is a splendid 10×50 binocular object; but IC 1805 and its small but somewhat brighter NW extension, NGC 896, requires supergiant glasses.

NGC 281: In terms of surface brightness, NGC 281, easily located $1^1/_2°$ due east of Alpha (α) Cassiopeiae, is far and away the brightest Cassiopeia H II region. Its integrated apparent magnitude, 8.0, is the same as that of IC 1848 in far eastern Cassiopeia; but IC 1848 spreads that light out over a 2°×1° area whereas NGC 281 is just $^1/_2°$ across. The problem is that at 10x or less the glare of NGC 281's mag 7.8 central star simply overpowers the nebula-glow; and even in 15×100 supergiant glasses the nebula, though not difficult to see as a moderately-bright medium-surface-brightness haze around the central star, does not call attention to itself as it

Photo 3.18 Emission nebula NGC 281 in Cassiopeia

Cassiopeia Bright Nebulae

Nebula	Type	RA (2000.0)	Dec	Dimensions	m_V*	*Spectrum	Distance	True Size
N281	EN	00^h53^m	+56 37′	35′×30′	7.8	06	7200 l-y	70×60 l-y
N896	EN	02 25	+61 34	20×10			7800	45×23
I 1795	EN	02 26	+62 04	40×15			7800	91×34
I 1805	EN	02 31	+61 27	145×100	8.10	04 If	7800	330×230
I 1848	EN	02 51	+60 26	120×55	7.06	06.5f	7800	272×125

Photo 3.19 The giant emission nebula IC 1805 in eastern Cassiopeia. North is to the left. The loose cluster Mel 15 is embedded in nebula haze at the center of the field. Toward the lower left of the field is the loose foreground cluster NGC 1027. The bright patch of nebulosity toward the upper left is NGC 896.

would if seen alone. In 15×100s a couple very faint star-glimmers are intermittently visible near the central star. Just west of the nebula is a field condensation of 8th and 9th mag stars, including a small equilateral triangle of mag 9–9$^1/_2$ stars on the nebula's NW edge.

The central star of NGC 281 is the multiple ADS 719 (HD 5005), the primary of which is the hot O6 object fluorescing the hydrogen gas of the nebula. This star's companions include a 10th mag component just 1.4" away in PA 82°, a 9th mag star 3.8" distant in PA 33°, and a mag 9$^1/_2$ star at 8.9" in 194°. On photos a concentration of very faint stars can be seen north and NW of ADS 719. The multiple and its faint entourage are catalogued as the open cluster IC 1590. This combination of a bright multiple attended by a cluster of much fainter stars appears at the center of many H II regions–including the Orion Nebula itself, within which the Theta (θ) Orionis multiple is surrounded by a cluster of faint nebular variables.

NGC 281 is estimated to be some 7,200 light-years away. Its true size consequently is at least 60 l-y. The nebula's 6° apparent distance from the galactic equator corresponds to a rather high 750 l-y off the Galactic plane. 2$^1/_2$° on the other side of the galactic equator from NGC 281 is the rather obscure Perseus Spiral Arm association Cassiopeia OB7. Thus the thickness of the Perseus arm here must be about 8$^1/_2$°~1070 l-y.

NGC 896 + IC 1805: Overshadowed by the glorious Perseus Double Cluster just a few degrees to their SW are the two huge but low-surface-brightness H II regions IC 1805 and IC 1848. They also suffer from the handicap of being in a very bright-star-poor area and thus difficult to find. Of the two, IC 1805 is the largest, measuring about 2$^1/_2$° E-W by 1$^1/_2$° N-S, but the lower in surface brightness. Perhaps the best way to find it is by first locating the large (20') and rather bright (magnitude 6.7) open cluster NGC 1027, 5° NE of the Double Cluster. NGC 1027 (described in more detail in the section on Cassiopeia in Chapter 2) is not physically involved with IC 1805, but is just 1$^1/_4$° due east of the central cluster of hot stars that is fluorescing the nebula, Melotte 15. Mel 15 is large but scattered, with nine mag 7.8 to 10.3 stars strewn over a $^1/_2$° area: it appears more like a field condensation than a true open cluster. However, in 15×100 supergiant binoculars IC 1805 is visible as a faint glow around Mel 15–not difficult, though amorphous. In 25×100s brighter patches of IC 1805 can be recognized in the area between Mel 15 and NGC 1027, and due north and due west of Mel 15.

About 1$^1/_2$° NW of the center of Mel 15 is a higher surface brightness patch of IC 1805 that has received its own NGC number, NGC 896. It is rather small, only 20'×10', and therefore, despite its relatively high surface brightness, is not visible in medium-sized binoculars. However, with the magnifying as well as the light-gathering power of 25×100s, NGC 896 is not difficult. It is particularly bright near the star at its eastern corner, from which it extends west and NW over a narrow fan-shaped area. To the east of NGC 896 across a thin but dark dust lane is another separately-designated section of IC 1805, the extremely low-surface-brightness IC 1795.

On Photo 3.19 it can be seen that, except immediately around Mel 15 itself, IC 1805 is brightest around its rim. This is the excited zone of collision between the hot expanding gas of IC 1805, and the cool, dense molecular cloud around it. The molecular dust and gas is particularly heavy east of Mel 15, where a "bay" of resistant dark matter juts in toward the cluster. The rim of this bay is the brightest outlying part of the nebula. Photo 3.19 also reveals that the interior of IC 1805 has more faint background stars than the region outside the rim. This is both because ionized gas like the hydrogen glowing in IC 1805 is transparent, and because the stellar winds and radiation pressure of the hot, luminous O-type stars in Mel 15 has "snowplowed" the dust near them out into the IC 1805 rim.

IC 1805 and IC 1848 are involved with the 7,800 light-year distant Cassiopeia OB6, a stellar association described in more detail under Melotte 15 in the section on Cassiopeia's open clusters in Chapter 2. The true size of IC 1805 is 330×230 l-y. which means that it is one of the largest H II regions in our quadrant of the Galaxy, inferior in size perhaps only to the Eta (η) Carinae Nebula, NGC 3372, in the far southern Milky Way.

IC 1848: The easiest emission nebula in Cassiopeia for binoculars is IC 1848, which is much larger than the bright though compact NGC 281 but has a higher surface brightness than the huge but tenuous IC 1805. IC 1848 has no conspicuous stars nearby but is not difficult to find because it happens to be 41/2° NE of the Perseus Double Cluster and the same distance north of Eta (η) Persei. Under good sky conditions it can be glimpsed with averted vision in just 8×40 binoculars as an amorphous area of haze. With 10×50s its rectangular shape becomes evident, measuring about 90'×45' E-W, and the scattering of stars across the nebula-glow is noticeably richer than the surrounding field (though that is not saying much). The magnitude 7.1 O6.5 star fluorescing IC 1848 is toward its western end. The nebula's appearance in supergiant binoculars is basically the same as in 10×50s: a roughly rectangular area of haze approaching 2°×1° E-W in size, over which is spread a dozen 7th, 8th, and 9th mag stars (sometimes catalogued as an open cluster under the same designation as the nebula). The true size of IC 1848 is about 270×125 l-y–not much less than that of IC 1805.

VI The Milky Way toward the Rim of the Galaxy

The galactic anticenter–the point exactly opposite the direction of the Center of the Galaxy–is $3^1/_2°$ east of Beta (β) Tauri on the Taurus/Auriga border. Thus from Perseus SE through Auriga, NE Taurus, SW Gemini/NE Orion (which share the same stretch of the Milky Way), Monoceros, and Canis Major we are in effect looking out toward the nearest arc of the rim of our Galaxy. Because of this, and because the outer arms of a spiral galaxy are typically less star-rich than the inner arms, the Milky Way from Perseus to Canis Major is nowhere particularly broad nor bright. In fact, because of the nearby dust clouds that cover most of Perseus and northern Taurus, in Perseus the Milky Way thins and dims almost to nothing.

Because we lie near the inner edge of our spiral arm, we look *back into* our arm when we look toward the Milky Way from Perseus to Canis Major. Fortunately the nearby gas and dust of our spiral arm does not exactly follow the central line of the Milky Way from Perseus to Canis Major or the Milky Way here would be even fainter than it is. The bulk of nearby interstellar material is along Gould's Belt, which arcs well below this stretch of the Milky Way and includes the Zeta (ζ) Persei Association, the Pleiades, the Orion Association, and the Canis Major Association. Some heavy dust clouds are just beyond the Pleiades; and many stars of the Orion and Zeta Persei associations are still embedded in the dust clouds from which they have only very recently (by astronomical time-scales) condensed.

Further back into our spiral arm, and along its arc out toward the Galactic exterior, the dust clouds are more scattered (true in general of the Galactic exterior), and between them are "windows" through which we have more distant views toward the rim of our Galaxy. One of these "windows" is in the Pentagon of Auriga (Alpha [α] - Beta [β] - Theta [θ] - Iota [ι] Aur + Beta Tau), and through it we see not only the 4,100 light-year distant (but not physically related) open clusters M36 and M37, and the 5,700 l-y distant M38, but also the 15,600 l-y distant NGC 1893 + IC 410 open cluster + emission nebula complex, which lies in an outer spiral arm (possibly a branch of the Perseus Arm). The star field in central Auriga is rich, by the standards of the Perseus-to-Canis Major stretch of the Milky Way, but does not have an especially bright background glow. Some sense of dark lanes snaking through a very pale haze can be gotten from careful looking at the field east and NE of M38; but the effect is not as pronounced as it is in the Sagitta/Vulpecula stretch of the Galactic Center interior of the Milky Way.

A couple visually interesting star arcs can be seen in the SW sector of the Pentagon of Auriga. The gentle curve of 19 + IQ + 16 Aur, with an 8th mag star between IQ and 16, is precisely paralleled just to its SW by the identically-long arc of 18 + 17 Aur + two 8th mag stars. In giant binoculars these two arcs are more identical than they look on star charts. However, there is no evidence that this double star-arc is an actual physical group. Also in this region, 1° due south of 14 Aur, is a remarkably smooth, south-opening, E-W curve of nine 8th-9th mag stars, with four or five mag 9–10 attendants scattered along on either side of the curve. This, however, has more the appearance of a mere chance alignment than does the splendid 19–18 Aurigae double star-arc.

The Auriga Window extends SE to the M35 region in the feet of Gemini. M35 itself is only about 2,200 light-years away. But just $^1/_2°$ to its SW is the partially-resolved little disc of the 13,000 l-y distant NGC 2158, a highly-evolved (3 billion year-old) open cluster that must be out very near the rim of our Galaxy's spiral disc (though absorption to the cluster is only 1.4 magnitudes). A scan west from the M35 region into NE Taurus brings you once more to the nearby Perseus/Taurus dust-cloud complex, obvious from the abrupt decrease in background star density. About $2^1/_2°$ ESE of M35, centered at the north vertex of an equilateral triangle with Eta (η) and Mu (μ) Geminorum, is a casually-concentrated star field dominated by a 1°-long NW-SE line of 5th and 6th mag objects. This loose stellar aggregation has been given an open cluster designation, Cr 89. And in fact it is the richest part of the 4,900 light-year distant Gemini OB1 association, which includes the mag 4.6 B2 Ia supergiant Chi-2 (χ^2) Orionis and, $1^1/_2°$ ENE of Chi-2, the high-surface-brightness emission nebula NGC 2174, an easy 10×50 binocular target.

However, the best binocular star field in the galactic anticenter part of the sky is off the Milky

Way proper. It is the Belt of Orion. All by themselves the three silver-blue supergiants that mark the Hunter's Belt–Delta (δ), Epsilon (ε), and Zeta (ζ) Orionis–would be a stunning binocular sight, for they flood the field with a frosty blue radiance. But they are also embedded in a large aggregation of 6th to 10th magnitude stars. This field, richest around Epsilon Orionis, has been catalogued as an open cluster, Cr 70. It is too large and too loose to be gravitationally bound; but it is part of the great Orion OB1 Association and therefore a true physical group. The Belt supergiants and the Belt Cluster are several million years old and therefore more evolved than the stars of the Sword to the south. Indeed, the populous star cluster embedded in the Orion Nebula (M42), which includes the two Theta (θ) Orionis multiples, is only a few hundred thousand years old. To get a good sense of the thickness of the interstellar material from which the Orion Association giants and supergiants have just recently condensed (and from which more Orion Association members are even now being created), scan NE from Zeta Ori toward the small but bright reflection nebula M78. The area around M78 is one of the regions along the Milky Way that have a peculiarly "murky" look in binoculars–not merely star-poor, but in fact *sooty.* (And they should look sooty, for a considerable amount of interstellar dust is in the form of carbon grains–soot.)

Meanwhile, SE from M35 along the galactic equator itself, the view out toward the Galaxy's rim becomes blocked by the dust clouds involved with the 4,900 light-year distant NGC 2174 emission nebula and its association Gem OB1, and with the 3,000 l-y distant "Christmas Tree" Cluster NGC 2264. But in central Monoceros we come to another window through our Orion-Cygnus Arm out toward the Galactic rim. The basic star field in the Central Monoceros Window is rather rich, its pale Milky Way background glow, garnished with scintillating partial resolution, generously over-scattered by a multitude of 8th to 10th mag foreground stars. And the Monoceros Window has two splendid open clusters for giant binoculars, both 2,900 light-years distant: NGC 2301, a well-resolved, peculiarly-structured group right in the middle of the Window; and the even better-resolved M50 on the Window's south edge. Both M50 and NGC 2301 are intermediate-age groups and therefore not true tracers of the Ori-Cyg Arm.

South of M50 is the 4,300 light-year distant Canis Major OB1 association (a different aggregation from the Canis Major Association proper, which includes most of the 2nd and 3rd magnitude stars in the southern part of the constellation). Its half dozen brightest members, 6th and 7th mag objects, are rather scattered, so the association does not give the star field south and SE of M50 much sense of concentration. However, CMa OB1's brightest nebulosity, the 2°-tall, reversed "S" of IC 2177, is just visible to careful looking in 10x50 binoculars. The Milky Way background glow is brighter, and the Milky Way faint-star field much richer, west of IC 2177 than to its east, where the bulk of the CMa OB1 dust lies. An attractive 2°-long, $^1/_2$°-wide star stream extends between (though it does not link) M50 on the NW and the large (20'), fairly rich and well-resolved, open cluster NGC 2353 on the SE. NGC 2353 is the core cluster of CMa OB1.

The final major "window" out toward the rim of our Galaxy is in central Puppis south of Rho (ρ) and Xi (ξ). Because galactic longitude 270°, the direction from which we have come in our orbit around the Center of the Galaxy, is on the Puppis/Vela border, when we look through the central Puppis Window we look not only *out* toward the rim of our Galaxy, but also *back* at an angle slanting through our Orion-Cygnus Spiral Arm. Nevertheless, the dust in our spiral arm is so thin in this direction that the total absorption out to the rim of the Galaxy here, 30,000 light-years away, is less than 3 magnitudes. Consequently the Milky Way in the Central Puppis Window is very star-rich in binoculars and richest-field telescopes (though its background glow is not exceptionally brilliant). It has been hypothesized that there is so little interstellar material toward central Puppis because a few hundred million years ago the complex of giant molecular clouds within which Gould's Belt was to form came from that direction and "scoured out" a tunnel through the interstellar medium as it passed.

The Central Puppis Window is also rich in open clusters, many visible with only 10×50 glasses. An easy emission nebula for 10×50 binoculars is the rather compact but bright H II region NGC 2467, located 1° ESE of Omicron (o) Puppis. This is a very remote object, 15,000 light-years away, and probably a distant tracer of our Ori-Cyg Spiral Arm.

VI.1 The Bright Nebulae of the Winter Milky Way

When we look toward the winter Milky Way we are looking out toward the rim of our Galaxy in a direction where the dense clouds of cool gas and dust from which stars form, and within which we find emission and reflection nebulae, are more scattered. However, along the winter Milky Way there are several nearby dust clouds containing several good binocular and richest-field telescope nebulae. One of these is the Orion OB1 complex, the stars and nebulae of which are described in a separate section at the end of this chapter. Another is the Perseus-Taurus Dark Cloud Complex with the reflection nebulae NGCs 1435 and 1579. Ori OB1 is centered some 1600 light-years away, with the Per-Tau DCC only about one-third that distance. Halfway between them is the Zeta (ξ) Persei Association, Perseus OB2, which has a small reflection patch visible in supergiant binoculars, NGC 1333, but features the large California Nebula, NGC 1499, just visible in 10×50 glasses. Two small emission nebulae in far NE Perseus, NGCs 1491 and 1624, are likely embedded in dust clouds involved with the 3,300 l-y distant Camelopardalis OB1 association.

Orion OB1, Perseus OB2, and the Per-Tau Dark Cloud Complex are all part of Gould's Belt, which here arcs well out of the winter Milky Way proper. Along the actual Milky Way from Auriga SE to Puppis bright nebulae are rather scattered. Several of them, however, are visible even in modest-size binoculars: the 15,600 l-y distant IC 410 in Auriga, NGC 2174 in extreme NE Orion, the Rosette Nebula NGC 2237 in Monoceros, the Seagull Nebula IC 2177 on the Monoceros/Canis Major border, and the 15,000 l-y distant NGC 2467 in the extremely transparent central Puppis window. The winter Milky Way also has three unusual bright nebulae: Hubble's Variable Nebula, NGC 2261 in Monoceros, small but of high surface brightness; the Flaming Star Nebula, IC 405 in Auriga, a supergiant binocular object illuminated by a "runaway" Orion Association member; and the Crab Nebula, M1 in Taurus, the easiest of all supernova remnants for small instruments.

The Crab Nebula, M1: Only two or three supernova remnants are visible even in supergiant binoc-

Bright Nebulae Along the Winter Milky Way

Nebula	Name	Type	RA (2000.0)	Dec	Dimen	m_V*	*Spectrum	Distance	True Size
M1	Crab	SNR	05^h34^m	+22° 01'	6'×4'	16v		6300 l-y	11.0×7.3 l-y
N1333		RN	03 29	+31 25	6×3	10.8		1600	2.8×1.4
N1435	Merope	RN	03 46	+23 47	30	4.18	B6 IVe	410	3.5
N1491		EN	04 03	+51 19	9×6	11.0	B0		
N1499	California	EN	04 01	+36 17	160×40	4.02	07.5 III	1300	60×15
N1579		RN	04 30	+35 16	12×8	12		450	5×3½
N1624		EN	04 40	+50 27	5				
N2174		EN	06 10	+20 30	40×30	7.55	06.5 V	4900	57×43
N2237	Rosette	EN	06 32	+05 03	80×60	7.25	03.5 V	4900	114×85
N2261	Hubble's Var	EN	06 39	+08 14	3½×1½v	10–12		3000	3×1½v
N2327		EN	07 04	−11 18	1½			4300	1.8
N2467		EN	07 53	−26 33	16×12	9.39	02-3 V	15,000	70×52
I 405	Flaming Star	EN+RN	05 16	+34 16	30×20	5.4-6.1	09.5 V	1500	13×8.7
I 410		EN	05 23	+33 31	40×30	9.04	04 V	15,600	181×136
I 2177	Seagull	EN	07 05	−10 38	120×40	6.21	06.5 V	4300	150×50

ulars, the easiest of them being the NGC 6992/5 arc of the Veil Nebula in Cygnus and the Crab Nebula, M1 in Taurus. Of the two the Crab is undoubtedly the more accessible to small instruments: it is a compact (6'×4') 9th magnitude object whereas the low surface brightness of NGC 6992/5 can make that nebula tricky unless skies are very clear and dark. M1 is visible as a tiny but fairly bright patch of haze in even 35mm binoculars if you have at least 10×: with 7× the Crab appears merely stellar. It is located just over 1° NW of Zeta (ζ) Tauri and about $^1/_2$° due west of a 6th mag field star.

The Crab Nebula is the still vigorously-expanding debris cloud of a supernova that occurred on July 4, 1054 A.D. and was recorded in Chinese (though not European) annals. It became sufficiently bright to be seen in full daylight–which means that it reached an apparent magnitude of at least $-4^1/_2$ or −5. It did not fade from view for almost two years.

The photographic appearance of the Crab is very peculiar: a smoothly-luminous squashed-S shaped glow within a netting of filamentary tendrils. (The squashed-S is visible in 12-inch telescopes; but the filaments can be glimpsed only with observatory instruments.) The remnant of the star that went supernova is equally peculiar: a 16th mag object near the center of the nebula that flashes on and off–"pulses"–in visible light, radio waves, and X-rays 30 times each second. The Crab Pulsar is a rapidly-rotating neutron star in which more than one Solar mass of material has collapsed into a sphere only a dozen miles in diameter. Its magnetic field is so intense that it channels all its radiation out in narrow beams from its magnetic poles and one of these beams (perhaps both) sweeps over our line of sight 30 times each second. The Crab Pulsar is also emitting large quantities of electrons travelling near the speed of light. As these electrons decelerate along the strong magnetic lines of force in the expanding supernova debris cloud, they emit radio waves, X-rays, and visible light, some of which interacts with the material in the debris cloud, causing it to glow.

The Crab Nebula is about 6,300 light-years distant, so its true size is about $11\times7^1/_2$ l-y. Its present expansion rate, 0.2" per year, implies an age of 750 rather than 950 years: but the expansion velocity of the debris cloud is actually *increasing* because of the energy input of the relativistic electrons from the Crab Pulsar. On the other hand the pulsar's period has slightly but measurably lengthened since its discovery in 1968 because the electrons are carrying away the neutron star's rotational energy. The original supernova might have reached an absolute magnitude of nearly −19, a luminosity of 3.3 billion Suns.

NGC 1333: The reflection nebula NGC 1333, located 3 $^1/_3$° WSW of Omicron (o) Persei, is typical of its class of object–small and faint. It is right at the limits for 25×100 supergiant binoculars, appearing as a small, circular, low-surface-brightness patch of haze around its magnitude 10.8 central star–it would be invisible either with less magnification or less aperture. In telescopes the nebula can be seen to be elongated 6'×4' NE-SW, its NE lobe having its own 12th mag central star.

NGC 1333 seems to be involved with the same complex of dust clouds which have given birth to the Perseus OB2 association (described under NGC 1499 below). It might be a bit beyond the core of Per OB2. The mag 10.8 and mag 12 illuminating stars of the nebula have themselves just been born as the consequence of the collision between two small molecular clouds, the nearer one, centered NE of the nebula, moving NE to SW, and the farther one, centered SW of the nebula, moving SW to NE. Infrared observations reveal just SW of NGC 1333 more than two dozen infrared sources: these are newly-born stars still buried in the dense dust of the molecular clouds. Eventually, when the dust clears, we should see here a small cluster of modest-mass, modest-luminosity stars.

The Merope Nebula, NGC 1435: The Perseus-Taurus Dark Cloud Complex covers north central Taurus, SE Perseus, and extreme SW Auriga. It is centered at about RA $4^h\ 30^m$, dec +25° just NE of the Pleiades. And its center is 450 light-years away just beyond the 410 l-y distant Pleiades. Indeed, at present the cluster seems to be passing through the nearer edge of the dust clouds. The reflection nebulae around the four "dipper" stars of the Pleiades, 17, 20, 23, and 25 Tauri, is Pleiades star light reflected off Tau DCC dust. The bright Pleiades *not* embedded in reflection patches, like 27 and 28 Tau at the end of the "dipper handle," must be slightly nearer to us than the "dipper bowl" stars and thus just outside the front edge of the Tau DCC. The brightest of the Pleiades reflection nebulae, NGC 1435 around and south of Merope = 23 Tau at the southern corner of the "dipper bowl," is a 10x50 binocular object. It is described in more detail in the section on the Pleiades under Taurus in Chapter 2. The Pleiades nebulosity cannot be dust left over from the giant molecular cloud in which the cluster was born at least 70 million years ago because stellar winds and radiation pressure from the brightest Pleiades stars would have blown away all such residual matter long ago.

NGC 1491 and NGC 1624: In extreme NE Perseus are two very small emission nebulae visible only with high-magnification supergiant binoculars: NGC 1491, located 1° NW of the 4th magnitude Lambda (λ) Persei, measures 9'×6' N-S; and NGC 1624, a few degrees ENE of the open cluster NGC 1545 (which is 2° due east of Lambda Per), is only 5' across. In 25×100 glasses NGC 1491 appears as a small circular patch with an 11th mag central star (telescopes show that the nebula is in fact elongated N-S and the mag 11.0 star is at the midpoint of its eastern edge), and NGC 1624 is a small spot of very low-surface-brightness haze. However, neither, despite their small size, are difficult. The fact that both are emission rather than reflection nebulae implies that their stars are somewhat hotter than the types of stars forming in the Taurus Dark Cloud Complex past the Pleiades, around which we find mainly just very faint reflection nebulosity. And indeed, the B0 illuminating star of NGC 1491 is a hotter, more massive type of object than the Tau DCC is capable of generating. If it has an absolute magnitude typical for main sequence B0 stars, around −4, the distance to NGC 1491 (compensating a perhaps too-high 5 mags for absorption by dust) is the same as that to the Camelopardalis OB1 association (described under NGC 1502 in Chapter 2), 3,300 light-years.

The California Nebula, NGC 1499: One of the nearest of the young stellar associations–indeed, nearer even than the Orion Association–is Perseus OB2, which includes Zeta (ζ), Omicron (o), Xi (ξ), and 40 Persei, as well as a scattering of fainter stars, 10° due north of the Pleiades. Per OB2 is centered 1100 or 1200 light-years away and is a compact, very young group whose members are still in, or very near, the giant molecular cloud of gas and dust within which they have just been born. The association is no more than 1 million years old. Its lucida, the magnitude 2.9 Zeta Per, is a B1 Ib supergiant with an absolute mag near −6 (22,000 Suns). The mag 4.0 Xi Per is a very hot O7.5 III giant with a bolometric absolute mag (total energy output) of −8.7 (a quarter-million Suns). Xi Per has a very high radial velocity, 59 kilometers per second in recession, and seems to be a "runaway" from the center of Per OB2, which is located to its south between Zeta and Omicron.

In its flight from the center of its association, Xi Per is encountering some denser interstellar matter. The ultraviolet radiation of the star is fluorescing the hydrogen gas of the large California Nebula, NGC 1499, a shallowly-curved arc of pale haze about $2^1/_2$° long and 1° thick centered just NE of Xi. The California has a very low surface brightness but can be seen in only 10×50 binoculars with averted vision when it is near the zenith of a dark-sky site: with such glasses even the nebula's arced-rectangle shape can be discerned. In 15×100 supergiant instruments the California is an amorphous area of haze extending from NW to NE of Xi Per: it is not necessarily difficult to see (again use averted vision), and it appears quite long and broad, but it lacks the sense of curved rectangularity it has in 10×50s. However, in 25×100s it is barely visible: its surface brightness is so low that 25× almost magnifies it out of existence. In fact it would not even be noticeable if you did not know it is there. The one part of the California improved by 25×100s is its NE rim, a thin streak of brighter nebula-glow reminiscent of the NGC 6960 arc of the Veil Nebula in Cygnus. This bright streak is the excited zone of collision where the California's hot hydrogen gas is expanding into the interstellar material around it.

Photo 3.20 The California Nebula, NGC 1499 in Perseus. The bright star toward the lower edge is Xi (ξ) Persei

Assuming NGC 1499 to be somewhat beyond the core of Perseus OB2 (the recessional velocity of Xi Per suggests that in the past million years it has moved 200 light-years out of the center of the association directly away from us), the true size of the California Nebula is 160×40 l-y.

NGC 1579: The Merope Nebula NGC 1435 is the reflection nebula around a star intruding into the Perseus-Taurus Dark Cloud Complex. However, star-formation is presently occurring in various parts of the Per-Tau DCC and some of these recently-born stars are still swaddled in small reflection and emission nebulae. The Per-Tau DCC is generating only modestly-luminous stars of a couple Solar masses or less, so the nebulae around these stars are likewise small and relatively faint. Most of them require the magnifying power of telescopes to be seen. However, NGC 1579, located 2° ENE of the 5th magnitude 54 Persei, is sufficiently large, 12'×8', that it can be glimpsed in 25×100 supergiant binoculars as a small shred of haze: it is amorphous, and of very low surface brightness, but immediately discernible. Assuming it to be 450 light-years away (the distance to the center of the Per-Tau DCC), the true size of NGC 1579 is just 5×3½ l-y.

NGC 2174: One of the brightest emission nebulae in Orion is as far from the M42 heart of the Orion OB1 association as it can get and still be within the constellation. It is the large (40'×30') and high-surface-brightness NGC 2174, centered 1½° ENE of Chi-two (χ^2) Orionis (which marks the end of Orion's club). The glow of NGC 2174 around its magnitude 7.5 central star can be seen (with averted vision) even in mere 40mm binoculars. In 10×50s the nebula appears as a perfectly circular disc around the central star.

NGC 2174 is not involved with the Orion OB1 complex at all. In fact it is 4,900 light-years distant, three times farther than the M42 core of Ori OB1, and is the showpiece H II region of its own association, Gemini OB1. Gem OB1 covers a 5°-long, 2°-wide area from Chi-two Ori on the SW to 8 and 10 Geminorum on the NE. The mag 4.63 Chi-two Ori, the lucida of the association, is a B2

Photo 3.21 Emission nebula NGC 2174 (bottom center) in NE Orion, and the field of the Gemini OB1 Association. The open cluster M35 is toward the upper right (NW). The two bright stars in the field are Eta (η, right center) and Mu (μ, left center) Geminorum. The arc of nebulosily ENE of Mu Gem is the supernova remnant IC 443

Ia blue supergiant with an absolute mag of –7.6 (a luminosity of 100,000 Suns). The second and third brightest stars of the association are the mag 5.75 and 6.26 3 and 8 Gem, which are B2.5 Ib and B3 Iab supergiants, respectively. The association has three distinct subgroups. Its youngest, most luminous members, Chi-two Ori and the extremely hot central star of NGC 2174, an O6.5 main sequence object with a total energy output of around a half million Suns each second, are at Gem OB1's SW end. At the association's NE end, around 8 and 10 Gem, is a loose cluster, catalogued as Collinder 89, of nine modest-luminosity main sequence and giant stars with spectral types around B0. (The mag 6.42 10 Gem is a B0 II giant with an absolute mag of –5.8.) In between (located NW and SW of Eta [η] Gem) are the association's very luminous but evolved red supergiants, BU, TV, and WY Gem, respectively M1 Ia, M1 Iab, and M2 Iab stars. Cr 89 is Gem OB1's oldest subgroup. Apparently star formation in the association has proceeded from NE to SW. The 5° apparent length of Gem OB1 corresponds to a true size of 430 l-y.

The Rosette Nebula, NGC 2237: The huge Rosette Nebula, which has an apparent diameter of over 1°–*twice* the apparent diameter of the Moon!–is one of those large, low-surface-brightness objects that are easier in 50 mm binoculars than in conventional medium-aperture telescopes. It is not at all difficult with 10×50 binoculars from a dark-sky site, and under ideal conditions can even be seen (with averted vision) in 40mm glasses at about 15×. With 15×100 supergiant instruments the Rosette, though very diffuse, and amorphous rather than circular in outline, is very distinct and covers a surprisingly large area around its $^1/_4$°-long central cluster, NGC 2244. The famous "hole" in the Rosette requires richest-field telescopes equipped with the lowest-power eyepiece available and a nebular or H-alpha filter.

The Rosette Nebula can be a bit difficult to find because it is in a bright-star-poor stretch of the Milky Way and its central cluster, though a 15'×5' NW-SE rectangle of eight 6th to 8th magnitude stars, is not sufficiently rich or concentrated that it stands out very well from the surrounding Milky Way field. The cluster and nebula are best located by star-hopping from Gamma (γ) Geminorum: first go 4° SSE to mag $3^1/_2$ Xi (ξ) Gem, then 3° south and slightly west to the 5th mag S Monocerotis (the lucida and northernmost star of the open cluster NGC 2264), then 3° SW to mag 4.5 13 Mon, and finally 2° due south to the Rosette complex.

The Rosette Nebula is one of the most photogenic objects in the entire sky. Its rose-red color comes from the 6562Å-wavelength hydrogen-alpha (Hα) line which is emitted by an electron as it cascades down a hydrogen atom's orbital shells after recombining with a free proton (an H II ion), emitting a characteristic wavelength of radiation at each jump. The electron was initially kicked off the proton by an ultraviolet photon from a nearby hot star. In the case of the Rosette, the ultraviolet photons that ionize the hydrogen gas in the nebula come from the O-type members of NGC 2244. The central "hole" of the Rosette is a true expanding hollow from which the gas and dust around NGC 2244 has been pushed by the strong radiation pressure of the O-type stars of the cluster: the intense radiation of these stars is accelerating the dust particles in the interstellar matter around them which then, because these particles carry electric charge, drags the electrically-charged ionized hydrogen gas with them.

The Rosette Nebula and Cluster is the heart of the 4,900 light-year distant Monoceros OB2 association. The 80'×60' photographic size of the nebula corresponds to true dimensions of 114×85 l-y–almost four times larger than the Orion Nebula, but barely a third the size of IC 1805 in Cassiopeia. The diameter of the central "hole," roughly 30 l-y, and the estimated radiation pressure of the stars of NGC 2244, implies that those stars began shining only some 300,000 years ago. The mass of the ionized gas in the nebula is about 10,000 Solar masses; but NGC 2237 is at the NE end of a 4°~325 l-y long giant molecular cloud which contains a quarter-million Solar masses. The Rosette Cluster, NGC 2244, and the Mon OB2 association are described in more detail in the section on the Monoceros open clusters in Chapter 2.

Photo 3.22 "Close-up" of dark lanes in the Rosette Nebula, NGC 2237 in Monoceros

Hubble's Variable Nebula, NGC 2261: A very small, but high-surface-brightness, nebula visible in 25×100 supergiant binoculars is the peculiar NGC 2261, "Hubble's Variable Nebula." This is a 10th magnitude, roughly triangular ($3^1/_2' \times 1^1/_2'$ N-S), patch of nebulosity that changes erratically in brightness, size, and internal detail. It is a combination emission/reflection feature illuminated by the nebular variable R Monocerotis, located within the nebula's southern tip. R Mon itself erratically varies between mags 10 and 12, and its spectrum has been classified anywhere from A to early G by different authorities. The variations of the nebula (and of R Mon) require telescopes to be followed, but NGC 2261 can be at least glimpsed with higher-magnification supergiant binoculars as a hazy spot $^3/_4$° SSW of the mag 7.1 star at the southern tip of the arrowhead-shaped open cluster NGC 2264, and $^1/_2$° WNW of a 6th mag field star (the latter easily spotted exactly 1° due south of S Mon, the 5th mag lucida of NGC 2264).

NGC 2261 is peculiar not simply because it is a "variable nebula," but also because its variations do not occur as a result of the light changes of its illuminating star: when R Mon brightens or fades, the nebula does not brighten or fade outward from the star with time (like a ripple on a pond) as would be expected. On the other hand, the changes of internal detail in the nebula occur too rapidly to be the consequence of actual physical motions of the nebula's gas and dust, which would need to be moving with velocities near the speed of light; so these changes must somehow be due to changing illumination from R Mon. Apparently what is occurring is a sort of "shadow-play": small dense clouds of dust drift near R Mon and cast shadows that radiate out through the nebula with the speed of light. Some of the light variations of the star must also be caused by this effect – that is, by obscuring dust clouds passing between the star and ourselves. R Mon is a strong emitter of infrared radiation, which implies that it is embedded in a dust cloud that absorbs much of the star's visible light and re-radiates it in cooler IR wavelengths. A "variable nebula" similar to NGC 2261 is the small comet-shaped NGC 6729 in Corona Australis (described in

Photo 3.23 Nebulosity around the open cluster NGC 2264 in Northern Monoceros. North is up. The bright star on the upper edge of the field is S Monoceroti. Directly below (south) of it in the left-center of the field is the Cone Nebula. Hubble's Variable Nebula, NGC 2261, is to the lower right (SW)

Section 1 above), which has the nebular variable R CrA in its head.

NGC 2261 is embedded in a small molecular cloud involved with the 3,000 light-years distant Monoceros OB1 association. (Mon OB1 is described in the discussion on NGC 2264 in the section on Monoceros in Chapter 2). It is located on the NE rim of a large ring of pale nebula-glow called the Monoceros Loop, the still-expanding shell of an old supernova. The interaction of this shell with the interstellar medium is stimulating star-formation at several sites around the Monoceros Loop, including the NGC 2261 molecular cloud.

NGC 2327: NGC 2327 is the brightest section of the Seagull Nebula, IC 2177, which is described in detail below.

NGC 2467: The region around and several degrees to the south of the magnitude 3.34 Xi (ξ) Puppis is one of the most transparent directions anywhere along the Milky Way. Consequently this is a very star-rich field for binoculars and RFTs. It also has quite a number of, mostly distant, open clusters, the best for binoculars and small RFTs being M93 $1^{1}/_{2}°$ NW of Xi and the quartet of NGCs 2527, 2533, 2567, and 2571 a few degrees south and SE of the 3rd mag Rho (ρ) Puppis.

Part of the reason for the transparency in this direction is that here we are looking out toward the rim of our Galaxy, and gas and dust is thinner toward the Galactic exterior than toward the Galactic interior. But another reason for the clear view here seems to be that the original giant molecular cloud from which formed the stellar groups around Gould's Belt–the Orion, Canis Major, and Scorpio-Centaurus associations, as well as the Pleiades, to name only the most conspicuous–came from this direction, sweeping up the gas and dust as it went.

Because there is comparatively little gas and dust in this direction, we cannot expect to see many bright emission and reflection nebulae toward central Puppis. However, there is one compact (about 10' across) but bright (mag 7.0) H II region here, NGC 2467, located about 2° SSE of Xi and 1° ESE of Omicron (o) Puppis. NGC 2467 is so compact, and its surface brightness so high, that with a casual scan in 10x50 binoculars it might be mistaken for a star. But once noticed its bright, fuzzy-edged disc is obvious, and averted vision reveals the nebula's mag 9.21 O6 V central star. But the principal source of the ultraviolet photons fluorescing NGC 2467 is an ultra-hot O2-3 V object, the mag 9.39 star just NE of the mag 9.21 nebula lucida.

The NGC 2467 complex is about 15,000 light-years away. Its full 16'×12' photographic size therefore corresponds to true dimensions of 70×50 l-y, though its bright interior is more like 35 l-y across. It is embedded in a very compact and isolated giant molecular cloud of gas and dust This molecular cloud is so compact that photos reveal, right on the galactic equator just $^{1}/_{2}°$ west of the nebula, four faint galaxies. NGC 2467 is the core of an association, Puppis OB2, the lucida of which is a mag 8.11 O7 star just SW of the nebula. This is a very complicated direction, with several young stellar groups along the same line of sight. Just over half the distance to NGC 2467 + Puppis OB2, about 8,200 l-y away, seems to be a thin association that is catalogued as Puppis OB1. And just NE of the nebula are two extremely remote, but intrinsically highly luminous, young open clusters: Haffner 18ab is a 2'-long, NW-SE line of mag 11–15 stars located 4' NE of NGC 2467's mag 9.21 lucida; and Haffner 19 is a 3'-diameter gathering of mag 11 and fainter stars a few minutes of arc due north of Haf 18ab. The integrated apparent mags of these two clusters are only 9.3 and 9.4, respectively; consequently they require moderately large telescopes to be seen at their best. But they both seem to be about 22,500 l-y distant and have absolute mags between −9 and $-9^{1}/_{2}$–greater than the absolute mags of the two components of the Perseus Double Cluster. They are about 44,300 l-y from the Center of the Galaxy and probably right on the outer rim of our Galaxy's spiral disc.

The Flaming Star Nebula, IC 405: Stellar associations and young open clusters are very active places. In both supernovae are frequent events (in terms of Galactic time scales). And both seem to contain a large proportion of double and multiple stars (the table of Orion Association members in the next section of this chapter gives a good example of

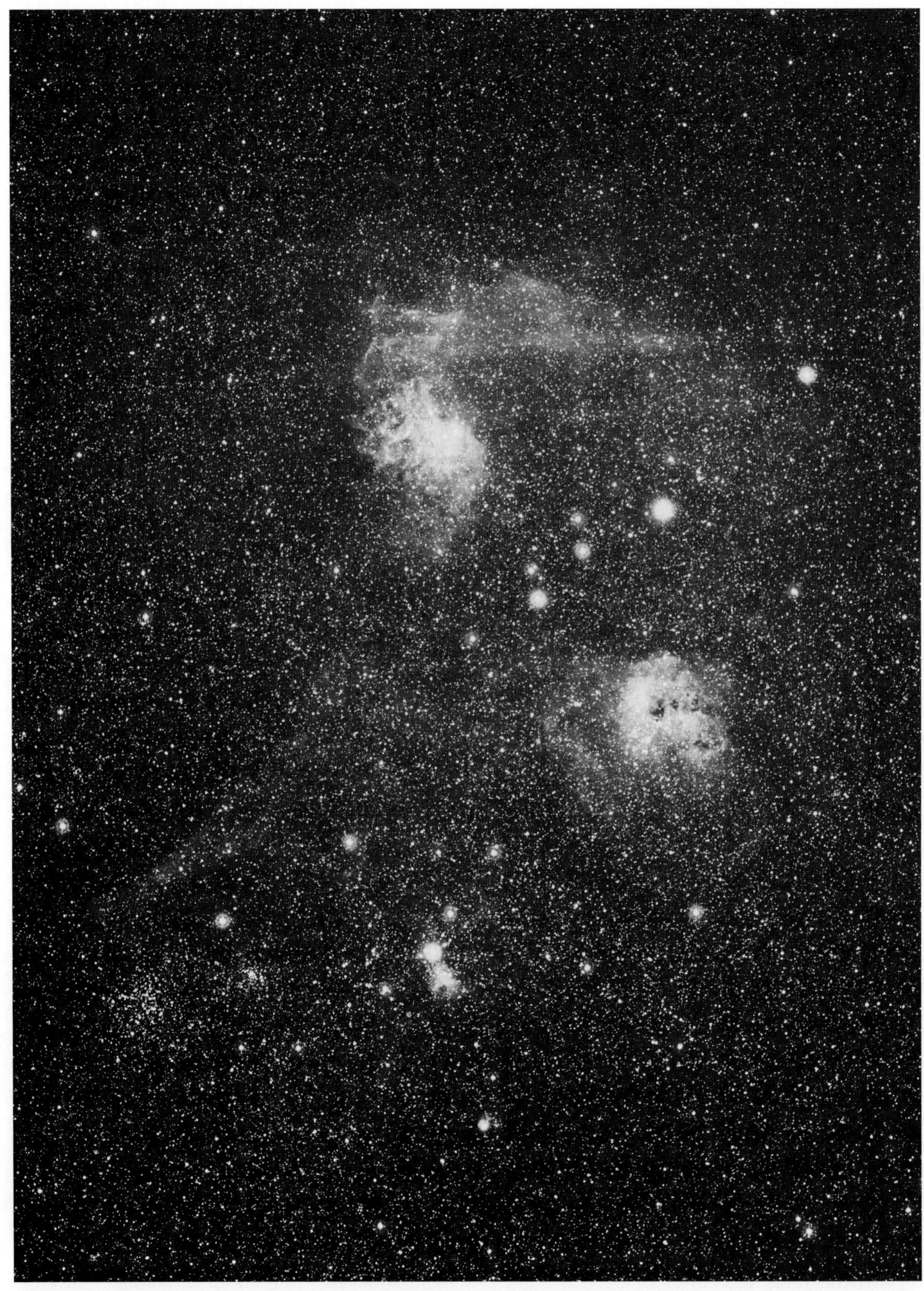

Photo 3.24 The Flaming Star Nebula, IC 405 (above center), and the distant emission nebula IC 410 (right center) in central Auriga. North is to the left. The large open cluster toward the lower left corner is M 38

that). If the primary of a double star goes supernova, the companion will suddenly find itself with nothing left around which to orbit and flies off into deep space on a straight line with its original orbital velocity. But sometimes when multiple stars pass sufficiently near each other to gravitationally interact the same thing happens: one of the components finds itself flung out at a high velocity into interstellar space. The latter situation is called the *dynamical ejection scenario* (DES), the former the *binary supernova scenario* (BSS). In either scenario the result is a *runaway star.*

Runaway stars are revealed by their high radial velocities and/or large proper motions, which imply high true space velocities. The direction of that space velocity points directly away from the star's parent group, which usually can be easily identified. There are three well-known runaways from the Orion Association: the magnitude 5.16 O9.5 V Mu (μ) Columbae, the B1.5 V 53 Arietis, and the variable (mags 5.4–6.1) O9.5 V AE Aurigae.

AE Auriga, located in central Auriga about 2° west of Phi (ϕ) Aur, has a true space velocity of 113 kilometers per second and probably was ejected from the M42 region of the Orion association by the dynamical ejection scenario. It is of particular interest among the Orion runaways because it is presently plowing through a rather dense cloud of interstellar gas and dust. The star is sufficiently hot to fluoresce the gas for a considerable distance around it and its blue light is being reflected from the thick dust it is encountering. The result is the combination emission/reflection nebula IC 405, the Flaming Star Nebula. The brightest part of the nebula, north and east of AE, measures about 30'×20' roughly SE-NW. However, it has extremely low surface brightness and cannot be seen with anything less than 15×100 supergiant binoculars, in which it appears only as a very pale amorphous glow around and north of AE. To confirm your sighting of it, compare the appearance of AE with that of the similarly-bright 16, 17, and 19 Aur just to its SE: the stars in that conspicuous star-arc should lack the dim halo that is around and north of AE. The area immediately west and SW of AE reveals the presence of thick obscuring clouds of dust by the region's poverty of background stars: its darkness is a striking contrast to the faint-star-rich background east of the 16 + 17 + 19 Aur star arc, and it has the "thick" or "murky" look of directions toward the nearby dust clouds (such as toward the Taurus Dark Cloud Complex beyond the Hyades and the Pleiades). The reason for the contrast is the passage through the region of the hot, luminous AE, which swept the dust away by radiation pressure as it fled north from Orion. As can be seen on Photo 3.24, the boundary between the cleared zone south of AE and the dust clouds to its SW is marked by very faint, very straight streaks of emission glow about 1½° long, oriented N-S in the direction of AE's motion.

IC 410: The rather large (20' in diameter) but very low-surface-brightness emission nebula IC 410 is around the large but loose open cluster of 9th magnitude and fainter stars NGC 1893, located in central Auriga 1½° SW of the 4th mag Phi (ϕ) Aurigae. The nebula can be glimpsed in 10×50 as well as in 100 mm supergiant binoculars as an amorphous area of haze overspreading the scatter of NGC 1893 stars. The NGC 1893 + IC 410 complex is extremely far from us, 15,600 light-years out toward the rim of the Galaxy, which in this direction must be only a few thousand light-years farther. The full 40'×30'photographic extent of IC 410 implies a true size of an extremely impressive 181×136 l-y–*six times* the true size of the Orion Nebula, M42. NGC 1893 and the association of which it is a part, Auriga OB2, are described in more detail in the section on Auriga's open clusters in Chapter 2.

The Seagull Nebula, IC 2177: Located right on the Monoceros/Canis Major border and centered about 3° ENE of the 4th magnitude Theta (θ) Canis Majoris, is the 2°-long, but very thin, N-S reversed "S" of the emission nebula IC 2177. Under dark-sky conditions, and with averted vision, not only the fact of IC 2177, but even its shape, can be glimpsed with just 10×50 binoculars. The problem with NGC 2177 is that it is in a bright-star-poor area and therefore difficult to locate. The best way to find it is first find the large and bright open clusters M50 and NGC 2353 (both of which are described in the section on Monoceros in Chapter 2). The northern end of IC 2177 is 2° due west of NGC 2353 and the same distance SSE of M50. The southern end of IC 2177 is $2\frac{1}{2}^{\circ}$ due east of Theta CMa. The open cluster NGC 2335 (also described in the section on Monoceros in Chap. 2) is on the northern tip of IC 2177, but is a rather difficult cluster even for giant binoculars. The brightest patch of IC 2177 has been separately catalogued as NGC 2327; it is just over $\frac{1}{2}^{\circ}$ due west of the only 5th mag star in the field, FN CMa, but is only $1\frac{1}{2}'$ across and therefore too small to be seen even in 25×100 super-giant glasses.

The main illuminating star of IC 2177 seems to be the mag 6.21 HD 54662, an O6.5 V object located $\frac{1}{2}^{\circ}$ west of NGC 2353 toward IC 2177. The intense radiation pressure of this hot, luminous star has cleared the area between it and the nebula of most of the gas and dust that was there, and even with binoculars it can be seen that there are more faint background stars east of IC 2177 than to its west. At least some of the glow of the nebula is from the collision of dust accelerated by the radiation pressure of HD 54662 with the dense molecular cloud west of IC 2177.

IC 2177 is about 4,300 light-years away and therefore about 150 l-y long and a maximum of 50 l-y wide. It is involved with the Canis Major OB1 association with HD 54662 and NGC 2353. CMa OB1 is described with NGC 2353 in the Monoceros section of Chapter 2.

The star patterns that form the constellations which we have inherited from the ancient Greeks are for the most part simply chance alignments of stars. However, a few of them are not accidents in the sense that most of the stars comprising them are physically related. The central five stars of the Big Dipper, for instance, are all members of the Ursa Major Moving Group centered some 71 light-years away. And most of the bright stars of Scorpius are members of the extremely long Scorpio-Centaurus Association.

The same is true of Orion. With the exception of Betelgeuse (Alpha [α] Orionis) and Bellatrix (Gamma [γ] Ori), the bright stars that constitute the figure of the Hunter are all members of the Orion OB1 association. Furthermore, the stars of the Belt of Orion, Delta (δ) + Epsilon (ε) + Zeta (ζ) Ori, are the heart of one of the major subgroups of Orion OB1; and the Orion Nebula, M42 + M43, with Iota (ι) Ori in the Sword of Orion are the most conspicuous components of the youngest Ori OB1 subgroup. The brightest Orion Association stars are listed in the accompanying table.

As the presence in the association of extremely luminous, rapidly-evolving O giants and B supergiants implies, Orion OB1 is extremely young. The Belt subgroup is at most a few million years old; and Iota and the Theta-one (θ^1) and Theta-two multiple stars buried in the glow of the Orion Nebula are probably just a few hundred thousand years old. Thus it is no surprise that much gas and dust is still within the association. Indeed, star formation is presently occurring in several regions of the Orion complex, in particular within the dust clouds behind the Orion Nebula and behind the small M78 + NGC 2071 nebulae $2^1/_2$° NE of Zeta Ori. The dense giant molecular cloud that extends from the Zeta Ori region NE to the M78 region is called Orion B, and the molecular cloud behind M42 is called Orion A. The presence of these star-obscuring dust clouds is

The Orion Association

Star	Apparent mags	Spectra	Distance	A_V	Absolute mags	Separation + PA
Rigel (β)	0.1var, 6.9	B8 Iae, B5 V	900 l-y	0.0	−7.1, −0.3	9" in 202°
Alnilam (ε)	1.70	B0 Ia	1600	0.2	−7.0	
Mintaka (δ)	1.9–2.1, 6.87	O9.5 II, B2 V	1600	0.4	−7.2, −2.2	53" in 0°
Saiph (κ)	2.04	B0.5 Ia	2100	0.2	−7.0	
Alnitak (ζ)	2.05, 4.2	O9.5 Ib, B0 III	1600	0.3	−6.7, −4.6	2.4" in 164°
ι	2.77, 7.3	O9 III, B7 IIIp	1500	0.2	−5.7, −1.2	11" in 141°
λ	3.54, 5.61	O8 III, B0 V	1800	0.4	−5.5, −3.4	4.4" in 44°
η	3.6var, 4.98	B1 IV + B, ?	940?		−3.7?, −2.3?	1.6" in 77°
σ	3.8, 7.5, 6.5	O9.5V+B3,B2,O9	1400*	0.2	−4.6	13" in 84°; 42" in 61°
λ Leporis	4.28	B0.5 IV	1900	0	−4.5	
υ	4.60	B0 V	1900	0.1	−4.3	
Σ 747	4.79, 5.7	B0 V, B1	2000	0.1	−4.3, −3.4	36" in 223°
θ^2 A, B, C	5.08	O9.5Vep	1600	0.6	−4.0	
	6.38	B0.5Vp			−2.6	52.5" in 92°
	8.1–9.8	B5			−1	129" in 97°
Trapezium =	5.14	O6ep	1600	1.1	−4.5	
θ^1 C,D,A,B	6.70	B0.5 Vp			−2.9	13.3" in 61°**
	6.7–7.7	B0.5 Vp			−2.6	13.7" in 311°**
	8.0–8.7	B			−2	16.8" in 342°**

* The mag 7.5 and 6.5 components are probably background stars not physically related to Sigma A.

** Distance and position angle measured from the bright C component.

especially obvious on photographs (see Plate XXXVII); but even in binoculars the area east and NE of Zeta Ori is peculiarly star-poor, and even "murky," in appearance.

The Orion complex is rich for the wide-field observer. First are its star colors. Rigel (Beta [β] Ori), Saiph (Kappa [κ] Ori), and the Belt supergiants all have a magnificent silver-blue glitter in binoculars. Though it is rather far from them even for 7× glasses, contrast their silver-blues with the ruddy-orange of the M2 Iab "red" supergiant Betelgeuse. These colors are better appreciated in binoculars than in telescopes, which gather so much light from 1st and 2nd magnitude stars that the star images are simply "over-exposed" on the eye's retina and their colors washed out into mere blue- and orange-tinted whites.

As the table of Orion Association members shows, the group is rich in double and multiple stars (which seems to be typical for OB associations). The best Orion double for small binoculars is the Σ747 pair 8' SW of Iota, two comparably-bright silver-blue early-B stars that are only 36" apart but an easy split even for 7× glasses. The 135"-wide Theta-one + Theta-two pair on the north edge of M42 is also an easy 7x split. 10x is sufficient to resolve the mag 6.9 component of Delta, which is almost 1' of arc due north of its primary, and the mag 6.5 companion of Sigma (σ), almost $^3/_4$' ENE of that system's lucida. (Though these stars are somewhat farther from their primaries than Σ747B is from Σ747A, there is a greater magnitude spread and consequently it is difficult or impossible to discern them from their primaries' glare with only 7×.) 10× is also sufficient to see the mag 6.4 and 8.1 components of Theta-two Ori, which are respectively 1' and 2' due east of their primary; but the bright background glow of M42 makes it a challenge.

The Orion complex is surprisingly poor in open clusters. However, low-power binoculars show encircling Epsilon, the middle star of the Belt, a rich field of 6th, 7th, and 8th mag stars. This field, about 2° in diameter, extends as far as Delta and Zeta at the ends of the Belt and is called the Belt Cluster and catalogued as Collinder 70. These stars are in reality physically related to the Belt supergiants: they formed in the same molecular cloud with the supergiants, the intense radiation pressure of which has cleared the area of most of the cloud's residual dust and gas. The only significant emission glow in the Belt is around Zeta, which is at the SW edge of the Orion B molecular cloud. The Belt Cluster is strictly a 7×–10× binocular field: it cannot take magnification. Its star-richness at low powers is enhanced by the icy silver-blue sheen of the three O/B supergiants.

Cr 70 is less a true open cluster than simply a rich association subgroup: its members are certainly not gravitationally bound. Other than the Trapezium cluster, most of which is still buried in Orion A dust behind M42, the only true open cluster in the Orion complex is the very large (25' long E-W) but very loose NGC 1981 at the north end of the Sword. This group has only about a dozen stars, three of mag $6^1/_2$–7 and several of mags 8–10, but is an attractive sight in low-power, large-aperture instruments like 11×80 giant binoculars. The whole Sword fits in the field of view of such glasses, so NGC 1981 tends to get overlooked because of the glories just to its south.

The centerpiece of the Sword of Orion–indeed of Orion as a whole–is the Orion Nebula, M42, one of the handful of emission nebulae that can be seen with the unaided eye. In 10×50 binoculars M42 covers an area equal to that of the full Moon. With averted vision the thick filament on the east side of the nebula (a conspicuous photographic feature) can be traced from the region of the Theta's south all the way to the western member of the wide E-W pair of 7th mag stars that are halfway to Iota at the tip of the Sword. From the Theta's the M42 glow extends west and WSW for a distance almost equal to the length of the Great Filament. Because the interior of the nebula SSW of the Theta's is somewhat fainter than is its eastern and western sides, in 10×50s M42 has something of a "double wing" look.

But the Orion Nebula's magnificent telescopic appearance begins to be evident in 15×100 super-giant glasses. In such instruments the nebula is *not* seen merely as a smooth glow paler in the interior and smoothly brightening out to its east and west wings. Instead it has the distinct impression of *texture*–of thin, dark tendrils snaking through the nebula-glow. This gives it a remarkably *living* look.

Photo 3.25 The Orion Nebula, M42. The 3rd magnitude Iota (ι) Orionis is at the lower (southern) edge of the field. On the upper edge is the NGC 1977 reflection nebula with its illuminating stars 42 (right) and 45 Orionis.

Bright Nebulae in Orion

Nebula	Type	RA (2000.0)	Dec	Dimensions	m_V*	*Spectrum	Distance	True Size
M42	EN	05^h35^m	−05°27′	65′×60′	5.14	O7 Vp	1600 l-y	30×28 l-y
M43	EN	05 37	−05 16	20×15	6.8	O7 V	1600	9.3×7.0
M78	EN+RN	05 47	+00 03	8×6	10.2,10.6		1600	3.7×2.8
N1973	RN	05 35	−04 44	5	7.1v	B3	1600	2.3
N1975	RN	05 35	−04 41	10×5	5.5	B2	1600	4.6×2.3
N1977	RN	05 35	−04 52	20×10			1600	9.3×4.6
N2024	EN	05 42	−01 05	30	2.05	O9.5 Ib	1600	14
N2064	EN+RN	05 46	00 00	$1\frac{1}{2}$×1			1600	0.7×0.5
N2067	EN+RN	05 46	+00 06	8×3			1600	3.7×1.4
N2071	EN+RN	05 47	+00 18	7×5	10.1		1600	3.3×2.3
I 434	EN	05 41	−02 24	60×10	3.8	O9.5 V	1400	24×4
I 2118	RN	05 35	−07 10	3°×1°	0.1 var	B8 Ia	900	48×16
Sh2-276	EN	Barnard's Loop		13°×5°			1600	390×140
B 33	DN	05 41	−02 28	6×4			<1400	2.4×1.6

The dark bay into the nebula from the NE reaches west almost all the way to Theta-one. The E-W line of the three bright components of Theta-two are well within the edge of the nebula-glow but neatly parallel the southern edge of the dark bay. Theta-one, the famous Trapezium quadruple star, though not resolved, is distinctly a disc rather than a single star-point. The faint (mag 9–11) stars, many of them very young nebular variables, that are part of the Trapezium Cluster are embedded in the nebula-haze and therefore do not call attention to themselves but should be looked for because their hard-edged images contrast markedly with the soft nebula-glow.

Just north of the Theta region of M42 is a lobe of the Orion Nebula that has its own mag 6.8 O7 illuminating star and has been separately catalogued as M43. It is about one-third the diameter of M42 and somewhat lower in surface brightness. With 15×100s M43 is very distinctive around its central star and appears clearly detached from M42; but it is rather difficult for 10×50s.

A true low-surface-brightness nebula in the Sword of Orion is NGC 1977, a 20′-long E-W reflection nebula around 42 and 45 Orionis, 5th mag stars several minutes apart E-W midway between M42 on the south and the open cluster NGC 1981 on the north. 42 and 45 Ori are early-B objects and therefore not sufficiently hot to fluoresce the hydrogen gas around them. However, they are rather bright stars and the interstellar dust around them is rather thick. Consequently they are embedded in a reflection haze just large and bright enough to be glimpsed with averted vision in 10×50 binoculars. In 15×100 glasses, NGC 1977 is pale and amorphous, but fairly extensive. NGC 1977 is the largest and brightest of a nest of reflection nebulae that includes the 10′×5′ NGC 1975 around a mag 5.5 B2 star to the NW of 42 Ori, and the compact NGC 1973 around a 7th mag double NNW of 42.

Thus the Sword of Orion offers the binocular observer in one 2° field of view, from north to south, an open cluster (NGC 1981), a reflection nebula (NGC 1977), a bright emission nebula (M43), an even brighter emission nebula (M42) with structural detail and its illuminating multiple star (the Theta group), a 3rd mag silver-blue O-type giant (the O9 III Iota), and a tight double of comparably-bright silver-blue early-B stars (Σ747). No other single field in the heavens can compete with the Sword of Orion for variety and celestial beauty.

However, the field of Zeta Ori at the SE end of the Belt of Orion is not far behind the Sword, at least for telescopes. Zeta itself is a brilliant silver-blue O-type supergiant (with a mag 4.2 companion for high powers in telescopes). Sigma to its SSW is a multiple with a hot blue O9 primary and so many components (one resolvable in 10× glasses) that it is almost more an open cluster. South of Zeta extends the very pale reef of nebula-glow, IC 434, upon which is silhouetted, 1° south of the star, the famous Horsehead Nebula, B33. Unfortunately the Horsehead is so small, 6'×4', and IC 434 is so faint that they require moderately-large telescopes and moderately-high magnifications to be seen. It's a balancing act: you need sufficient magnification to make B33 large enough to be seen; but too much magnification will wipe IC 434 out of existence, and the Horsehead can be seen at all only by contrast with the background glow of IC 434.

But centered just $^1/_4$° NE of Zeta is a very large (30' diameter) and bright emission nebula, NGC 2024, which can be glimpsed with 50mm optics-providing the power is at least 10× to get the nebula sufficiently far from the glare of Zeta. With 10x50s NGC 2024 is distinct, but it appears basically just as a lobe of haze extending a little NE out of the glare of Zeta. In telescopes NGC 2024 reveals a very conspicuous 3'-wide lane of obscuring matter, curving from south to NW, reminiscent of the lane of obscuring matter that bisects the Lagoon Nebula, M8 in Sagittarius. Except for the glare of Zeta, the NGC 2024 lane would be visible in 15×100 binoculars. Both NGC 2024 and IC 434 seem to be fluoresced by Zeta.

Considerably smaller, measuring only 8'×6', and fainter than NGC 2024 is the combination emission + reflection nebula M78, located $2^1/_2$° NE of Zeta. However, M78 is much easier to see than NGC 2024 because the latter has to compete with the glare of Zeta. In 10×50 glasses M78 is a small, moderately high surface brightness, hazy-edged disc with a 10th mag star at its center. In 15×100 supergiant binoculars the central star of M78 intermittently resolves into mag 10.2 and 10.6 components, separated by 51" in PA 202° (NNW-SSE). M78 is the brightest of a group of small, newly-emerged emission + reflection nebulae around modest-luminosity early-B stars that have just formed in this region of the Orion B giant molecular cloud. Though a little smaller (7'×5') than M78, NGC 2071, 15' to the larger nebula's NNE, can be seen in 10×50 instruments as a tiny shred of haze.

The edge of the Orion B giant molecular cloud is only about 1° NE of M78. It is bordered by a very faint emission glow which photographs reveal to be part of a huge semicircle of nebulosity that begins $3^1/_2$° SE of Gamma Ori and arcs SE, south, and then west around the Belt and Sword of Orion, ending near Rigel. This is *Barnard's Loop* (discovered by the early 20th century Milky Way photographer E. E. Barnard). Its brightest segment, NE of M78, is catalogued as Sharpless 2-276. Barnard's Loop seems to be an expanding shell driven by the radiation pressure of the Belt supergiants and by shock waves from past supernovae in the Belt and Sword. It is reported to be visible from high-altitude observing sites with the unaided eye assisted by a nebular or Hα filter. Assuming Barnard's Loop to be centered about 1600 light-years from us, its 13° N-S diameter corresponds to a true size of 360 l-y and its full arc length is 575 l-y.

A possible extension of Barnard's Loop is the 3°-long NNE-SSW patch of reflection haze IC 2118, the Witch Head Nebula, centered $2^1/_2$° NW of Rigel (see Plate XLI.) It seems to be reflecting the light of Rigel, which, as a B8 star, is too cool to fluoresce the gas around it but so luminous, 60,000 times brighter than the Sun, that it could easily produce a reflection nebula even at the projected distance of IC 2118, 30–40 l-y. The Witch Head Nebula is reported to be visible from high-altitude sites with 10×50 binoculars.

Chapter 4
Galaxies and Galaxy Groups

Perhaps the greatest surprise awaiting the wide-field observer are the galaxies visible in low-power instruments. The sheer *numbers* of galaxies that can be seen even with standard binoculars–more than six dozen in 10×50s–would be sufficiently astonishing all by itself; but in fact with large binoculars the *structural details* of literally *scores* of galaxies can be glimpsed. Of course you must look carefully from a dark-sky site to discern these details; but once you know what to look for, it is perfectly easy to distinguish a giant elliptical galaxy, for example, from a face-on Sb spiral by the differences between their eyepiece images. Indeed, galaxies offer the wide-field observer much, *much* more structural detail than planetary nebulae and globular clusters–and a lot more of them can be seen. Only open clusters, emission nebulae, and the Milky Way itself are better targets for wide-field instruments than galaxies.

Like globular clusters, in the eyepiece most galaxies display a tenuous halo surrounding a core that contains a stellar nucleus. The main difference between a globular image and a galaxy image is that the features of the latter (excluding the stellar nucleus, if present) are indescribably more tenuous and ambiguously-edged than the former. However, galaxies also come in a striking variety of shapes, from perfectly circular face-on spirals and E0 ellipticals to highly-elongated edge-on spirals. Indeed, the hauntingly aethereal sliver of such edge-on spirals as NGC 4565 in Coma Berenices, NGC 891 in Andromeda, and NGC 5907 in Draco make them among the most memorable sights in binoculars or telescopes of any size. Nothing else in Nature can compare to the ghostly streak of an edge-on spiral galaxy.

Galaxy-searching with giant binoculars and RFTs is aesthetically superior to galaxy-searching with even much larger conventional telescopes in two ways. First, because such instruments *are* wide-field, frequently more than one galaxy can be seen in the same field of view. A few among many examples are M65 + M66 + NGC 3628 in Leo, M81 + M82 in Ursa Major, M31 + M32 + NGC 205 in Andromeda, and M84 + M86 in the core of the Coma-Virgo Galaxy Cluster. Indeed, galaxies are so dense in the core of the great Coma-Virgo Cluster that often a half-dozen systems can be seen at once. M87, M89, M90, and M58, for example, all conveniently fit in the same richest-field view along with fainter attendants (the exact number of which seen will depend upon the aperture and power of your binocular or RFT).

The second advantage of wide-field galaxy-observing is that very often galaxies in or around the same field of view will have contrasting structural appearances (hence true structural differences). Among the most dramatic structural contrasts are between the M81 + M82 pair in Ursa Major, between the M102 and NGC 5907 pair in Draco, and among the M65 + M66 + NGC 3628 trio in Leo. But even when the structural differences are less flagrant–as between M105 + NGC 3384 in Leo, or among the giant ellipticals in the core of the Coma-Virgo Cluster–the visual effect is quite beautiful.

In this chapter the galaxies visible with binoculars and wide-field telescopes are for the most part divided into the actual physical groups or clusters of which they are members. The largest and brightest galaxies are naturally the nearest galaxies and therefore members of the Local Galaxy Group with our Milky Way and its two Magellanic satellites. Most of the brightest Local Group members happen to be in the autumn sky, as is the next nearest galaxy gathering, the Sculptor Group.

Just a little beyond the Sculptor group are the Ursa Major (M81) and Centaurus (NGC 5128) galaxy groups in the spring skies. The spring constellations as a whole are especially rich in galaxies. Indeed, the spring heavens has something of a "milky way" of galaxies that begins in Ursa Major around the Big Dipper and extends SE through Canes Venatici, Coma Berenices, and western Virgo, ending west of Spica. The densest aggregation of spring galaxies off this galactic "milky way" is in Leo, which has three galaxy groups with members visible in wide-field instruments: the M65/M66 Group has two members which can be seen with just small binoculars; the M95/M96 Group has five members that can be glimpsed with 10×50 glasses; and the distant NGC 3607 Group has two or three members within the reach of supergiant instruments.

Numerous systems in the spring "milky way" of galaxies from the Big Dipper SE to Spica can be seen with 10×50 or larger glasses. The densest galaxy-aggregation in this galaxy-stream, and a particularly fine galaxy-field for supergiant binoculars, is in southern Coma Berenices and NW Virgo. This region, the "Realm of the Galaxies," is the Coma-Virgo Galaxy Cluster, the core of the Coma-Virgo, or Local, Supercluster, which includes not only the galaxies and galaxy groups in the spring "milky way" of galaxies, but also the Local and Sculptor galaxy groups. The brightest members of the Coma-Virgo Galaxy Cluster will be described in one section. Then the other bright galaxies in the spring galactic "milky way" will be discussed under their separate constellations: Ursa Major, Canes Venatici, and Coma Berenices. Finally, the brighter spring galaxies off the axis of the Coma-Virgo Supercluster will be covered: three in Leo, three in far eastern Virgo, and then eight scattered around the other spring constellations.

The richness of galaxies and galaxy groups in the spring constellations dramatically contrasts with the poverty of nearby galaxies and galaxy groups in the autumn skies, with have just some Local Group members and Sculptor. This is because we are near the outer edge of our Local Supercluster: when we look toward the Coma-Virgo Galaxy Cluster in southern Coma and NW Virgo, we are looking into the core of the Coma-Virgo Supercluster; but when we look toward the autumn skies we are looking out past the sparse Sculptor Galaxy Group into the relatively galaxy-poor spaces between the Local Supercluster and its distant neighbors (see Figures 4.1–4.3.) The final section of this chapter describes autumn galaxies outside the Local and Sculptor groups, which for the most part are either isolated ungrouped systems, or bright galaxies on the nearest edges of other superclusters.

Brightest Members of the Local Galaxy Group									
Galaxy	Constel	RA (2000.0)	Dec	m_V	Dimensions	Type	Color Index	Distance	M_V
M31=N224	And	00^h43^m	+41°16′	3.4	178′×63′	SA(s)b	0.92	2.4×10^6l-y	-21.1
Milky Way						Sbc			-20.6
M33=N598	Tri	01 34	+30 39	5.7	73×45	SA(s)cd	0.55	2.7	-18.9
LMC	Dor	05 24	-69 45	0.4	650×550	SB(s)m		0.16	-18.1
IC 10	Cas	00 20	+59 18	11.3	7.3×6.4	SB(s)m		4.1	-17.6
M32=N221	And	00 43	+40 52	8.1	8×6	dE2	0.94	2.4	-16.4
SMC	Tuc	00 53	-72 48	2.2	350	Im		0.16	-16.2
N205	And	00 41	+41 41	8.5	17×10	dE5p	0.84	2.4	-16.0
N3109	Hya	10 03	-26 09	9.8	16×3	Sm		4.1	-15.8
N185	Cas	00 39	+48 20	9.2	$14^1/_2\times12^1/_2$	dE3p	0.90	2.2	-15.3
N147	Cas	00 33	+48 30	9.5	15×9.4	dE4	0.94	2.2	-15.0
N6822	Sgr	19 45	-14 48	8.8	19×15	IB(s)m	0.70	1.6	-14.4
IC 1613	Cet	01 05	+02 07	9.2	$20\times18^1/_2$	Im		2.5	-14.4

Few galaxies are loners: almost all are members of groups of at least a dozen systems. The Milky Way is no exception. It is the second brightest in a group of three dozen, mostly dwarf, galaxies. The brightest member of the Local Galaxy Group is the Andromeda Spiral M31. The most luminous members of the Local Group, including all those visible with any type of binoculars or small richest-field telescope, are listed in the accompanying table. The three brightest systems in the Local Group are all large spirals: the Andromeda Galaxy M31, the Milky Way, and the Triangulum Galaxy M33. The Local Group also has three Magellanic spirals (irregular systems with incipient spiral structure)–the Large Magellanic Cloud (LMC), NGC 3109, and IC 10–and four dwarf ellipticals–M32, and NGCs 147, 185, and 205. However the majority of our Local Group members are either dwarf irregulars resembling, but even smaller than, the Small Magellanic Cloud (SMC), or extremely loose, low-luminosity dwarf spheriodals (none of which are visible in wide-field instruments). The Local Group does not have any lenticular (S0) or giant elliptical (E) systems: the nearest of these types of galaxies are in the Centaurus (NGC 5128) Galaxy Group. The total mass of the Local Group is estimated to be 3.7 *trillion* Solar masses. Half of the total luminosity of the Local Group comes from the Andromeda Spiral alone.

Four Local Group galaxies can be seen with the unaided eye: the Large and Small Magellanic clouds in the far southern heavens, and the Andromeda and Triangulum spirals in the northern sky. Easy for 10×50 binoculars are the M32 and NGC 205 satellites of M31; more difficult are the NGC 185 satellite of M31 and Barnard's Galaxy NGC 6822 in Sagittarius. Possible supergiant binocular galaxies include the NGC 147 satellite of M31, NGC 3109 in Hydra, and the large, but low-surface-brightness irregular system IC 1613 in Cetus. Other Local Group galaxies have too low surface brightness to be glimpsed in any size binoculars or RFTs: their apparent sizes are plenty large enough, but their stars are too faint and too spread out to create a perceptible haze.

The Local Galaxy Group has a pronounced binary structure. At one end is our Milky Way, its two Magellanic satellites, and several low-luminosity dwarf spheriodal systems. Our nearest galactic neighbor is the Sagittarius Dwarf, a dSph system with an integrated absolute magnitude of just -14.0 (a luminosity of 33 million Suns), which we see through our Galaxy's central bulge. It is only 80,000 light-years away from the Solar System, and so near the Milky Way that it will probably be "eaten" by it in the astronomically near future. The bright Sagittarius globular M54 is in fact a Sgr Dwarf cluster rather than a member of our Milky Way's family of globular clusters.

At the other end of the Local Galaxy Group is M31, its four dwarf elliptical satellites (M32 and

Photo 4.1 The Andromeda Galaxy and its satellites M32 and NGC 205. North is up. The star on the left (east) edge of the field is Nu (ν) Andromedae

NGCs 147, 185, and 205), M33, and several more dwarf spheriodal systems. The great Andromeda Spiral **M31** (Plates XLII, XLIII, XLIV; Photo 4.1) is easily seen on reasonably dark nights as a NE-SW streak of haze just west of the 4th mag Nu (ν) Andromedae. (Strange to say, M31 does not seem to have been noticed by the ancient Greek or Babylonian astronomers, though they were otherwise attentive watchers of the skies.) The integrated apparent mag of M31 is 3.4; but its apparent size depends critically upon the darkness of its sky background and upon the size of the instrument with which you view it. In supergiant glasses it can appear (with averted vision) almost 4° long and 1° wide. The galaxy appears so elongated because it is a spiral tilted only about 15° from fully edge-on. The nearer side is the NW, and even in binoculars it is obvious that this edge of M31 is more distinct, less ambiguously-bordered, than the SE side. The reason is clear from photographs, which reveal dark dust lanes along the NW side of the galaxy: they are features exactly like the Great Rift in the Milky Way, except viewed from a much greater distance.

The tripartite structure typical of spiral galaxies–a spiral disc around a large central bulge containing a bright, concentrated nucleus–is clear in the 10×50 binocular view of M31. In the center of the highly-elongated spiral plane of M31 is a pale circular glow about 20' across, as ambiguously-edged as the SE side of the galaxy's spiral disc but dramatically brightening toward its interior. This is the system's central bulge. At its center is the tiny, bright, but not quite stellar image of the bulge's nucleus. The distance to the Andromeda Galaxy, derived from its Cepheid variables, is 2.4 million light-years. Thus the full diameter of M31's spiral disc is over 150,000 l-y (our Milky Way's disc, for comparison, is less than 100,000 l-y across), and its bulge's diameter is roughly 12,000 l-y. Like our Galaxy's, the M31 nucleus is only a couple dozen l-y across but contains millions of stars.

In supergiant binoculars the main structural features of M31 are extremely distinct and provide a splendid example of the typical eyepiece appearance of even very distant Sb spirals. The highly-elongated spiral disc fades ambiguously towards its NE and SW extremities, which lengthen almost disconcertingly with averted vision. The bulge is very bright and strongly circular, but without distinct edges.–Strange how an object can be both circular and yet lack an actual outline! M31's bulge is a good example of how ambiguously-bordered this feature of spiral galaxies really is; and the way *both* bulge and spiral disc fade into the basic background sky glow demonstrates convincingly that we see galaxies only by contrast with the sky background–and that there is a lot more of any individual galaxy than we actually see, no matter how good the sky background.

The nucleus of M31 in supergiant binoculars is a bright, but not perfectly stellar, dot. In distant Sa and Sb systems the nucleus does indeed appear star-like. The difference in distinctness between the NW and SE long sides of the M31 spiral disc is very pronounced in supergiant glasses. Indeed, 15×100s give the impression that the inner of the two major dust lanes along that side of the galaxy can be seen. However, a bit more magnification is required for a certain sighting, and in 25×100s the inner dust lane is difficult but unquestionable with averted vision.

One of the interesting differences between the Milky Way and the M31 halves of the Local Galaxy Group is that the brightest satellites of the Milky Way are two Magellanic irregular systems with young star populations and vigorous star-formation complexes whereas all the bright satellites of M31 (excluding the rather distant M33) are dwarf ellipticals with evolved-star populations and comparably little gas and dust. Three of M31's elliptical satellites are visible in 10×50 binoculars, **M32**, **NGC 205**, and **NGC 185**. The last is 7° due north of the big spiral in southern Cassiopeia. However, M31, M32, and NGC 205 all easily fit in the same wide-field view. M32 is quite bright, magnitude 8.1, but so compact, only about 7' across, and concentrated that it appears almost stellar. It is just south of the M31 bulge, with a 9th mag star between it and the bulge's nucleus. M32 is tricky to spot because it is an almost circular dE2 system and fades very rapidly outward from its stellar core. By contrast the elongated dE5 NGC 205, 35' NW of the M31 bulge, though slightly fainter (mag 8.5) than M32, is easier to see because of the distinctly oval outline of its image. Unfortunately, the surface brightness of NGC 205 is so low that it disappears at 7×, which does not sufficiently darken the sky background. NGC 185, located 1° due west of Omicron (o) Cassiopeiae, is a dE3 system almost as large and elongated as NGC 205, but only one-half as bright and therefore marginal for 10×50s. In 15×100 supergiant glasses NGC 185, though appearing as an amorphous, very low-surface-brightness patch of haze requiring averted vision, is not difficult once noticed because of its size. In 15×100s M32 remains a very small disc, almost just a dot with a stellar core: though it is bright, you really have to *look* for it. But in these instruments NGC 205 is even more obvious than in smaller glasses because of its elongation. It is distinctly brighter toward the middle, but does not appear to have an elongated core or a stellar nucleus. The fourth M31 dwarf elliptical satellite, the elongated dE4 **NGC 147**, is about the same size as NGC 185 just over 1° to its east, but 0.3 mag fainter. It is difficult even for supergiant binoculars, requiring ideal sky conditions.

The integrated absolute magnitude of M31 is -21.1, a luminosity of 24 billion Suns. (In reality the galaxy's overall luminosity is much higher because this number does not compensate for the fact that we see M31 almost edge-on and thus it suffers considerable dimming from *internal absorption* by its dust lanes.) The absolute magnitudes of M32, NGC 205, and NGC 185 are respectively -16.4, -16.0, and -15.3, luminosities of 300, 230, and 110 million Suns. The high color index of M31, +0.92–it is as red as its M32 and NGC 205 elliptical satellites–is because the Andromeda Spiral is a bulge-dominated system with at present only a modest rate of star formation, hence of the production of luminous blue supergiants, occurring in its spiral arms.

M33 (Plate XLV, Photo 4.2), the Triangulum Galaxy, located 15° SE of M31, is a large, low-surface-brightness object that spreads the light of a mag 5.7 star out over an area larger than the full Moon. When it is near the zenith of extremely dark sites it can be glimpsed with the unaided eye. M33 has such low surface brightness because it is a very loose-armed Scd spiral which we view almost face-on–its slightly oval shape (oriented NE-SW) is from the slight inclination of its spiral disc to our line of sight. In 15×100 supergiant binoculars M33 appears as a large NE-SW glow, brightening slightly toward the interior, at the center of which is an intermittently-resolved stellar nucleus. Except for the nucleus, the galaxy image is featureless–*strangely* featureless for something so large! However, some internal structure can be made out in 25×100 instruments: the bulge becomes visible as a faint little circular patch, almost lost in the surrounding galaxy glow; and the tiny, but bright, fuzzy dot of the NGC 604 star cloud in one of the galaxy's outer arms can be glimpsed SSW of the nucleus. An arc of a half dozen faint stars curves toward the star cloud from the west.

The distance to M33, estimated from the brightness of its Cepheid variables, is 2.7 million light-years, confirming the long-held belief that it lies a little beyond the Andromeda Spiral. The true separation between the two galaxies is not quite 1 million l-y. M33 is an almost bulgeless system: on photos its spiral arms can be traced nearly into the nucleus. The galaxy's rather blue color index, +0.55, is a consequence of the vigorous star-formation presently occurring in its spiral arms. M33's integrated absolute mag, -18.9, corresponds to a luminosity of 3.2 billion Suns.

The final Local Group galaxy that can be glimpsed with standard 10×50 binoculars is the faint (mag 8.8) but large (19'×5') NGC 6822, a dwarf irregular system in extreme NE Sagittarius. It is a very low-surface-brightness object, and inconveniently far south for mid-northern observers. Fortunately it is in a well-marked location: it is 80' NE of 55 Sagittarii, with a mag $5^1/_2$ star midway between 55 Sgr and the galaxy. In 10×50s under excellent sky conditions NGC 6822 appears as a small N-S smudge of light. However, with 25×100s it is not even difficult, a N-S patch of haze discernably brighter toward the middle. In fact, the bar-like character of the galaxy's core, the feature which gives it an "IB" rather than a simple "I" morphological classification, is evident. The Cepheid-derived distance of NGC 6822 is 1.6 million l-y: thus it is ten times farther from us than the Magellanic satellites of our own Milky Way, but just two-thirds the distance of the Andromeda Spiral. Its integrated

Photo 4.2 The Triangulum Galaxy M33 and Comet Tempel-Tuttle (lower right; parent comet of the Leonid meteor shower). Photographed on January 30, 1998

absolute mag of −14.4 converts to a luminosity of nearly 50 million Suns.

Two challenge Local Group objects for super-giant binoculars are the Magellanic spiral **NGC 3109** in Hydra and the dwarf irregular **IC 1613** in Cetus. Both have integrated apparent mags of about $9^1/_2$, but are rather large in apparent size and therefore of extremely low surface brightness. NGC 3109 is, moreover, very thin (we see it almost edge-on) and even farther south than NGC 6822, with a declination of −26° 09'. If all that wasn't enough, both galaxies are in bright-star-poor regions and thus as difficult to locate as they are to see: NGC 3109 is about $7^1/_2$° NW of the 4th mag Alpha (α) Antliae and 2° east of a 5th mag star; and IC 1613 is a little over 12° due north of Eta (η) Ceti 0.8° north of the 6th mag 6 Cet and the same distance almost due west of the 6th mag 29 Cet. The Cepheid variables in IC 1613 imply that it is about 2.5 million l-y distant. Its integrated absolute mag is therefore about the same as that of NGC 6822, −14.4. NGC 3109 is rather remote, 4.1 million l-y away, and therefore its membership in the Local Group has been questioned.

II Nearby Galaxy Groups

II.1 The Sculptor Galaxy Group

One of the most star-poor regions of the heavens is south of the constellation Cetus. This area was outside any of the ancient constellation figures and first named *l'Atelier du Sculptor,* "the Sculptor's Studio," in the mid-18th century by the French astronomer Lacaille. The single most important fact about this region of the sky is that it contains the south galactic pole (located 9° due south of Beta [β] Ceti). This means that our view here is at right angles to the dust-rich Milky Way and that, in the absence of any nearby dust clouds, we should have a clear view into deep intergalactic space. If there are any bright galaxies in this direction, we should be able to see them.

And in fact toward Sculptor lies the nearest galaxy group to us outside our Local Group. The Sculptor Galaxy Group is centered about 8.2 million light-years distant, but its individual members lie between 4 and 13 million l-y away. Its five brightest systems are all loose-armed spirals and therefore, except for the nearly edge-on NGCs 55 and 253, of very low surface brightness. However, their integrated apparent mags are 9.1 and brighter and all five can be glimpsed with 10×50 binoculars. The main problem is elevation: all five galaxies are between −20° and −40° declination and therefore none of them ascend very high above the horizon haze for observers in Europe and much of the United State.

The largest and brightest Sculptor Group member is **NGC 253**, easily spotted 7$^1/_2$° due south of Beta Ceti. It is almost $^1/_2$° long but only a few minutes of arc wide because we view it almost edge-on. However, it has an integrated apparent mag of 7.6 and is obvious even with 7×35 glasses as a skinny gash of light oriented NE-SW, a close E-W pair of relatively bright (mag 8$^1/_2$) field stars near its southern tip. In 15×100 supergiant binoculars the gash thickens into a streak with a bright, extremely elongated, core, and is so conspicuous that it can be seen in a scan before the two mag 8$^1/_2$ stars (which, however, add to the excellent aesthetics of the sight) are spotted. In 25×100s NGC 253 appears to its full photographic length and its core becomes a bright elongated oval and its halo a somewhat thicker oval, but very tenuous and much extended at the ends. NGC 253 is about 9 million l-y distant, which implies that it has an integrated absolute mag of −19.8, a luminosity of 7 billion Suns. However, this is a minimum value: the galaxy's full face-on luminosity would be about twice as great and therefore its true brilliance comparable to our own Milky Way's. It is the nearest *starburst galaxy:* systems in which much enhanced rates of star-formation are occurring either on a galaxy-wide scale, or in large regions of the disc and/or bulge. Most starburst galaxies have bluer color indices than NGC 253's +0.85; but, because of its edge-on orientation, its starlight is reddened as well as dimmed by the dust in its spiral plane.

NGC 55 and **NGC 300** both have integrated apparent mags of 8.1 and would be easy 10×50 binocular galaxies. But they are very far south, with declinations of −39° 11′ and −37° 41′, respectively, and consequently can be seen even with supergiant glasses only from the southern half of the United States. NGC 55, located 3° NNW of Alpha (α) Phoenicis, is as long as NGC 253, but even thinner. NGC 300, in a very star-poor area 8° south and slightly east of NGC 253 and about 2$^1/_2$° ENE of the wide 6th mag pair Lambda-one (λ^1) + Lambda-two

Brightest Sculptor Group Galaxies

Galaxy	RA (2000.0)	Dec	m_V	Dimensions	Type	Color Index	Distance	M_V
N 55	00^{h}15^m	−39°11′	8.1	30′×6.3′	SB(s)m	0.50	4.2×10^{6}l-y	−17.6
N 247	00 47	−20 46	9.2	19×5.5	SAB(s)d	0.56	10	−18.2
N 253	00 48	−25 17	7.6	30×7	SAB(s)c	0.85	9	−19.8
N 300	00 55	−37 41	8.1	20×13	SA(s)d		7	−18.7
N 7793	23 58	−32 35	9.2	10.5×6.2	SA(s)d	0.54	13	−18.9

Sculptoris, is a nearly-round oval because we see it nearly face-on. This gives it an extremely low surface brightness. NGC 55 is the more conspicuous of the two galaxies because of its bright needle-like appearance. The apparent magnitudes of the Cepheid variables in NGC 300 imply that the system is 7 million l-y away, slightly nearer than NGC 253. NGC 55, however, is only about half as far as NGC 253, 4.2 million l-y distant. NGC 55 and most of the other bright Sculptor Group galaxies have relatively blue color indices, +0.56 to +0.50. This is because their central bulges are very small and because star-formation is currently very vigorous in their spiral arms. Recent observations of NGC 300 have identified no less than 58 very young, very luminous Wolf-Rayet stars in the central 6.8'×6.8' of the galaxy. WR stars are only found in associations and clusters of hot, luminous blue giants and supergiants.

Though only of mag 9.2, both **NGC 247** and **NGC 7793** can be detected with 10×50 binoculars. NGC 247, just over the border in southern Cetus $4^1/_2$° due north of NGC 253 and $2^1/_2$° south and slightly east of Beta Ceti, is as thin as NGC 253 but only two-thirds as long. In 10×50s it appears as an intermittently visible N-S smudge of low-surface-brightness glow (it has strangely low surface brightness for an edge-on system). 25×100s improve the view to a N-S oval of pale haze steadily, no longer intermittently, visible. NGC 7793, $2^1/_2$° SSW of the 5th mag Zeta (ζ) Scl and about 1° SE of a 6th mag star, is an almost round oval about one-half the size, and therefore only one-fourth the area, of the likewise nearly circular NGC 300. However, it can be seen in 25×100 supergiant binoculars when it is only about 10° above the horizon (on dark, clear nights). NGC 7793 looks like NGC 300 because it is exactly the same sort of galaxy–a loose-armed Scd system which we see almost face-on. Its luminosity, 3 billion Suns, is just about the same as that of NGC 300: it appears only half as large and over one mag fainter simply because it is almost twice as far, 13 million l-y. NGC 247 is somewhat nearer than NGC 7793, 10 million l-y, but intrinsically only half as luminous.

11.2 The Ursa Major/M81 Galaxy Group

Ursa Major is one of the most galaxy-rich constellations in the sky. Its greatest galaxy concentration is in and around the "Bowl" of the Big Dipper. This galaxy concentration is the NW end of the great "milky way" of galaxies that extends SE through Canes Venatici, Coma Berenices, and western Virgo. However, Ursa Major's brightest galaxies are in more remote parts of the constellation: M101, the Pinwheel Galaxy, is located 5° NNE of the star at the end of the Dipper "Handle" (Eta [η] Ursae Majoris); and the M81 + M82 galaxy pair lies 10° NW of the northwestern star in the Dipper "Bowl" (Alpha [α] UMa).

M81 and **M82** (Plate XLVI, Photo 4.3) are only about 40' apart due N-S, M81 being the southern member of the pair. They are in a rather bright-star-poor part of the north circumpolar sky: the nearest 4th magnitude star to them is Lambda (λ) Draconis at the end of the Dragon's tail 8° to their east. They would therefore be very difficult to find but for the hazy glow of M81, a mag 6.9 object with a full photogra-

Brightest Ursa Major/M81 Group Galaxies

Galaxy	RA (2000.0)	Dec	m_V	Dimensions	Type	Color Index	Distance	M_V
M81=N3031	09^h56^m	+69°04'	6.9	21'×10'	SA(s)ab	0.95	$12×10^6$l-y	-21.1
M82=N3034	09 56	+69 41	8.4	9×4	I0	0.89	12	-19.6
N 2403	07 37	+65 36	8.5	25×13	SAB(s)cd	0.47	10.5	-19.0
N 2976	09 47	+67 55	10.2	5.0×2.8	SAc pec	0.68	12	-17.8
N 3077	10 03	+68 44	9.8	5.5×4.1	I0 pec	0.76	12	-18.2
N 4236	12 17	+69 28	9.6	21×7.5	SB(s)dm		12	-18.4
IC 342	03 47	+68 06	8.4	22	SAB(rs)cd		12	-19.6

Photo 4.3 The galaxy pair M81 (right) and M82 (left) in northern Ursa Major. North is left and west is up

phic size of about 20'×10' NNW-SSE. However, despite its size and brightness M81 can be tricky to spot in 7×50 or even 10×50 glasses: much of its light is concentrated in its compact core, which appears stellar at 7× and nearly so at 10×. But when you do spot the two galaxies in 10×50s, they present a strange spectacle: M81 has an extensive, low-surface-brightness aura enveloping a small but bright oval core, and M82 is a thin (9'×4') mag 8.4 spindle of light oriented ENE-WSW at right angles to its neighbor. In 7× binoculars the two systems' orientations, but not the character of their images, can be detected.

With 15×100 supergiant binoculars M81 displays an intense stellar nucleus within the large, bright central bulge which fades subtly into a vast, tenuous oval halo lacking distinct edges (the darker the sky, the more of M81's halo you will see). The galaxy looks exactly like what it is: a bulge-dominated Sab spiral slightly inclined to our line of sight. M82, by contrast, has neither core nor nucleus: it is a bar of uniform surface brightness with strangely sharp edges–very unlike the edges of the vast majority of galaxy-images.

Even in the absence of other information we would suspect that something unusual is going on with the M81 + M82 pair–in part because of the violent way the bar of M82 slashes across the orientation of the M81 oval, in part because of the bizarre appearance of M82, almost unique among galaxies visible in binoculars. M82 is a starburst galaxy in which star-formation has been occurring at an accelerated rate for perhaps a half billion years. Most of the present starburst activity in the galaxy is near its center but hidden to our sight by massive clouds of gas and dust. However, the vigorous star-formation within the galaxy can be detected in radio and infrared wavelengths because hot young stars ionize the gas still around them, which emits radio waves, and the dust in which they are still embedded absorbs their visual light, re-radiating it at infrared wavelengths. And some of the strong radio emission from M82 must be from the frequent supernovae that no doubt have been occurring in it during the astronomically recent past. Apparently all this activity was initiated some 500 million years ago by a close encounter of M82 with the massive

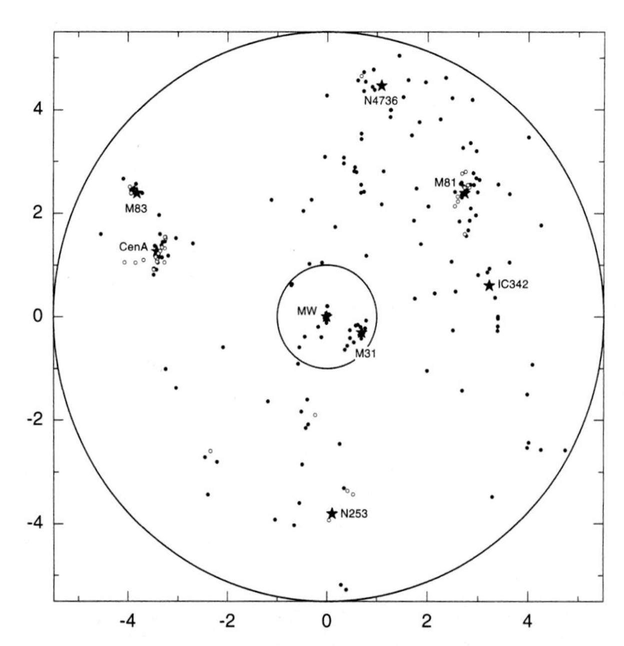

Figure 4.1 Nearby galaxy groups. The Milky Way ("MW) is at the center. This view is "above" the plane of the Local Supercluster. The Coma-Virgo Cluster is located beyond the top of the figure. The scale around the edge is in intervals of one megaparsec (3.26 million light-years).
(I. D. Karachentsev, et al., AA 398, 479; Fig. 7.)
Courtesy M. E. Sharina

M81. The latter seems not to have suffered much from the event: it is an almost classic two-armed grand-design spiral in which the neutral hydrogen (H I) gas is densest along, and just within, the optical arms–just where it should be in a well-behaved spiral system. The most unusual feature of M81 is its central bulge, which is abnormally extensive, populous, massive, and bright even for a galaxy with such luminous and well-defined spiral arms.

M81 and M82 are estimated, on the basis of the apparent magnitudes of the Cepheid variables in the big spiral, to be 12 million light-years distant. M81 has a luminosity of about the same as the Andromeda Galaxy, 23 billion Suns. M82 is very bright for its size (as would be expected from a starburst galaxy), 5 billion Suns. This galaxy pair is the heart of a large, but rather scattered, galaxy group that extends from the edge-on Sd spiral NGC 4236 near Kappa (κ) Draconis through the M81 + M82 pair at least as far west as NGC 2403 in eastern Camelopardalis and perhaps all the way to the face-on Scd system IC 342 near the Camelopardalis/Cassiopeia border.

NGC 2403 is almost as bright as M82, with an integrated apparent mag of 8.5. It is, however, considerably larger, its photographic measurements being 25'×13' NW-SE. NGC 2403 is a loose Scd spiral which is much tilted to our line of sight. The elongation of the galaxy's image makes it easier to spot; which is fortunate because it is in a very star-poor area 14° WSW of M81 + M82, the nearest "bright" star to it being the 6th mag 51 Camelopardalis a degree to the east. In 10×50 binoculars NGC 2403 displays a small, much-elongated bright core within a compact much-flattened oval halo which, because of its low surface brightness, appears a good deal shorter than its 25' photographic length. 100mm supergiant glasses reveal that the NW-SE oval patch of the galaxy is between a close E-W pair of mag $9^1/_2$–10 stars. The apparent mags of the Cepheids in NGC 2403 imply that the system is 10 million light-years from us and thus slightly nearer than the M81 + M82 pair, from which it lies around 3 million l-y. Its integrated absolute mag is therefore −19.0, a luminosity of 3.3 billion Suns. NGC 2403 has the very bluish color index of +0.47 because it, like M82, is a starburst galaxy. However, the cause of its accelerated rate of star-formation is not immediately obvious.

Two faint 10th mag satellites of the massive M81 + M82 pair are detectable in supergiant binoculars. The mag 10.2 **NGC 2976**, which measures just 5'×3', is located only $1^1/_4$° SW of M81. It is a very low-surface-brightness dwarf Sc spiral that can be seen in 15×100 glasses, but only as a tiny amorphous smudge. A faint field star is near the galaxy's center. The integrated absolute mag of NGC 2976, −17.8 (a luminosity of about one billion Suns), is 0.3 mag less than that of Large Magellanic Cloud. Just 0.4 mag brighter than NGC 2976 (and therefore intrinsically just about as bright as the LMC) is **NGC 3077**, located $^1/_2$° ESE of M81, an 8th mag star just to its NW. This is a peculiar system with the smooth texture of elliptical galaxies but an irregular outline. Like M82, its structure has probably been influenced by the gravitational field of the massive M81. NGC 3077 is about the same size as NGC 2976 but of little greater surface brightness. However the glare of the 8th mag star makes it a challenge.

Almost as difficult as NGC 2976 and NGC 3077 is **NGC 4236**, located at the extreme eastern end of the Ursa Major Group. This is a highly-tilted Sd spiral that measures an impressive 21'×7½' but has an integrated apparent mag of just 9.6, only 0.2 magnitude greater than the much more compact NGC 3077, and consequently has extremely low surface brightness. Its N-S orientation, but not much else about the galaxy, can be glimpsed in 15×100 glasses. NGC 4236 is, however, in a well-marked location 1½° WSW of Kappa (κ) Draconis and just south of a 7th mag field star.

IC 342, located in far west-central Camelopardalis 3° south of Gamma (γ) Cam, is so distant from the M81 + M82 pair, about 6 million light-years, that it perhaps is best thought of as the lucida of its own separate galaxy group (which would include the dwarf spirals NGC 1560 and NGC 1569 as well as the heavily dimmed and reddened systems Maffei I and Maffei II, partially obscured by the Milky Way dust between the Perseus Double Cluster and the Cassiopeia emission nebula IC 1805). IC 342 is a very loose-armed Scd spiral that we see face-on. It therefore is another very low-surface-brightness galaxy: it shines with the light of a mag 8.4 star spread out over a 20' diameter area. Nevertheless IC 342 is not difficult in 10×50 binoculars (under dark skies, and using averted vision). It is of course easier in 100 mm supergiant glasses, but its appearance is the same as with smaller instruments: an amorphous/circular patch of extremely pale haze behind for or five intermittently-resolved field stars. These stars are outliers of the Perseus/Cassiopeia stretch of the Milky Way, for IC 342 is only about 10° off the galactic equator. But because it lies behind the fringes of the Milky Way, it is dimmed a difficult-to-estimate amount by our Galaxy's dust and its distance and integrated absolute mag are consequently somewhat uncertain. The absolute mag of −19.6 given for IC 342 in the table of Ursa Major Group members is a minimum value uncorrected for extinction: the reality is probably at least −20½.

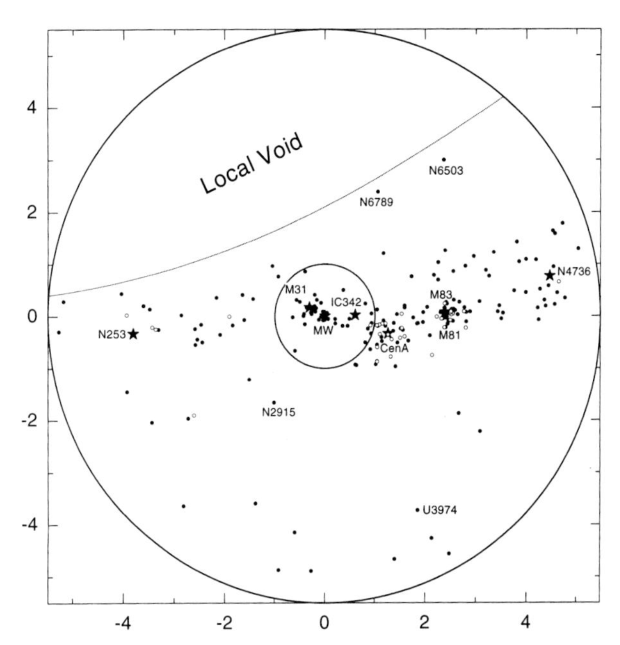

Figure 4.2 The distribution of nearby galaxy groups from an "edge-on" perspective—that is, in the plane of the Local Supercluster, with the Coma-Virgo Cluster to the right (beyond the NGC 4736 = M94 group). The NGC 253/Sculptor Group is "behind" us with respect to the Coma-Virgo core of our Local Supercluster. The Local Void is a peculiarly galaxy-empty space "above" our region of the Local Supercluster. From this perspective the IC 342 and M81 groups are in the foreground of the Local Group, and Cen A = NGC 5128 + M83 behind. (I. D. Karachentsev, *op. cit.*) Courtesy M. E. Sharina

II.3 The Centaurus/NGC 5128 Galaxy Group

The Ursa Major Galaxy Group is at the extreme NW end of the spring "milky way" of galaxies that extends SE from Ursa Major through Canes Venatici, Coma Berenices, and western Virgo. Past the extreme SE end of that celestial river of galaxies is the third nearest galaxy aggregation to us, the

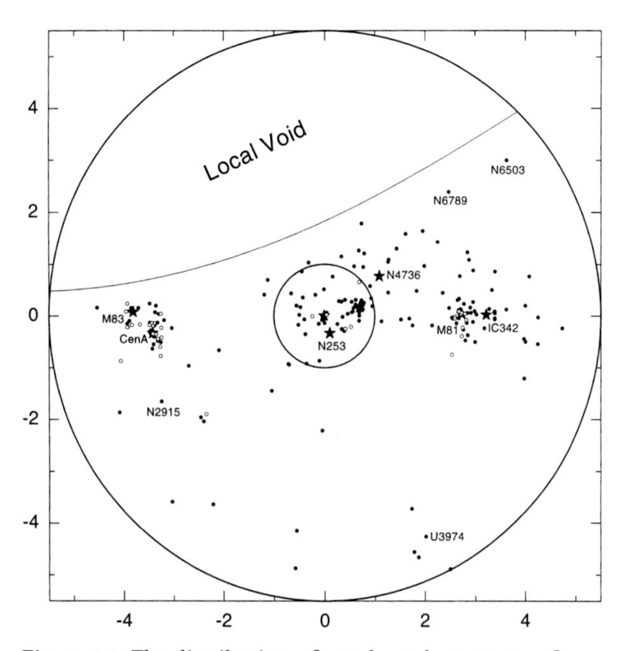

Figure 4.3 The distribution of nearby galaxy groups from outside the rim of the Local Supercluster looking "in." From this perspective we look through the NGC 253/Sculptor Galaxy Group and our Local Group (the Milky Way is the star-image at the center) in toward the Coma-Virgo Galaxy Cluster (the members of which are not plotted). This is basically the view we have when we look at the spring constellations from Ursa Major (to the right, at the M81 Group) SE through Canes Venatici, Coma Berenices, and Virgo to Centaurus (to the left, at the M83 group). (I. D. Karachentsev, *op. cit.*) Courtesy M. E. Sharina

Centaurus Galaxy Group, a 30°-long chain that includes NGC 5068 in far southern Virgo, M83 in the tail of Hydra, and NGCs 5253, 5102, 5128, and 4945 down into central Centaurus. Like the Local Galaxy Group, the Centaurus Group seems to have a binary structure: the giant radio galaxy NGC 5128 and two other bright galaxies, NGC 5102 and NGC 4945, are 12 million light-years away from us; and M83 and NGC 5253, 2° to its south, are perhaps 15 million l-y distant.

The Centaurus Galaxy Group is of special interest to astronomers because it contains the nearest explosive giant galaxy, NGC 5128, and because it contains the nearest lenticular (type SO) systems, NGC 5128 itself and NGC 5102. However, both are atypical lenticulars: NGC 5128 is suffering some cataclysm in its deep interior connected with the broad band of obscuring matter that slashes across the galaxy's circular and otherwise smoothly-luminous disc; and NGC 5102 has an unusually blue color and an unusually great amount of neutral hydrogen gas for a lenticular. NGC 5253 near M83 is also peculiar: it combines the smooth outer profile of an evolved-star-populated elliptical galaxy with a core that resembles a Magellanic irregular (like the Small Magellanic Cloud) rich in dust, gas, and young-star complexes; and it has suffered two extremely luminous supernovae in the past century.

M83, located 7° WSW of the 3rd magnitude Pi (π) Hydrae, itself has some peculiarities. At first glance it is simply an Sc spiral which we see almost face-on: but it has three arms, its spiral disc is warped (one edge bent "down" and the diametrically-opposite edge bent "up"), its central bulge contains *two* nuclei (one at the bulge's center of light, the other at the galaxy's center of spiral symmetry), and it has had five supernovae in only 70 years–twice the supernova rate of our own Milky Way (though dust prevents us from seeing most of our Galaxy's super-

Photo 4.4 The explosive galaxy NGC 5128 in the Centaurus Galaxy Group

Brightest Centaurus /NGC 5281 Group Galaxies								
Galaxy	RA (2000.0)	Dec	m_V	Dimensions	Type	Color Index	Distance	M_V
M83=N5236	13^h37^m	−29°52′	7.6	15.5′×13′	SAB(rs)c	0.89	15×10^6l-y	−20.8
N 4945	13 05	−49 28	8.8	23×6	SA(s)d:	0.66	12	−19.2
N 5102	13 22	−36 38	8.8	9.8×4.0	SA(s)0^- pec	0.70	12	−19.2
N 5128	13 25	−43 01	6.7	31×23	S0 pec	0.98	12	−21.3
N 5253	13 40	−31 39	10.2	5.1×2.3	Im: pec		15	−18.2

novae). All these abnormalities are thought to be the consequence of a past interaction of M83 with NGC 5253, which presently (if both systems are precisely 15 million l-y from us) is only 260,000 l-y from its large neighbor. At declination −30° M83 is the farthest north of the brighter Centaurus Group members. It is easily spotted with 10×50 binoculars from as far north as latitude 50° (given dark skies) because it has rather high surface brightness: its integrated apparent mag is 7.6, but its diameter a rather modest 15′. M83's high surface brightness in contrast to other face-on Sc systems (such as IC 342 in the Ursa Major Group, which has a diameter of over 20′ but an integrated apparent mag of just 8.4) is the consequence of the vigorous star-burst activity presently occurring in its spiral arms. The total luminosity of M83, 18 billion Suns, is a little greater than that of our Milky Way.

However, the other Centaurus Group galaxies are a problem for mid-northern observers because they are so far south and, except for **NGC 5128**, so faint. NGC 5128 (Photo 4.4) has an integrated apparent mag of 6.7 and a diameter approaching $^1/_2$°, but culminates at least 20° above the horizon only for observers in southern Texas and Florida. Of the other Centaurus Group galaxies, **NGC 5253** is only 2° farther south than M83 but just of mag 10.2 and 5′×2$^1/_2$′ in size and therefore would be a challenge for 25×100 supergiant glasses even from the southern hemisphere, where it can be looked for near the zenith. **NGC 5102** is twice as large as NGC 5253, 10′×4′, and four times brighter, mag 8.8; but it is at declination −36$^1/_2$°, and just NE of the mag 2.8 Iota (ι) Centauri, the glare of which increases the difficulty of spotting the galaxy. Finally, the mag 8.8 **NGC 4945** is a very long and thin (23′×6′) low-surface-brightness Sd galaxy highly-inclined to our line of sight. It would not be difficult for 10×50s, but is at declination −50°. However, NGC 4945 is easily located if you are far enough south: it is just $^1/_2$° NW of the mag 3$^1/_2$ Xi-two (ξ^2) Cen and $^1/_4$° due east of Xi-one.

III Galaxy Groups in Leo

The spring "milky way" of galaxies that begins in and around the Big Dipper in Ursa Major and extends SE through Canes Venatici, Coma Berenices, and western Virgo misses the constellation of Leo, which lies to its SW and west. Nevertheless Leo has three galaxy groups with systems visible in wide-field instruments. The M65/M66 and M95/M96 galaxy groups are the two halves of the Leo I Galaxy Cloud. This is a major substructure of the Coma-Virgo Supercluster that is almost 40 million light-years from us but 15 million l-y in front of, and at a direction slightly over, the supercluster's core in southern Coma Berenices and NW Virgo. The third Leo galaxy aggregation, the NGC 3607 Group, is very distant (54 million l-y) but also part of the Coma-Virgo Supercluster. Its two or three brightest members are right at the limits for supergiant binoculars.

III.1 The M65/M66 Galaxy Group

One of the best galaxy trios for binoculars in the sky is the M65 + M66 + NGC 3628 group 2$^1/_2$° SSE of Theta (θ) Leonis. All three systems are Sa or Sb spirals which we see not far from edge-on. Thus, though their apparent magnitudes are from 8.9 to 9.5, they are a bit of a challenge for standard binoculars. Fortunately, they are not difficult to find: the mag 9.3, 8'×1$^1/_2$' M65 is $^3/_4$° ESE of the 4th mag 73 Leo, with mag 8.9 8'×2$^1/_2$' M66 another 20' in the same direction. NGC 3628, which measures 14'×4' but is only a mag 9.5 object, is 35' due north of M66 and 20' NE of a 7th mag star. This 7th mag star is also 20' NNE of M65 and in looking for NGC 3628, the most challenging of these three galaxies, it helps to think of it as forming a flattened isoceles triangle with the 7th mag star and M65.

M66 (Photo 4.5) is sufficiently bright and thick that it can be glimpsed in 7×50 binoculars as a tiny NNE-SSW oval. However 10×50s are required to see the NNW-SSE streak of the fainter and thinner M65 (Photo 4.5), which has an unusually small bulge for an Sa galaxy. The E-W patch of **NGC 3628** can also be discerned in 10×50s, but, because of its very low surface brightness, extremely clear skies are necessary. In 15×100 supergiant glasses NGC 3628 is surprisingly easy: indeed, it seems to gain more with the increase in light-gathering power and magnification than M65 and M66. A NW-SE pair of mag 9$^1/_2$–10 stars NW of the stellar nucleus of M66 can be seen pointing directly at it, the nucleus and the two stars being evenly spaced.

The M65/M66 Galaxy Group, estimated from the apparent magnitudes of the Cepheids in M66, is 39 million light-years away. If M65 and M66 are precisely the same distance from us (rather unlikely: one of the galaxies is probably a few hundred thousand light-years beyond the other), their true separation would be only around a quarter million light-years. M65, M66, and NGC 3628 are all very luminous spirals. In fact, the integrated absolute mags given for them in the accompanying table are too low because these values are not corrected for the fact that these systems are highly inclined to our line of sight and therefore significantly dimmed by the dust in their own spiral planes–and photos reveal that M66 has a great deal of dust. But even without that correction, M66 has an integrated absolute mag of –21.4, 0.3 mag brighter than the Andromeda Galaxy, and M65 and NGC 3628 are not much fainter.

III.2 The M95/M96 Galaxy Group

In the bright-star-poor region between Regulus (Alpha [α] Leonis) on the west and Theta (θ) and

Brightest M65/M66 Group Galaxies

Galaxy	RA (2000.0)	Dec	m_V	Dimensions	Type	Color Index	Distance	M_V
M65=N3623	11^h19^m	+13°05'	9.3	8'×1$^1/_2$'	SAB(rs)a	0.92	39×10^{6}l-y	–21.0
M66=N3627	11 20	+12 59	8.9	8×2$^1/_2$	SAB(s)b	0.73	39	–21.4
N 3628	11 20	+13 36	9.5	14×4	Sb pec	0.80	39	–20.8

Photo 4.5 The galaxy pair M65 (right) and M66 (left) in Leo

Iota (ι) Leonis on the east is the M95/M96 Galaxy Group. Its members are scattered between the 5th magnitude 52 Leo on the north and 53 Leo on the south. (52 and 53 Leo are a little more than halfway from Regulus to Theta and Iota Leo, respectively.) The group includes the tightly-wound spirals M95 and M96, the giant ellipticals M105 and NGC 3377, and the lenticular NGC 3384. In fact, M105 and NGC 3377 seem to be the nearest giant ellipticals to our Local Galaxy Group, and NGC 3384 (with the possible exception of NGC 3115 in Sextans) the nearest "normal" lenticular S0 system since the lenticulars NGC 5128 and NGC 5102 in the Centaurus Galaxy Group are decidedly abnormal.

Because they are ellipticals, lenticulars, or bulge-dominated early-type spirals, the bright galaxies of the M95/M96 Group all present the observer with the same type of image: an intense, even stellar, core within an ambiguously-bordered, low-surface-brightness halo. This makes them difficult for 7×–10× instruments: 15×–20× is necessary to darken the sky background sufficiently to provide contrast for the subtle glow of the galaxy halo.

The one M95/M96 Group galaxy reasonably easy for 10×50 binoculars is the compact (6'×4') mag 9.2 Sab spiral M96, located exactly $1^1/_2$° NNE of 53 Leo. However, it appears only as a tiny fuzzy spot–

Brightest M95/M96 Group Galaxies

Galaxy	RA (2000.0)	Dec	m_V	Dimensions	Type	Color Index	Distance	M_V
M95=N3351	10^h44^m	+11°42'	9.7	4.4'×3.3'	SB(r)b	0.79	37×10^6 l-y	-20.5
M96=N3368	10 47	+11 49	9.2	6×4	SAB(rs)ab	0.86	37	-21.0
M105=N3379	10 48	+12 35	9.3	2.0	E1	0.94	37	-20.9
N 3377	10 48	+13 59	10.4	4.1×2.6	E5–6	0.84	37	-19.8
N 3384	10 48	+12 38	9.9	5.5×2.9	SB(s)0$^-$:	0.93	37	-20.3

bright enough, perhaps; but tricky if you haven't done a lot of galaxy-searching. **M95**, located about 40' west and slightly south of M96, is even smaller (4.4'×3.3') and fainter (mag 9.7) and in 10×50s looks merely stellar (in effect it can be seen but not identified). However, in 15×100s the galaxy pair is a fine sight. M96 appears as a compact but rather high surface brightness glow around a strong stellar core and is quite easy. M95 looks as large as M96 but is paler and lacks an observable stellar core. A 10th mag star is just to its west.

An even tighter M95/M96 Galaxy Group pair for supergiant binoculars is **M105 + NGC 3384**, located about 40% of the distance from 52 to 53 Leo. M105 is a bright (mag 9.3) but *very* compact (only 2' diameter) E1 system; and NGC 3384, just a few minutes of arc to its NE, is a faint (mag 9.9), elongated ($5^1/_2$'×3') lenticular. In 15×100 instruments both galaxies have intense stellar nuclei within compact amorphous halos. NGC 3384 is decidedly fainter; but neither galaxy is all that difficult. In telescopes the 12th mag Sc spiral NGC 3389 is visible a few minutes SE of NGC 3384.

The fifth member of the M95/M96 Galaxy Group that can be glimpsed in supergiant binoculars is the very faint (mag 10.4), very small (4'×$2^1/_2$') E5-6 elliptical **NGC 3377**, conveniently located only about 20' SE of 52 Leo. In 15×100s NGC 3377 is distinctly fainter than even NGC 3384, and has a less intense nucleus than NGC 3384 or M105; but its halo appears as large (though the halo's elongation is not evident).

The Cepheid variables in M96 imply a distance to that galaxy of 37 million light-years. Thus the M95/M96 Group might be centered a couple million light-years nearer than the M65/M66 Group. (But even Cepheid-based distances of galaxies this far away are subject to considerable uncertainty.) Thus the 1° apparent distance between the M95 + M96 and the M105 + NGC 3384 subgroups corresponds to a true distance of 700,000 l-y; and the $2^1/_2$° from M95 on the south to NGC 3377 on the north is over 1.5 million l-y.

M105, as probably the nearest giant elliptical galaxy to us, has been the target of numerous studies. To a casual inspection it seems very ordinary: its surface brightness, hence star density, decreases smoothly in all directions from its nucleus; its color index, +0.94, is typical for giant ellipticals and its color changes very little from its center to its periphery; it rotates very slowly around its minor axis (unlike spirals, the discs of which spin very rapidly); and it contains very little gas and dust, and shows no overt sign of any recent disturbing interaction with the nearby NGC 3384. However, deep within its core is a dust ring (with an apparent size of only 1.5") that is rotating around an axis highly tilted to the galaxy's overall axis of rotation. And around this nuclear molecular ring is a cloud of ionized gas rotating around an axis tilted in yet a *third* direction. Now, as the common bicycle demonstrates, the law of conservation of angular momentum is harsh and uncompromising: it takes a tremendous amount of energy to tilt a spinning disc out of its plane of rotation. The origin of three separate planes of rotation in the core of such an otherwise well-behaved galaxy as M105 is very much a puzzle.

III.3 The NGC 3607 Galaxy Group

Located $2^1/_2$° SE of Delta (δ) Leonis is a small group of faint, compact galaxies which, though estimated to be 54 million light-years from us, has a couple members that can be glimpsed with 100 mm binoculars. The brightest member of the group is the len-

Brightest NGC 3607 Group Galaxies

Galaxy	RA (2000.0)	Dec	m_V	Dimensions	Type	Color Index	Distance	M_V
N 3605	11^h17^m	+18°01'	12.3	1.2'×0.6'	E4–5		54×10^6l-y	−18.7
N 3607	11 17	+18 03	9.9	4.6×4.1	SA(s)0°:		54	−21.1
N 3608	11 17	+18 09	10.8	2.7×2.3	E2		54	−20.2
N 3626	11 20	+18 21	11.0	2.6×1.8	(R)S(rs)0+	0.83	54	−20.0

ticular **NGC 3607**. Though neither very bright (magnitude 9.9) nor large ($4^{1}/_{2}$'×4'), this galaxy is not difficult for 15×100 glasses. It has an intense stellar core within a moderately large but very low surface brightness amorphous/circular halo. Two 9th mag stars centered approximately $^{1}/_{2}$° due north of the galaxy are oriented N-S pointing right at it.

Also due north of NGC 3607, but centered only 6' away from it, is the very faint (mag 10.8) and very small (2.7'×2.3') E2 elliptical **NGC 3608**. Under ideal conditions this galaxy can be glimpsed in 15×100s as a tiny, very faint haze just discernably brighter in its center. Another elliptical system, the E4-5 **NGC 3605**, is just 2.7' WSW of NGC 3607. However this galaxy is so small, just $1^{1}/_{2}$'×1', that even in richest-field telescopes it will appear only as a 12th mag "star."

Finally, $^{3}/_{4}$° ENE of NGC 3607 is the fourth major member of the group, **NGC 3626**, a faint (mag 11.0) and small ($2^{1}/_{2}$'×2') lenticular. Like NGC 3607 itself, NGC 3626 is $2^{1}/_{2}$° from Delta Leo; but it is a bit easier to locate because it lies exactly on the line from Delta to the $3^{3}/_{4}$° distant 71 Leo. In 15×100 glasses NGC 3626 displays an intense, but not stellar core: it lacks the hard-edged quality of the core of NGC 3607. The galaxy's halo is small and of low surface brightness. NGC 3626 is the middle object in a WNW-ESE line of five rather evenly-spaced "stars," the brightest of which is at the line's SE end.

V The Coma–Virgo Galaxy Cluster

The richest concentration of galaxies visible in binoculars or telescopes anywhere in the sky is in a 15°-diameter area along the Coma Berenices/Virgo border centered about midway between Denebola (Beta [β] Leonis) and Vindemiatrix (Epsilon [ε] Virginis). This area has sixteen Messier-numbered galaxies, half of which are brighter than magnitude 9.5. And a total of sixteen Messier or NGC galaxies in the region are brighter than mag 10.0. But none of these have apparent sizes as large as 10′. However, this means that many of these systems have rather high surface brightness. Consequently, though they might be a bit small for 7× binoculars, most of the region's Messier galaxies can be glimpsed in 10×50 binoculars; and literally dozens of the Messier and NGC galaxies in the area are visible, if you know where to look, in supergiant binoculars and small RFTs.

The galaxy concentration in southern Coma Berenices and NW Virgo is called the *Coma-Virgo Galaxy Cluster* (or often simply the *Virgo Cluster*). It is the galaxy-dense core of the larger aggregation termed the *Local Galaxy Supercluster*, toward the outer edge of which lies our Local Group. Essentially all the systems in the spring "milky way" of galaxies from the Big Dipper in Ursa Major on the NW down through Canes Venatici, Coma Berenices, and western Virgo to the Centaurus Galaxy Group on the SE are in the Local Supercluster. The flattening of the supercluster into a disc implies that the galaxies in it are in orbits around the supercluster's center of gravity in the Coma-Virgo Cluster (just as the individual stars in the flattened disc of the Milky Way orbit around the Galactic Center in our Galaxy's bulge).

The Coma-Virgo Cluster is dominated by giant ellipticals and lenticulars. It has its share of giant spirals, especially Sb and Sbc systems like M58 and M100; but there is some suspicion that these gas-and-dust-rich spirals are in fact either somewhat nearer, or somewhat farther, than the E and S0 rich core of the Cluster. The idea is that the close encounters between galaxies that would be frequent in such a crowded environment would have long ago stripped the gas and dust out of the arms of any spiral galaxies there, thus hastening their evolution into gas-and-dust-poor, old-star-populated, E and S0 systems.

The distance to the center of the Coma-Virgo Cluster is a key to cosmology, because it can be used as a *standard candle* for estimating the distances to more remote clusters of giant elliptical and lenticular systems, which in turn will provide an idea of the distribution of galaxies in particular, and of matter as a whole, in our Universe. And an estimate of the amount of matter in the Universe will tell us whether or not the Universe is sufficiently dense to reverse the presently-observed expansion–the "flying apart" of galaxies–that has resulted from the Big Bang.

Consequently a reliable distance to the center of the Coma-Virgo Cluster, which is in the vicinity of the huge ellipticals M84, M86, and M87, has long been sought. For some time it was thought to be around 40 million light-years. Then estimates, based particularly upon the apparent magnitudes of the Type Ia supernovae observed in Coma-Virgo galaxies, were raised to 65–70 million l-y. But Type Ia supernovae do not seem to always reach the same peak brilliancy, and in the past decade the Hubble Space Telescope and larger ground-based telescopes have permitted Cepheid variables to be identified in several Coma-Virgo giant spirals. The mean distance of six Coma-Virgo spirals with observed Cepheids turned out to be 50.2 ± 1.6 million l-y. However, if these systems are really in front of the giant ellipticals and lenticulars that dominate the core of the Coma-Virgo Cluster, this is a minimum. Unfortunately the Type II Cepheids that must be in the evolved-star populations of the Coma-Virgo E and S0 giants are on the average $2^1/_2$ magnitudes fainter than the classical Cepheids that are found in the arms of spiral galaxies and therefore cannot be observed. The value now most commonly used for the distance to the center of the Coma-Virgo Cluster is 54 million l-y (corresponding to a distance modulus of 31.0 magnitudes). Assuming this distance to the center of the Local Supercluster, the observed motions of its galaxies imply that its total mass is 1.2×10^{15} = 1,200 trillion times the Sun's (about 0.3% of which is in the Local Galaxy Group).

For the binocular and richest-field telescope observer, the two practical problems posed by the Coma-Virgo Cluster are the compactness of its galaxies and the lack of bright stars between Denebola and Vindemiatrix to use as celestial guideposts. The absolute minimum magnification for galaxy-searching in the Coma-Virgo Cluster is 10×. Because almost all these galaxies are ellipticals, lenticulars, or bulge-dominated Sb spirals, they basically appear merely as tiny amorphous or circular patches of smooth, low-surface-brightness haze within which is a stellar core. Greater magnification makes the patch of haze easier to see both by spreading it out a little, and by darkening its sky background. There is a significant gain from 10× to 15× in the number of Coma-Virgo galaxies identifiable.

Because of the lack of bright stars in the region, wide-field instruments do have one decided advantage over conventional telescopes for galaxy-searching in Coma-Virgo: you can use the few 5th, 6th, and 7th mag field stars that are in the region to find your way around. (And, frankly, much of the pleasure of looking at Coma-Virgo galaxies is simply in seeing such distant objects, for there is no real gain in the visual character of their image with increase in aperture: the image is merely larger and brighter, not more detailed.) The best approach to the M84 + M86 heart of the Coma-Virgo Cluster is from the NW. About 7° due east of Denebola is the 6th mag star 6 Comae Berenices. M84 is precisely 3° SE of 6 Comae and there happens to be a 7th mag star almost exactly halfway between 6 Comae and the galaxy (the star is slightly SW of the midpoint of the line joining them).

Brightest Members of the Coma-Virgo Galaxy Cluster

Galaxy	RA (2000.0)	Dec	m_V	Dimensions	Type	Color Index	Distance	M_V
M49=N4472	12^h30^m	+08°00'	8.4	9'×7.5'	E2	0.96	54×10^6l-y	−22.6
M58=N4579	12 38	+11 49	9.4	5.5×4.5	SAB(rs)b	0.82	65	−22.0
M59=N4621	12 42	+11 39	9.6	5×3.5	E5	0.94	54	−21.4
M60=N4649	12 44	+11 33	8.8	7×6	SO_1	0.97	54	−22.2
M61=N4303	12 22	+04 28	9.7	6×5.5	SAB(rs)bc	0.53	36	−20.4
M84=N4374	12 25	+12 53	9.1	5	E1	0.98	54	−21.9
M85=N4382	12 25	+16 28	9.1	7.1×5.2	SA(s)0+p	0.89	54	−21.9
M86=N4406	12 26	+12 57	8.9	7.5×5.5	E3	0.93	54	−22.1
M87=N4486	12 31	+12 24	8.6	7.0	E+0−1p	0.96	54	−22.4
M88=N4501	12 32	+14 25	9.6	7×4	SA(rs)b	0.73	58	−21.6
M89=N4552	12 36	+12 33	9.8	4.0	SO_1		54	−21.2
M90=N4569	12 37	+13 10	9.5	9.5×5	SAB(rs)b	0.72	29	−20.3
M91=N4548	12 35	+14 30	10.2	5.4×4.4	SB(rs)b	0.81	49	−20.6
M98=N4192	12 14	+14 54	10.1	9.5×3.2	SAB(rs)ab	0.81	47	−20.6
M99=N4254	12 19	+14 25	9.9	5.4×4.8	SA(s)c	0.57	42	−20.6
M100=N4321	12 23	+15 49	9.3	7×6	SAB(s)bc	0.70	47	−21.4
M104=N4594	12 40	−21 37	8.0	9×4	SA(s)a	0.98	31	−21.9
N 4261	12 19	+05 49	10.4	3.5×3.1	E2−3		110	−22.1
N 4365	12 24	+07 19	9.6	5.6×4.6	E3	0.96	54	−21.4
N 4429	12 27	+11 07	10.0	5.6×2.6	SA(r)0^+	0.98	42	−20.5
N 4435	12 28	+13 05	10.8	3.2×2.0	SB(s)0°	0.92	54	−20.2
N 4438	12 28	+13 01	10.2	8.9×3.6	SA(s)0/a pec	0.85	40	−20.2
N 4458	12 29	+13 15	12.1	1.5	E0−1			
N 4461	12 29	+13 11	11.2	3.7×1.4	SB(s)0^+:		39	−19.1
N 4473	12 30	+13 26	10.2	3.7×2.4	E5	0.96	54	−20.8
N 4526	12 34	+07 42	9.7	7.1×2.9	SAB(s)0°:	0.96	54	−21.3
N 4636	12 43	+02 41	9.5	7.1×5.2	E0−1	0.94	54	−21.5
N 4699	12 49	−08 40	9.5	4.4×3.2	Sab:			
N 4710	12 50	+15 10	11.0	3.9×1.2	SA(r)0^+?		75	−21.6

M84 and **M86** are only about 20' apart E-W. Both are giant ellipticals, but M84 is a nearly circular E1 system and M86 a slightly elongated E3 galaxy. They are both about mag 9.0 and only 5' across but not difficult to spot in 10×50 binoculars *if* you keep in mind that what you are looking for is essentially a 9th mag "star" embedded in a small pale halo. However, the two are much more conspicuous in 15×100 supergiant glasses, which enlarges and brightens them sufficiently that you can see they have a triple structure: an extremely faint, ambiguously-bordered outer halo envelopes a rather broad inner halo of moderate surface brightness, at the center of which twinkles a stellar nucleus. Both galaxies have integrated absolute magnitudes around −22, a luminosity of over 50 billion Suns. Like the other Coma-Virgo giant ellipticals, M84 and M86 are well-populated with globular clusters, each galaxy estimated to have around 3000.

If you can see M84 and M86, you will have no trouble spotting the mag 8.6 E0-1 giant elliptical **M87**, $1^1/_2$° to their ESE. In 15×100 glasses this galaxy has a tripartite structure like that of M84 and M86: a very extensive, very, very faint outer halo contains a broad inner halo of moderate surface brightness centered by an intense stellar nucleus. The most intriguing feature about the appearance of M87 is its blazing stellar nucleus: contrast it with the equally stellar, but less intense, nuclei of M84 and M86.

Thus even binoculars hint that there is something special about M87. Even before the galaxy was discovered to be a strong radio source, photos had revealed a peculiar "jet," especially bright in blue light, pointing out of its nucleus. Hale Telescope plates showed that this jet consisted of discreet but very closely-spaced, blobs of light. And the system is a strong X-ray emitter. The best explanation for all these bizarre features is that at the very center of M87 is a massive black hole into which matter from a dense ring is vortexing, with the generation of large amounts of radiation at all wavelengths and the occasional ejection at high velocities of blobs of matter at right angles to the vortexing ring. M87 is one of the two or three most luminous Coma-Virgo Cluster galaxies, its integrated absolute magnitude of −22.4 corresponding to the brilliance of 80 billion Suns. It is also extraordinary for its extremely large population of globular clusters, estimated to be 13,000. One theory is that many of these globulars are actually the nuclei of dwarf elliptical systems which have been stripped of their halos by the massive M87, the cannibalized halo stars becoming part of M87's own halo and the nuclei going into orbit around the giant galaxy's center of globular clusters. At a distance of 54 million light-years, the $1^1/_2$° apparent separation between M84 and M87 corresponds to a true separation of (at least) 1.4 million l-y.

The M84 + M86 pair is at the west end of a $1^1/_2$°-long group of galaxies that curves ENE and NE into Coma Berenices. Several of these systems can be glimpsed with supergiant binoculars (though it is a good idea to search for them only after you have gotten accustomed to the character of galaxy images). Just over 20' ENE of M86 is the 4'-wide NNW-SSE pair of S0 galaxies **NGC 4435** and **NGC 4438**. The SSE system, NGC 4438, is brighter (mag 10.2 vs mag 10.8) and larger (9'×$3^1/_2$' vs $3^1/_2$'×2') than its companion and consequently easier to see. In 15×100 glasses it appears as a NNE-SSW gash of pale haze with a stellar nucleus: it is faint, but under good sky conditions, and with averted vision, neither it nor its orientation should be doubtful. However, the other galaxy, NGC 4435, is so small and faint that even under ideal viewing conditions with supergiant glasses it will be seen only as a very, very faint, ambiguously-shaped, halo around a stellar nucleus. Though apparently so near each other, these two lenticulars are in reality very far apart: NGC 4435 is in the core of the Coma-Virgo Cluster 54 million light-years away; but NGC 4438 is thought to be well in front of the Cluster core just 40 million l-y from us.

About 20' ENE of the NGC 4435 + NGC 4438 pair, and a few minutes of arc NE of a 9th mag star, is the small (4'×$1^1/_2$') and extremely faint (mag 11.2) S0 system **NGC 4461**. This is a challenge object for supergiant binoculars and small RFTs. If seen, it will appear as a patch of haze, very tiny but seemingly oriented N-S around a stellar nucleus that is just south of an 11th mag star (nucleus and star will be about equally bright). However, even *more* of a challenge is the elliptical galaxy **NGC 4458** just west of the 11th mag star: it is a mag 12.1, 1½' diameter, object that *might* appear as an amorphous patch of haze to an experienced observer looking with good glasses during an ideal night. Assuming that it really is in the Coma-Virgo Cluster, NGC 4458 has the rather modest integrated absolute mag of −17.9 and therefore is even less luminous than the Large Magellanic Cloud. Its neighbor, NGC 4461, is estimated to be 39 million l-y away and consequently on the near edge of the Cluster.

Finally, much easier than NGCs 4458 and 4461 is **NGC 4473**, located 20' NE of NGC 4461 just over the border in Coma Berenices. (It is exactly 1° NE of M86.) This is an elongated E5 giant elliptical measuring only 3½'×2½' and just mag 10.2. It appears merely as a stellar nucleus in a halo so faint, small, and amorphous that it does not give even the impression of circularity, much less of any real elongation.

The remainder of the major Coma-Virgo Cluster galaxies will be described in order of Messier number or, for those without an M designation, by NGC number:

M49: The E2 giant elliptical M49 is one of the two or three intrinsically largest and brightest of the Coma-Virgo galaxies. In fact, its absolute magnitude, −22.6 (a luminosity of 90 billion Suns) is 0.2 mag greater than that of the exploding radio giant M87. M49 is about 5° due south of the M84 + M86 core of the Coma-Virgo Cluster at the center of its own group, Subcluster B, which is somewhat less dense in giant galaxies than Subcluster A with M84, M86, M87, M58, M59, M60, and several other systems sufficiently bright to have Messier numbers. In apparent as well as absolute terms M49 is large (9'×7½') and bright (mag 8.4), and it is easily seen even with just 10×50 binoculars. (Look for it not quite 4° SW of the 4th mag Rho [ρ] Virginis almost midway between a 1½°-wide, NW-SE pair of 6th mag stars.) In 15×100 glasses the galaxy displays a bright, compact core surrounded by a very faint, ambiguously-bordered halo. In fact, the compact core of M49 has sometimes been interpreted to be a bulge and the galaxy accordingly classified as a lenticular rather than as an elliptical system.

M58: Perhaps the most luminous of the giant spirals in the Coma-Virgo Cluster is the Sb system M58, located almost 2° ESE of M87 (and exactly 2° NNW of the 4th magnitude Rho [ρ] Virginis. The galaxy is just east of the southern member of a 20'-wide, N-S pair of mag 7 stars.) It is a mag 9.4 object measuring 5½'×4½', and in 15×100 supergiant binoculars has the features typical of an Sb system seen nearly face-on: a conspicuous stellar nucleus embedded within a condensed core (the central bulge) enveloped by a diffuse low-surface-brightness halo (the spiral disc). M58 is estimated to be 65 million light-years away and therefore a little beyond the core of the Coma-Virgo Cluster. Its absolute mag is therefore −22.0, a luminosity of 55 billion Suns–*four* times that of our Milky Way Galaxy, itself a very respectable spiral.

M59: The E5 giant elliptical M59 is almost exactly 1° east and slightly south of M58 and 1½° due north of Rho Vir. A 9th mag star is just a few minutes to its NW. It is both slightly fainter (mag 9.6) and slightly smaller (5'×3½') than M58, but no more difficult to see. However, it lacks its spiral neighbor's stellar nucleus and bright core: its glow merely brightens a little toward its center.

M60: The lenticular system M60, located only about 20' ESE of M59 (M58, M59, and M60 lie along a virtually straight 1½°-long WNW-ESE line), is one of the largest (7'×6') and brightest (mag 8.8) of the Coma-Virgo galaxies. It is not difficult to see in 10×50 binoculars as a tiny disc of haze; but it is not easy to find because of its nearly starless region. The nearest naked-eye star to it is the 5th mag 34 Vir 1° to its ENE; however the best way to locate it is by "galaxy-hopping" from the M84 + M86 pair ESE to M87 and on to the M58 + M59 + M60 line. In 15×100 glasses M60 displays a compact, relatively faint core (the galaxy's central bulge) within a large but dim halo. It seems to be the most luminous of the Coma-Virgo giant lenticulars, its integrated

Plate XXXIII

Plate XXXIV

Plate XXXV

Plate XXXVI

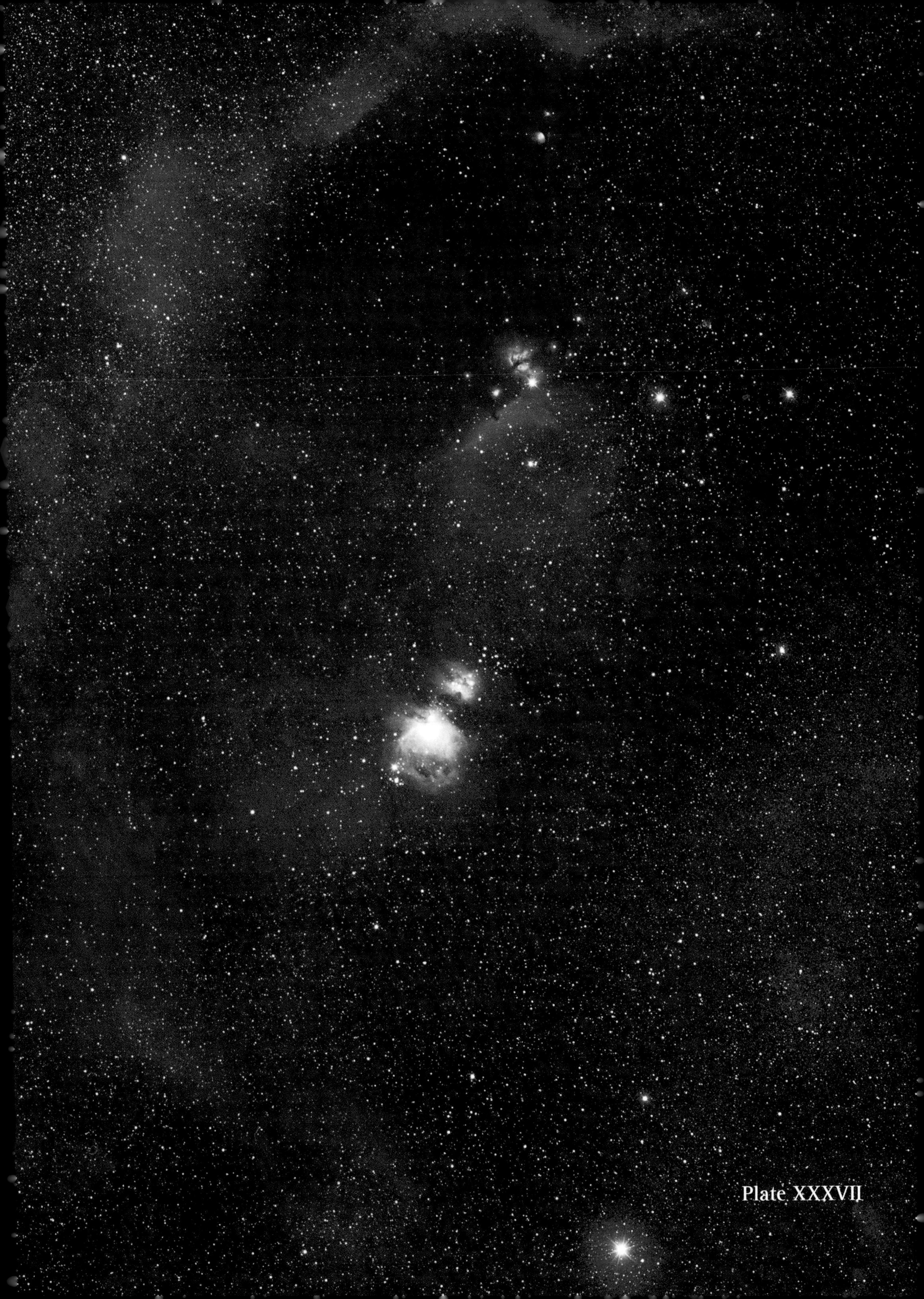

Plate XXXVII

Plate XXXVIII

Plate XXXIX

Plate XL

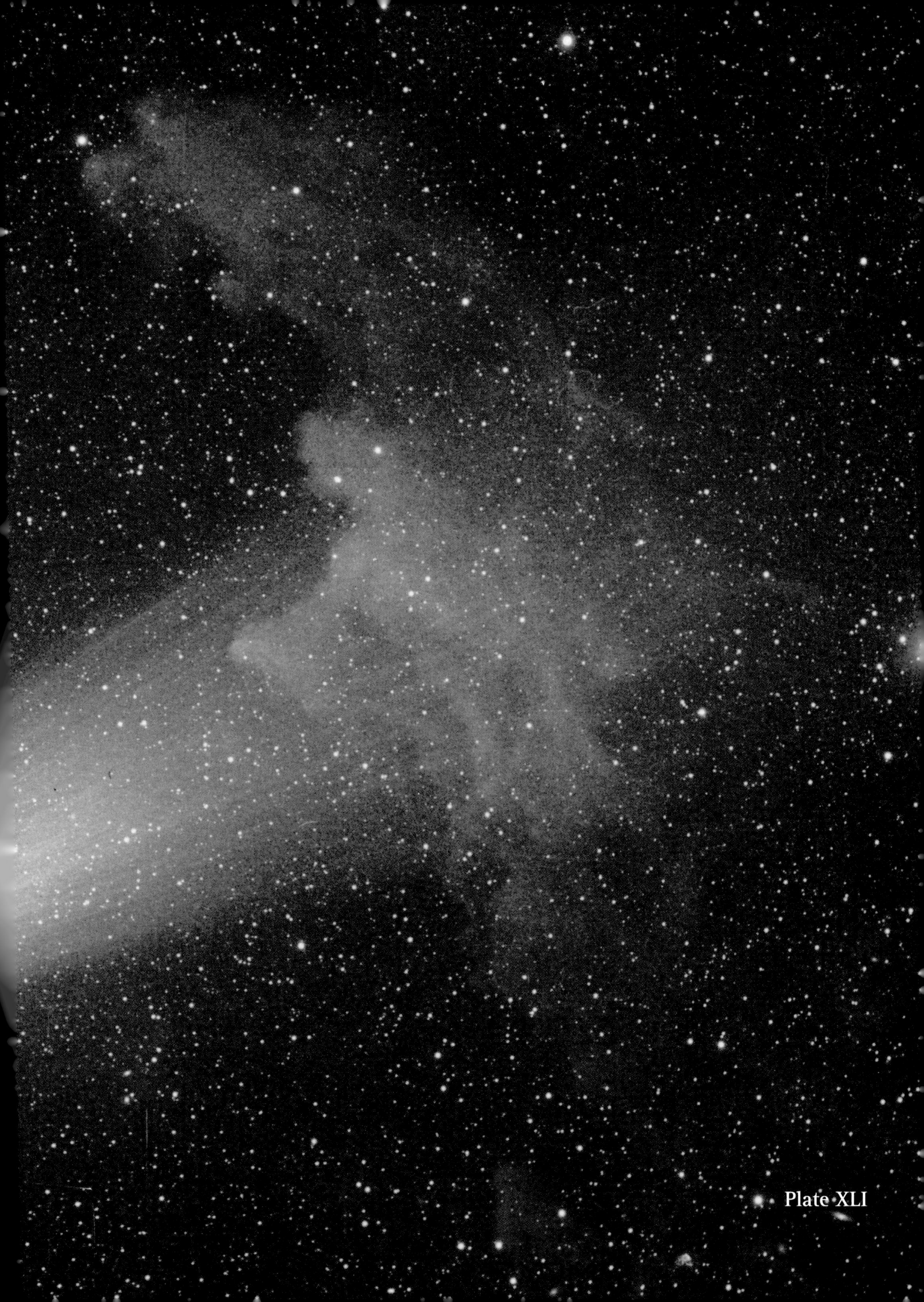

Plate XLI

Plate XLII

Plate XLIII

Plate XLIV

Plate XLV

Plate XLVI

Plate XLVII

Plate XLVIII

absolute mag of –22.2 corresponding to a luminosity of over 60 billion Suns. In telescopes the low-surface-brightness glow of the faint face-on Sc spiral NGC 4647 can be glimpsed just $2^1/_2$' NW of M60. But because there is no structural distortion in either galaxy, they must not be sufficiently near each other to be a true gravitationally-interacting pair.

M61: The Sbc spiral M61 is about 10° south of the M84 + M86 core of the Coma-Virgo Cluster 1° NNE of the 4th magnitude 16 Virginis. (The galaxy is about 60% of the way from 16 Vir to the 5th mag 17 Vir.) In 15×100 supergiant binoculars it appears as an amorphous/circular low-surface-brightness patch with a very faint stellar nucleus: it looks like the nearly face-on, loose-armed, small-bulged spiral it is. Despite its low surface brightness (the light of a mag 9.7 star spread over a 5' diameter area), it is surprisingly easy, almost conspicuous, even in 10×50 glasses, and requires no searching. M61 seems to be about 36 million light-years away, only two-thirds the distance to the core of the Coma-Virgo Cluster.

M85: The lenticular galaxy M85 is in Coma Berenices $4^1/_2$° due north of the M84 + M86 pair. It is one of the more difficult Coma-Virgo Messier galaxies to find and to see in wide-field instruments. It is in a rather isolated position, with nothing better nearby to help locate it than the 5th magnitude 11 Comae $1^1/_4$° to the WSW and the 5th mag 24 Comae $2^1/_4$° east and slightly north. And the galaxy is not as conspicuous as its stats–mag 9.1; size 7'×5'–imply. Even in supergiant binoculars it appears only as an amorphous halo with a stellar nucleus behind a couple field stars, the glare of which partially obscures the galaxy-glow and makes its nucleus hard to identify.

M88: Another difficult Coma-Virgo Cluster Messier galaxy to find is the Sb spiral M88, which is 2° NE of the M84 + M86 pair just over the border in southern Coma Berenices. It is not as isolated from the rest of the cluster as M85; but it is in an even more star-poor area, with no star even of the 6th magnitude anywhere near it. And the closest Coma-Virgo Messier galaxy to it, M91 1° to the east, is the faintest Messier galaxy in the Cluster. About the only thing the observer can do is estimate 2° NE from M84 + M86, and hope for the best. A 7th mag star is about 40' WNW of the galaxy. M88 is rather faint, mag 9.6, but sufficiently large, 7'×4', that it is not difficult to see in supergiant binoculars (when you know where to look), appearing as a small but distinct NW-SE oval of haze with a stellar nucleus. It thus looks how a large-bulged Sb spiral tilted to our line of sight should look. However, the galaxy appears not merely elongated, but strongly and smoothly oval. Photos reveal that its spiral structure is highly symmetrical, with multiple arms swirling out of the small bulge in an even better "whirlpool" than that of M51. It is this spiral symmetry which gives the galaxy its smooth profile in the eyepiece.

M89: Though rather faint (mag 9.8) and small (only 4' in diameter), the face-on lenticular galaxy M89 is easily seen in large binoculars just over 1° east and slightly north of the bright M87. It has a conspicuous stellar nucleus within a compact but only moderately low-surface-brightness halo. M89's nucleus is an interesting contrast to that of M87 because it is so much less "blazing" and intense. And its halo is more compact than that of M87, not as "spreading." M89, as a lenticular system, has a broad bulge (the nucleus of the binocular image) and a compact, relatively star-dense inner halo (the halo of the binocular image. The galaxy's outer halo is too faint to be seen in the eyepiece.)

M90: About 40' NNE of M89, midway between M89 and a conspicuously-solitary 7th magnitude star, is the Sab spiral M90. This is one of the larger Coma-Virgo galaxies, measuring about $9^1/_2$'×5'. But its integrated apparent mag is only 9.5 so its surface brightness is very low. Nevertheless it is not particularly difficult to see in 15×100 supergiant binoculars as a thin, nearly N-S, oval, though it displays neither core nor nucleus. The galaxy-image is elongated because M90 is a spiral highly tilted to our line of sight. However, its image, like that of M88, is not merely an elongated gash of light, but a smoothly-flattened and evenly-outlined oval. This is because the galaxy's outer arms are very smooth-textured (and thick). Photos also show, however, that its inner arms are mottled with dark dust clouds (which is why the galaxy's binocular image lacks a bright core). M90 is also peculiar for having a radial velocity in *approach:* it is one of only a half-dozen galaxies outside of the Local Group with true space motions that over-ride the cosmological expansion of the Universe. All six galaxies lie toward the Coma-Virgo Cluster, which demonstrates how extreme galaxy orbital velocities can be around the deep gravitational well which is at the center of the Cluster. Also, M90 is only 29 million light-years away, less than half the distance to the Coma-Virgo Cluster, and consequently subjected to less than half the cosmological expansion velocity of galaxies at the Cluster core. *The Hubble Constant*–the average increase in the observed recessional velocity of galaxies with increase in distance from us–is estimated to be around 50 kilometers per second for every megaparsec (= 3.26 million light-years) of distance.

M91: The Sb spiral M91 is both difficult to find and difficult to see. It is difficult to find because it is in extreme southern Coma Berenices with nothing conspicuous nearby, only the faint galaxy M88 1° to its west (and slightly south) and a mag 7 star 1° to its SE. It is difficult to see because it is a mag 10.2 object that measures about $5^1/_2' \times 4^1/_2'$: thus it is approximately the same size as the Sb spiral M58 to its south but only one-half as bright. Even in 15×100 supergiant binoculars M91 appears only as a very small, extremely low-surface-brightness patch of haze, with a stellar core perhaps visible at moments of better seeing.

M98: Though conveniently located exactly $^1/_2°$ due west of 6 Comae, the nearly edge-on Sb spiral M98 is extremely difficult to see even in supergiant binoculars. It not only is just mag 10.1 and thus almost as faint as the challenging M91, it is very thin, $9^1/_2' \times 3'$. Even 15×100 glasses show M98 (after considerable looking) only as an amorphous smudge of pale haze with no hint of the galaxy's strong NNW-SSE orientation. M98 seems to be 47 million light-years away, slightly nearer than the core of the Coma-Virgo Cluster, and is one of the half dozen galaxies toward the Cluster which have radial velocities in approach.

M99: Another Coma-Virgo giant spiral near 6 Comae Berenices is the nearly face-on Sc system M99, located 50' SE of 6 Comae and just SW of a 7th mag field star. M99 is easier to spot than M98 because it is 0.2 magnitude brighter and round ($5^1/_2' \times 5'$), not thin. Its surface brightness is sufficiently high that its tiny disc can be glimpsed even in 10×50 binoculars. With 15×100 glasses the disc enlarges into a small, dim, amorphous/circular patch of haze with a tiny, almost stellar, core. M99 is 42 million light-years distant on the near edge of the Coma-Virgo Cluster and has an integrated absolute mag of –20.6, a luminosity of about 15 billion Suns–comparable to that of our own Milky Way.

M100 (Photo 4.6): One of the most photogenic of all grand-design spirals is M100, a nearly face-on Sbc system. It is easily found about $1^1/_2°$ NE of 6 Comae Berenices. (Two 6th mag stars oriented SW-NE are between M100 and 6 Comae, approximately pointing at the galaxy.) M100 is fairly bright (mag 9.3) and large (7'×6') and not difficult to spot with 10×50 binoculars. In 15×100 glasses it appears as an amorphous/circular patch of haze with a tiny pinprick nucleus–the image that would be expected from a face-on, small-bulged Sbc system. The patch's surface brightness is quite low, but the galaxy is immediately visible. The distance to M100, derived from its Cepheid variables, is 47 million light-years. Its integrated absolute mag is therefore –21.4, a luminosity approaching 33 billion Suns.

M104, the Sombrero Galaxy: The photogenic M104, though certainly a member of the Local Supercluster, is not in the Coma-Virgo Cluster proper. It is located in far southwestern Virgo $11^1/_2°$ due west of Spica (Alpha [α] Virginis) on the Virgo/Corvus border. No bright stars are nearby, but the galaxy is easy to find because it is at the SW vertex of an equilateral triangle with Chi (χ) and Psi (ψ) Vir. However, it is not as easy to spot in small binoculars as its numbers–mag 8.0; size 9'×4'–imply because it is a borderline Sa/S0 spiral/lenticular galaxy with a very bright, but compact, bulge, and

Photo 4.6 The giant spiral M100 in the Coma-Virgo Galaxy Cluster.

its disc is almost edge-on and partially obscured by the famous dust lane. Thus in 10×50 glasses M104 looks smaller than its official 9'×4', appearing only as a tiny (but bright) E-W sliver of light. However, the surface brightness of its elongated core is so high that it takes magnification very well, and the galaxy is in fact easier at the higher powers of 40mm zooms than it is with 10×50s. In 15×100 supergiant instruments M104 has a tiny, bright E-W core within a compact, rapidly-fading, halo. Just to the galaxy's west is a well-resolved multiple star; and conspicuous to its SW is the bright field star that is equally conspicuous on photos. This isolated galaxy somehow looks more distant than the giant ellipticals and spirals in the Coma-Virgo Cluster to the north; but that seems to be an optical illusion because the most recent estimates place M104 only 31 million light-years away (half previous estimates). Nevertheless the galaxy has the very respectable integrated absolute mag of –21.9, a luminosity of 50 billion Suns and twice that of the Andromeda Spiral.

NGC 4261: A challenge galaxy for supergiant binoculars and small richest-field telescopes in the M61 region is the 3½'×3', magnitude 10.4, NGC 4261. This E2-3 giant elliptical is located 1° NW of 17 Virginis, with 8th mag field stars ¼° to its NE and NW. In 15×100 glasses the galaxy will appear, at best, as a very tiny smudge of haze with a very faint stellar nucleus. However, it is worth the effort; for NGC 4261 is *extremely* distant–perhaps as much as 110 million light-years away and therefore twice as far as the center of the Coma-Virgo Cluster. The galaxy's integrated absolute mag is around –22, about that of M84 and M86. NGC 4261 is the lucida of a distant cluster called the Virgo W Galaxy Cloud.

NGC 4365: Located 1½° WSW of M49 is NGC 4365, one of the brighter members of the M49-dominated Coma-Virgo Subcluster B. This E2 giant elliptical is, even in comparison with the Messier-numbered Coma-Virgo galaxies, rather large (5½'×4½') and bright (magnitude 9.6). 15×100 binoculars show a strong stellar nucleus in a low-surface-brightness, fuzzy-textured, circular patch. The patch is neither terribly small nor too difficult to detect. NGC 4365 has an integrated absolute mag of –21.4, corresponding to the brightness of 33 billion Suns. Nevertheless it is 1.3 mags = 3 times less luminous than the huge M49 near it.

NGC 4429: On the southern fringes of the core of the Coma-Virgo Cluster about 1½° SE of M87, and the same distance WNW of the 6th magnitude 20 Virginis, is the lenticular galaxy NGC 4429. This system is just 2' SSW of one 9th mag star and a few minutes north of another. The glare of the nearer star makes the galaxy a bit of a challenge even in supergiant binoculars; but it can be glimpsed as a very small, very faint halo around a very faint non-stellar core. NGC 4429 has an integrated apparent mag of 10.0 and is well-elongated, measuring 5½'×2½'; but higher magnifications than those provided by wide-field instruments are necessary to get the galaxy sufficiently far from the 9th mag star that its elongation can be detected. NGC 4429 seems to be about 42 million light-years away and thus a little nearer than the core of the Coma-Virgo Cluster.

NGC 4526: About 1° ESE of M49 is another member of Coma-Virgo Subcluster B, the lenticular system NGC 4526. This is one of the brighter Coma-

Virgo non-Messier galaxies, with an integrated apparent magnitude of 9.7. It is also quite large and elongated, measuring 7'×3'. NGC 4526 is extremely easy to locate because it is between a $^1/_4$°-wide, ENE-WSW pair of mag $6^1/_2$–7 stars. But with supergiant binoculars it is very easy to see anyway–a rather bright halo surrounding a small core in which is a sharp stellar nucleus. However the elongation is not evident. The integrated absolute mag of NGC 4526 is –21.3, a luminosity of nearly 30 billion Suns.

NGC 4636: 3° west and slightly south of Delta (δ) Virginis, and $1^1/_2$° SW of 35 Vir, is an outlying giant elliptical member of the Coma-Virgo Cluster, the E0-1 system NGC 4636. This galaxy is comparatively bright and large (mag 9.5; size 7'×5'), and with supergiant binoculars is immediately visible at the NW vertex of an equilateral triangle with two 8th mag stars. The galaxy has a tiny, sharp stellar core within a rather large amorphous/circular, moderate-surface-brightness halo. NGC 4636 is brighter and larger than some of the Messier-numbered Coma-Virgo galaxies and consequently should be visible in higher-power 50mm glasses. Its integrated absolute mag is –21.5, a luminosity of over 30 billion Suns.

NGC 4699: One of the brighter galaxies in Virgo outside the Coma-Virgo Cluster is NGC 4699, some 20° south of the core of the cluster and $4^1/_2$° NE of M104. It lies $1^1/_2$° NNE of Psi (ψ) Virginis, and 1° east and slightly south of a 7th magnitude star. NGC 4699 has an integrated apparent mag of 9.5 but is rather small, only about $4^1/_2$'×3'. Thus its surface brightness is moderately high. In 15×100 binoculars it displays a circular, moderately-bright halo that is compact and conspicuous, within which is a very faint stellar nucleus. NGC 4699 in fact looks exactly like a face-on Sab spiral galaxy should.

NGC 4710: A challenge galaxy for supergiant binoculars is the lenticular system NGC 4710, located on the far NE outskirts of the Coma-Virgo Cluster 1° almost due north of 29 Comae Berenices. It is very faint, only mag 11.0, and very short and thin, measuring just 4'×1.2', but conveniently located several minutes of arc due east of a 9th mag star. In 15×100 supergiant binoculars under dark skies NGC 4710 can be discerned as an intermittently-visible stellar nucleus within a compact, low-surface-brightness halo. The halo's elongation is not evident.

V Other Galaxies of the Spring Skies

The Coma-Virgo Cluster is the richest concentration of galaxies in the "milky way" of galaxies that stretches from the Ursa Major/M81 Group on the NW through Canes Venatici, Coma Berenices, and Virgo to the Centaurus/NGC 5128 Group on the SE. However, the area immediately NW of the Coma-Virgo Cluster, from the Coma Star Cluster NW through Canes Venatici to the Big Dipper, is also well-populated by bright galaxies, including eight with Messier numbers. These galaxies are all members of the Local Supercluster and distributed either in larger aggregations resembling the Leo I Galaxy Cloud or in smaller groups like the Local Galaxy Group. But because our line of sight in this direction angles through the interior of the Local Supercluster, some of these galaxy aggregations are superimposed upon one another and therefore difficult to disentangle. In this section the bright galaxies in Ursa Major (outside the M81 Group), Canes Venatici, and northern Coma Berenices will be described by constellation. At the end of the section other bright galaxies of the spring–Local Supercluster members, but in directions somewhat off the plane of the Supercluster–will be discussed in west-to-east order.

V.1 Ursa Major

The Great Bear has three Messier-numbered galaxies in addition to M81 and M82, the lucidae of the Ursa Major Galaxy Group: the very bright Pinwheel Galaxy M101, and two late tack-ons to the Messier catalogue, M108 and M109. M101 is an easy small binocular target; but M108 and M109, which are among the faintest of the Messier galaxies, both require magnification to be seen–M108 because it is so thin, and M109 to get it away from the glare of the 2nd magnitude Gamma (γ) Ursae Majoris. M101, M108, and M109 all are around the Big Dipper. But in the extreme western and southwestern parts of Ursa Major are four more binocular galaxies: NGC 2841 and NGC 3184, found among the stars marking the feet of the Bear, are accessible to 10×50s; but NGC 2768, toward Omicron (o) Ursae in the Bear's head, is best looked for with large binoculars, and its companion, the extremely faint NGC 2742, is a challenge object for supergiant instruments.

M101, the Pinwheel Galaxy (Plate XLVII, Photo 4.7): The Pinwheel Galaxy is one of those large, low-surface-brightness objects just made for binoculars. It spreads the light of a magnitude 7.9 star out over an area more than 20' in diameter and thus defies conventional telescopes. However, it is *easily* seen in 10×50 binoculars or with the higher powers of 40 mm zooms (the higher power darkening the sky background sufficiently to provide good contrast for the tenuous galaxy glow). M101 is at a well-marked spot 5° NNE of Eta (η) Ursae Majoris at the end of the "Handle" of the Big Dipper and the same distance due east of Zeta (ζ) Ursae. However, it is not necessarily as easy to find as this sounds because directions this near the north celestial pole can be confusing. (Remember that *north* is always toward Polaris and *west* the direction the stars move around Polaris.) Moreover, it can be difficult to adjust from looking at point-source star-sparks to trying to see a low-surface-brightness object like M101: you must keep reminding your eyes that they are searching for something two-thirds the diameter of the full Moon but shining with only about one-fifty-millionth as much light! However, in 100 mm

Ursa Major Galaxies Outside the M81 Group

Galaxy	RA (2000.0)	Dec	m_V	Dimensions	Type	Color Index	Distance	M_V
M101=N5457	14^h03^m	+54°21'	7.9	22'	SAB(rs)cd	0.45	23×10^6l-y	−21.2
M108=N3556	11 11	+55 40	10.0	8×1	SB(s)cd	0.66	~40	$-20^1/_2$
M109=N3992	11 58	+53 23	9.8	7×4	SB(rs)bc	0.77	46	−20.8
N 2742	09 08	+60 29	11.4	3.0×1.6	SA(s)c:	0.62	88	$-20^1/_2$
N 2768	09 12	+60 02	9.9	6.6×3.2	S0/E6:	0.97	88	−22.0
N 2841	09 22	+50 58	9.2	6.8×3.3	SA(r)b:	0.87	46	−21.4
N 3184	10 18	+41 25	9.8	7.8×7.2	SAB(rs)cd	0.58	36	−20.4

Photo 4.7 The Pinwheel Galaxy, M101 in Ursa Major

supergiant binoculars M101 is immediately visible as a big amorphous/circular patch of pale haze, slightly brighter through the central region. Both in photographic and binocular appearance the Pinwheel is the prototype loose-armed, small-bulged Sc galaxy seen face-on.

M101 is estimated to be about 23 million light-years distant. Its integrated absolute mag is therefore −21.2, a luminosity of 25 billion Suns. Its color index, +0.45, is very blue for a spiral galaxy because of the vigorous star-formation presently occurring in its majestic spiral arms. On photos, however, it can be seen that the Pinwheel's arms do not arc out evenly, but have some sharp bends. This, and the starburst activity in the arms, are due to the Pinwheel's gravitational interaction with the Scd system NGC 5474, an 11th mag satellite galaxy only about $^1/_2$° to the giant spiral's south. The M101 Galaxy Group has several other dwarf spiral and irregular members. It has been suggested, however, that the M101 Group might be part of a larger galaxy cluster that includes M51, the Whirlpool Galaxy, and M63 in Canes Venatici. If M101 and M63 are both 25 million l-y from us, the 15° apparent separation between them corresponds to a true separation of roughly $6^1/_2$ million l-y.

M108 (Photo 5.3): M108, one of the posthumous additions to Charles Messier's catalogue of nebulous objects, is not difficult to find, for it is located $1^1/_2$° ESE of Beta (β) Ursae Majoris, the 2nd magnitude star that marks the SW corner of the "Bowl" of the Big Dipper. A 7th mag star is halfway between the galaxy and Beta Ursae. M108 is, however, fairly hard to see because it is faint, only mag 10.0, and very thin, measuring just 8'×1'. It is beyond conventional binoculars. But it is not difficult with 15×100 supergiant glasses, in which it appears as a thick E-W streak, peculiarly ragged in outline and texture, a couple faint stars superimposed upon the galaxy-image. It looks wider than its 1' catalogue value. The supergiant binocular appearance of the galaxy faithfully reflects its true nature, for M108 is a loose-armed, tiny-bulged Scd system seen nearly edge-on is and mottled with dark dust clouds and bright star-formation complexes. The galaxy is estimated to be about 40 million light-years away and thus has the same absolute mag as our Milky Way, $-20^1/_2$. The Owl Nebula, M97, one of the largest of the planetary nebulae, is just 1° SE of M108 and therefore in the same wide-field view. (M97 is described in Sect. 3 of Chap. 5.)

M109 (Photo 4.8): M109, like M108, is a late addition to the Messier catalogue. It is both a little brighter (by 0.2 magnitude) and a little larger (measuring 7'×4') than M108, but more difficult to see in wide-field instruments because it is only 40' SW of the 2nd mag Gamma (γ) Ursae Majoris. M109 can be glimpsed in 15×100 supergiant binoculars with some effort: its ENE-WSW elongation is evident and a stellar nucleus intermittently visible. Just to the galaxy's SW is a 10th mag field star. M109 definitely needs higher powers to get it out of the glare of Gamma Ursae. It is estimated to be 46 million light-years distant and intrinsically about as luminous as M108 and our Milky Way.

NGC 2742 + NGC 2768: The giant S0 lenticular or highly-elongated E6 elliptical galaxy NGC 2768 is located in west central Ursa Major exactly 5° east and slightly south of the 3rd magnitude Omicron (ο) Ursae Majoris. It is $1^1/_2$° SSW of 16 UMa. NGC 2768 is rather faint, mag 9.9, and not large, only $6^1/_2$'×3',

Photo 4.8 The barred spiral galaxy M109 near Gamma (γ) Ursae Majoris (the bright star to the upper right)

and thus requires giant binoculars or small RFTs to be seen with certainty. In 15×100 supergiant glasses it displays a moderately-bright, non-stellar core (the galaxy's central bulge) within a compact, moderately faint halo, and its E-W elongation is intermittently visible. NGC 2768 is extremely distant, 88 million light-years, and extremely luminous, its integrated absolute mag of –22.0 being fully equal to that of the M84 + M86 giant ellipticals at the center of the Coma-Virgo Cluster. It is the lucida of a galaxy group that includes the Sc spiral NGC 2742, a tiny ($3^1/_2$'×$1^1/_2$'), extremely faint (mag 11.4) object right at the limit for 100 mm aperture. In 15×100 instruments NGC 2742 is only a very, very faint, intermittently-visible "star": there is no haze around it and therefore no galaxy-glow to give it away. It can be certainly identified only because of its location a few minutes of arc SE of a mag $7^1/_2$ field star. If NGC 2742 is indeed at the same distance as NGC 2768, its integrated absolute mag is around $-20^1/_2$, about the same as that of M108, M109, and our own Milky Way.

NGC 2841: The brightest galaxy in Ursa Major after M81, M82, and M101 is the magnitude 9.2 NGC 2841, an Sb spiral located 20' SE of the 6th mag star 3 Lyncis, which is just south of the midpoint of the E-W line joining Theta (θ) and 15 Ursae Majoris. An 8th mag star is a couple minutes NE of the galaxy and another 8th mag star just a couple minutes NE of 37 Lyn, the three stars and galaxy forming a very distinctive parallelogram, the galaxy at the parallelogram's southern corner. NGC 2841 is fairly small, $6^1/_2$'×3', but its surface brightness is quite high. In 10×50 binoculars it appears as a small, moderately high-surface-brightness patch of haze elongated NE-SW, with a rather bright core. The galaxy image upgrades considerably with 15×100 glasses: a very faint stellar nucleus can be glimpsed in the core; the core itself is distinctly oval; and the halo appears not smooth and symmetrical, but peculiarly "ragged." The stellar nucleus and oval core are characteristic of the eyepiece image of bulge-dominated Sb spirals (including the Andromeda Galaxy). But the ragged appearance of NGC 2841 is a consequence of the extreme flocculence of its spiral features. Indeed, photographically NGC 2841 is the archetypal *flocculent spiral:* it has no true spiral arms, just a multitude of short arcs oriented along a spiral design that is more suggested than actually traceable. Four supernovae have been observed in NGC 2841. The fourth, in 1999, suggested a distance of 46 million light-years to the galaxy. This is rather less than previous estimates, but still implies a very respectable integrated absolute mag for the galaxy of –21.4.

NGC 3184: The face-on Scd spiral NGC 3184 is easily located about $^3/_4$° due west of the 3rd magnitude Mu (μ) Ursae Majoris. A 7th mag star is just to the galaxy's west. NGC 3184 is rather faint, only mag 9.9, but over 7' in diameter. Despite its very low surface brightness, characteristic of face-on late-type spirals, the galaxy can be glimpsed with averted vision in 10×50 binoculars as a tiny smudge of haze. With 15×100s a stellar nucleus is intermittently visible at the center of the galaxy's extremely low-surface-brightness amorphous/circular patch. Visually NGC 3184 is a smaller version of IC 342 of the Ursa Major Galaxy Group; and in fact it is exactly the same type of galaxy also seen face-on but from three times the distance, 36 million light-years. The absolute mag of NGC 3184 is therefore about –20.4, twice the luminosity of IC 342.

V.2 Canes Venatici

Though not large in area, Canes Venatici has not less than four Messier-numbered galaxies, all bright magnitude $8^1/_2$ objects: M51, M63, M94, and M106. The showpiece object among its galaxies is M51, the Whirlpool: but the constellation's other three Messier systems are also easy medium-binocular targets. Of the NGC galaxies in Canes Venatici, two are interesting 10×50 binocular sights: NGC 4631, large and bright but hard to find; and NGC 5195, small and faint but easy to find because it is the northern lobe of M51. Three of Canes Venatici's other NGC galaxies are not difficult for large binoculars and small RFTs: NGC 4490, easily found NW of Beta (β) Canum Venaticorum; and the NGC 5005 and NGC 5033 pair SE of Alpha (α) CVn.

Canes Venatici Galaxies

Galaxy	RA (2000.0)	Dec	m_V	Dimensions	Type	Color Index	Distance	M_V
M51=N5194	13^h30^m	+47°12'	8.4	11'×7'	SA(s)bc pec	0.60	29×10^6 l-y	-21.4
M63=N5055	13 16	+42 02	8.6	10×6	SA(s)bc	0.72	24	-20.7
M94=N4736	12 51	+41 07	8.2	7×3	(R)SA(r)ab	0.75	15	-20.2
M106=N4258	12 19	+47 18	8.4	19×8	SAB(s)bc	0.69	23	-20.8
N 4490	12 31	+41 38	9.8	6.4×3.3	SB(s)d pec	0.43	26	-19.7
N 4631	12 42	+32 32	9.2	15.5×3.3	SB(s)d	0.69	23	~-20
N 5005	13 11	+37 03	9.8	5.8×2.8	SAB(rs)b	0.80		
N 5033	13 13	+36 36	10.2	10.5×5.1	SA(s)c	0.55	39	-20.7
N 5195	13 30	+47 16	9.6	6.9×4.6	I0 pec	0.90	29	-20.2

M51, the Whirlpool Galaxy, and NGC 5195 (Plate XLVIII, Photo 4.9): The classic *grand-design* spiral galaxy is the magnificent M51, the Whirlpool, located in extreme NE Canes Venatici about $3^1/_2$° SW of Eta (η) Ursae Majoris. The grand-design pattern which is so conspicuous on visual-light photos of the galaxy is even *more* uniform in the red-light, 6562 Å wavelength, H-alpha line radiated by the giant emission nebulae around the young-star complexes that define the true light ridge of spiral arms. Nevertheless, M51's two arms do not arc out evenly from the central bulge to the galaxy's periphery but have two or three abrupt changes of direction–"elbows." These irregularities have been blamed upon the gravitation influence of the peculiar elliptical or lenticular galaxy on the north edge of the Whirlpool, NGC 5195. This system is indeed a satellite of the big spiral, but somewhat beyond it: dust from one of M51's arms is conspicuously silhouetted upon the smaller galaxy's disc. The Whirlpool's grand-design pattern itself might be the consequence of the gravitation of NGC 5195. The star-richness of M51's arms, and the present starburst activity along those arms, are almost certainly the consequence of a recent interaction between the systems in which the gravitational field of NGC 5195 not only provoked wide-spread contraction of giant molecular clouds in the big spiral, but matter from the smaller galaxy was lost to the spiral, providing yet more fuel for star-formation.

Photo 4.9 The Whirlpool Galaxy, M51 in Canes Venatici, the classic grand-design spiral galaxy

Though bright, magnitude 8.4, M51 is not all that large, 11'×7', and therefore surprisingly difficult, especially in 7× binoculars, with which you must look carefully to discern the galaxy's fuzzy-edged "dot." And matters are not helped by the lack of bright stars nearby to use as guides. M51 can be especially frustrating to look for in small-field telescopes that lack good setting circles. However, with 10×50 binoculars the Whirlpool not only is easy to see, it appears pear-shaped, a small lobe projecting from it toward the north. That lobe, a mag 9.6 object measuring $6^1/_2$'×$4^1/_2$', is NGC 5195. In 15x100 supergiant instruments both the M51 spiral and its NGC 5195 companion display conspicuous stellar nuclei. The spiral portion not only is visibly elongated N-S (the direction in which the gravitational pull of the satellite is distorting it) but granular rather than smooth in surface texture. This is the beginning of the resolution of the spiral arms: M51 is perhaps the only galaxy in which some perception of spiral pattern is possible in wide-field instruments; and it is possible simply because of how extremely luminous and thick its grand-design arms actually are.

The Whirlpool Galaxy is estimated to be 29 million light-years distant. Thus its integrated absolute mag is -21.4, a luminosity of over 30 billion Suns. Its NGC 5195 companion has an absolute mag of -20.2 (10 billion Suns) and therefore, though "just" a satellite of the big spiral, is almost as brilliant as our Milky Way.

M63: The Sbc spiral M63 is located just over 1° due north of 19 Canum Venaticorum, the northernmost star in the small asterism named "Asterion" by Hevelius, the inventor of this constellation. (Asterion, which includes 19, 20, and 23 CVn, is 5° NE of Alpha [α] CVn and an attractive little group for 7× binoculars.) M63 is a magnitude 8.6, 10'×6' object and therefore only about 0.2 mag fainter and 1' smaller than M51. It is easily visible with 10×50 binoculars as a thin E-W streak with a tiny core (which corresponds to the system's compact bulge and bright inner spiral arms). A mag $8^1/_2$ field star is near the galaxy's NW extremity. In 15×100 glasses M63 enlarges into an E-W gash or bloated streak of light with a bright core containing a true stellar nucleus. The galaxy glow can be traced almost all the way to the mag $8^1/_2$ star and gives the strong impression of being ragged in texture rather than smooth. (Contrast the raggedness of M63 with the smoothness of M94.) This is a correct impression: M63, like NGC 2841 in Ursa Major, is a fully flocculent spiral galaxy, with a multitude of short spiral arcs rather than true spiral arms. M63 is estimated to be 24 million light-years distant.

M94: The brightest galaxy in Canes Venatici is not the famous M51, but the magnitude 8.2 M94. However, M94 is significantly smaller than the Whirlpool, 7'×3'. Though this means that it has even higher surface brightness than M51, it makes M94 a bit tricky to spot in 7× binoculars, with which it will appear virtually stellar. Fortunately the galaxy is at an easily-identified spot, for it is exactly 3° NNW of Alpha (α) Canum Venaticorum and the same distance almost due east of Beta (β). In 15x100 glasses M94's image takes on some character: it displays a dim, ambiguously-edged outer halo surrounding a tight, unusually bright inner halo, within which is a bright non-stellar core. The galaxy appears strongly circular, like a globular cluster, but is structurally very different: contrast it with M3 in far SE Canes Venatici 6° due east of Beta Comae Berenices, a globular with a large, bright, dense core and an extensive low-surface-brightness halo but no hint of a sharp boundary between the two as we see between the core and inner halo, and between the inner and outer halos, of M94. On photos M94 shows an extremely large, star-rich, central bulge into which curves a multitude of tight, smooth, and bright spiral fragments. The outer spiral arms are faint, but also tightly wound. And encircling the galaxy is a large, very faint ring (hence the "(R)" in its morphological classification). M94 is only about 15 million light-years away. It and M64 in Coma Berenices are the brightest members of the nearby Canes Venatici I Galaxy Cloud, most of the other systems of which are dwarf spirals and irregulars.

M106: Probably the easiest galaxy in Canes Venatici, particularly for wide-field instruments, is the late addition to the Messier catalogue, M106. This Sbc spiral is about as bright as the three other Messier galaxies in the constellation, magnitude 8.4, but much larger, measuring 19'×8'. It is easily seen even with 7×50 binoculars as a NW-SE spindle of light. The galaxy's one problem is that it is in the bright-star-poor NW quadrant of Canes Venatici and consequently can be tricky to find. The nearest reasonably conspicuous star to it is the 4th mag Chi (χ) Ursae Majoris, 6° due west. Not quite 2° due north of the galaxy is the 5th mag 3 Canum Venaticorum. But M106 is so large and so bright that with anything larger than 10×50s you need only a general idea of its location to see it any way. In 15×100s the galaxy displays a bright but substellar nucleus within a bright N-S oval core surrounded by a much fainter halo. A faint star is near the halo's southern tip. On photos the halo can be seen to be elongated NNW-SSE a bit out of alignment with the system's bright interior. M106 is estimated to be about 23 million light-years distant and therefore has an integrated absolute mag of –20.8, slightly greater than that of our Milky Way. The nucleus of the galaxy is unusually brilliant, and is an X-ray and radio source because of explosive activity presently occurring there. M106 is one of the nearest of the AGN, *active galactic nucleus,* systems.

NGC 4490: Though rather faint, magnitude 9.8, and rather small, $6^1/_2'\times3^1/_2'$, the Sd spiral NGC 4490 is one of the easier galaxies in Canes Venatici to find because it is conveniently located just $^3/_4°$ NW of Beta (β) Canum Venaticorum. It is a bit of a challenge with 10×50 binoculars, but in 15×100 glasses is immediately visible as a little gash, fairly short and thick, pointed SE at Beta Canum. The gash contains an intermittently-visible stellar nucleus and has a peculiarly ragged texture. Photos in fact reveal that NGC 4490 is mottled with dense dust clouds and bright star-formation complexes. Because of the rigorous star formation presently occurring in it, NGC 4490 has a very blue color index, +0.43, and is classified as a starburst galaxy. This starburst activity probably was provoked in the astronomically-recent past by a close encounter between NGC 4490 and its satellite, the faint NGC 4485 just to its NNW. NGC 4490 is about 20 million light-years away and, with NGC 4631 10° to its SSE, one of the bright members of the Canes Venatici II Galaxy Cloud, which is at about the same distance as the M101 + M51 Galaxy Cluster but in the background of M94, M64, and the Canes Venatici I Cloud.

NGC 4631 (Photo 4.10): As long and thin as M106, but only half as bright, is the edge-on spiral NGC 4631 in extreme southern Canes Venatici. It is a challenge object for 10×50 binoculars–in part because it is so thin; in part because of the lack of any bright stars nearby to use as signposts. The best way to find it is by looking just over half-way from Alpha (α) Canum Venaticorum to Gamma (γ) Comae Berenices, the 4th mag star at the northern

Photo 4.10 The edge-on spiral galaxy NGC 4631 in southern Canes Venatici. Just to the north of its bulge is its small satellite galaxy, NGC 4627

corner of the triangular Coma Star Cluster (though Gamma Comae is not a true Coma Cluster member). NGC 4631 is actually rather easy with 100 mm supergiant glasses, in which it appears as a long, finely-tapered E-W spindle around a thin, moderately-bright core. Not even the extremities of the spindle are difficult to see (with averted vision). Very likely the full photographic extent of the galaxy can be traced. No stellar nucleus is visible: it is blocked by the dust in the galaxy's spiral disc. Because of its edge-on orientation, the distance of NGC 4631 is somewhat uncertain; but the best guess is that it is about 23 million light-years away and a member of the Canes Venatici II Cloud with NGC 4490.

NGC 5005 and NGC 5033: About 3°–4° SE of Alpha (α) Canum Venaticorum is a pair of 10th magnitude galaxies for giant and supergiant binoculars. The Sb spiral NGC 5005, just over 3° SE of Alpha Canum between a NW-SE pair of mag $6^1/_2$–7 stars, is a mag 9.8 object that is elongated about $5^1/_2$'×$2^1/_2$' (NE-SW) because it is highly tilted to our line of sight. In 15×100 glasses it displays a compact, rather bright, almost (but not quite) stellar core within a compact amorphous/circular halo. The galaxy's orientation is not evident. Its bright core is because of its unusually star-rich and luminous bulge. About 40' to the SE of NGC 5005, and 20' due south of a 6th mag star, is another highly-tilted, loose-armed spiral, the Sc system NGC 5033. This galaxy is much larger than NGC 5005, $10^1/_2$'×5' (N-S), but 0.4 mag fainter and therefore of even lower surface brightness. NGC 5033 in fact is rather difficult even in supergiant binoculars: it appears only as a stellar nucleus within a *very* low-surface-brightness amorphous halo. The faintness of the halo is a result of the galaxy's rather thin, under-luminous spiral arms.

V.3 Coma Berenices

Southern Coma Berenices shares the galaxy-dense Coma-Virgo Cluster with NW Virgo. However, northern Coma is also well-populated with galaxies, including one easy for small binoculars (M64), and three possible for medium binoculars and easy in giant instruments (NGCs 4494, 4559, and 4565). Two of its galaxies are distinctive: M64 is called the "Black-Eye Galaxy" because of the exceptionally thick and dark lane of dust superimposed upon its bright inner disc and bulge; and NGC 4565 is one of the largest, brightest, and most-photographed of the precisely edge-on spiral galaxies. NGC 4565 with NGC 4494 and NGC 4559 are the brightest members of the 31 million light-year distant NGC 4565 Galaxy Group of the Coma I Galaxy Cloud.

M64, the Black-Eye Galaxy: The large ($9^1/_2$'×$5^1/_2$') and bright (magnitude 8.5) M64 is easily found 1° ENE of the 5th mag 35 Comae Berenices. Because of its high surface brightness, the galaxy's E-W orientation is readily visible in 10×50 binoculars. With 15×100 glasses M64 has a very distinctive and individual appearance. It is fairly fat, elongated E-W but not thin N-S. However, it is not oval in profile: the northern edge is distinctly flatter, less curved, than the southern. The galaxy has no halo, just a fairly high surface brightness extended core (within which is a sharp stellar nucleus) with a sharp drop-off in brightness toward the periphery. M64's binocular appearance faithfully reflects its structural reality. The flattened northern edge is from the heavy dust lane after which the galaxy has been named. (Thus the northern edge of the galaxy's disc is the nearer side to us, the dust lane being silhouetted against the system's bright bulge and inner spiral arms.) The extended core is because of M64's unusually thick, star-rich, and smoothly-luminous

Bright Coma Berenices Galaxies Outside the Coma-Virgo Cluster

Galaxy	RA (2000.0)	Dec	m_V	Dimensions	Type	Color Index	Distance	M_V
M64=N4826	12^h57^m	+21°41'	8.5	9.3'×5.4'	(R)SA(rs)ab	0.84	15×10^6 l-y	−19.9
N 4494	12 31	+25 47	9.8	4.6×4.4	E1-2	0.88	31	−20.1
N 4559	12 36	+27 58	10.0	12.0×4.9	SAB(rs)cd	0.45	31	−19.9
N 4565	12 36	+25 59	9.6	14.0×1.8	SA(s)b?	0.84	31	−20.3

inner spiral arms–unusually thick and bright and smooth even for an early-type spiral. M64 is about 15 million light-years away and consequently has an integrated absolute mag of –19.9, a luminosity of 8 billion Suns. (The galaxy's full face-on luminosity would be about twice as much.) M64 and M94 in Canes Venatici are both early-type Sab spirals with unusually smooth spiral arms; but the other galaxies in the CVn I Cloud are loose-armed, small-bulged Sc and Sd spirals and ragged Magellanic irregulars.

NGC 4494: The E1-2 giant elliptical galaxy NGC 4494 is 40′ east and slightly south of the magnitude $5^1/_2$ star 17 Comae Berenices (a true member of the Coma Star Cluster). This system is sufficiently large, $4^1/_2$′ in diameter, and bright, mag 9.8, that it could be glimpsed with 10×50 binoculars were it not for the glare of the mag $8^1/_2$ star just 5′ to the galaxy's NNE. However, in 15×100 glasses NGC 4494 is easily seen as a stellar core within a compact, but rather bright and surprisingly high surface brightness, halo. It is not at all difficult; but the galaxy's compactness, and the glare of the neighboring star, makes it rather easy to scan right over. The integrated absolute mag of NGC 4494, assuming a distance to the Coma I Cloud of 31 million light-years, is –20.1, a luminosity of 9 billion Suns. The distance of the Coma I Cloud is somewhat uncertain: recent estimates put it at 52 million l-y. But if the lower distance to the Cloud is correct, NGC 4494, and not M105 in the M95/M96 Galaxy Group in Leo, would be the nearest giant elliptical galaxy to us.

NGC 4559: In an area poor in even faint stars 2° east and a little south of Gamma (γ) Comae Berenices is the very large and much elongated (12′×5′) but faint (magnitude 10.0) NGC 4559. Despite its unpromising numbers, this galaxy is readily visible in 15×100 supergiant binoculars as a surprisingly large, moderate-surface-brightness amorphous patch. It does not, however, display a stellar core or any sense of elongation. NGC 4559 is a highly-tilted (hence its elongation) Scd spiral (thus the lack of a stellar core). It has the exceptionally blue color index of +0.45, equivalent to that of M101 in Ursa Major. This indicates vigorous star-formation is presently occurring in its spiral arms. The integrated absolute mag of –19.9 cited for the galaxy in the table is not corrected for its high inclination to our line of sight and consequently for the significant amount NGC 4559 is dimmed by its own dust.

NGC 4565: The magnificent, precisely edge-on, spiral galaxy NGC 4565 is located $1^1/_2$° due east of 17 Comae Berenices. Though it is quite bright, magnitude 9.6, and extremely long, 14′, it is a challenge object for medium binoculars because it is so thin, just 1′ wide. Nevertheless it can be glimpsed in 10×50 glasses (under good sky conditions, and with averted vision) as a needle-like streak pointing NW at Gamma (γ) Comae. In 15×100s the galaxy is much easier, and during moments of superior seeing appears much as it does in the photographs: a very long sliver of light, at the midpoint of which is a slight swelling containing a non-stellar core. Indeed, NGC 4565, with its astonishingly thin and fragile streak of gentle radiance, its subtle bulge-swelling, and its twinkling core, is one of the most awe-inspiring sights in the sky: in supergiant binoculars it looks almost *living*. The most recent distance estimate to NGC 4565 and the Coma I Galaxy Cloud, 52 million light-years, is rather more than the older estimate of 31 million l-y. At 52 million l-y, the integrated absolute mag of the galaxy would be an impressive –21.2, a luminosity of 25 billion Suns. But this is for the galaxy *in the orientation we see it* and not corrected for the large internal absorption caused by its famous dust lane: the true face-on absolute mag of NGC 4565 would be much higher–perhaps *too* high, favoring a lower distance to the galaxy and the Coma I Cloud. The effect of the dust lane is evident in the galaxy's color index, +0.84, which is almost as red as the color index of the giant elliptical member of the Coma I Cloud, NGC 4494. At 52 million l-y the 14′ apparent length of NGC 4565 implies a true disc diameter of 210,000 l-y–possible; but again a rather high value. At 31 million l-y the galaxy would be 120,000 l-y long and have an (uncorrected) integrated absolute mag of –20.3 (11 billion Suns).

V.4 Miscellaneous Spring Galaxies

In this section will be described the brightest spring galaxies outside the "milky way" of galaxies from Ursa Major to Centaurus–excluding the galaxy groups in the constellation Leo, which have their own section earlier in this chapter. All these galaxies are members of the Local Supercluster; but several of them, including NGC 2683, NGC 2903, NGC 3115, and NGC 6503, seem to be true ungrouped individualists that are not members of any smaller galaxy aggregation. The brightest system in this miscellany is the magnitude 8.9 NGC 3115, which has the additional distinction of being the nearest true lenticular S0 galaxy to our Milky Way. But NGC 3115 is very small. An easier medium-binocular target is the mag 9.0 Sbc spiral NGC 2903 in Leo. However, visually the most intriguing galaxies in this collection are the nearly edge-on NGC 2683 in Lynx and the *precisely* edge-on NGC 5907 in Draco, an astonishingly long and thin needle of elusive light that rivals in aesthetic impact the more famous NGC 4565 of Coma Berenices.

NGC 2683: The small off-Milky Way constellation of Lynx does not offer observers much. However, it does have one good large-binocular galaxy, the reasonably bright (magnitude 9.8) but very thin ($8^1/_2$'×$2^1/_2$') NGC 2683. This is an Sb spiral oriented almost exactly edge-on. Such galaxies are beautiful sights in the eyepiece, but can be difficult to see. And NGC 2683 can be difficult to find as well because it is in a bright-star-poor part of the sky: the nearest conspicuous star to it is the 3rd mag Alpha (α) Lyncis 5° east and slightly north. The best way to find the galaxy is through Cancer, for it is located 1° due north of the 6th mag Sigma-one (σ^1) Cancri and 1° NW of the 5th mag Sigma-two Can. Another 6th mag star is about $^1/_4$° WSW of the galaxy. NGC 2683 is so thin that that it is only intermittently visible with 10×50 binoculars. However, it is surprisingly easy with 100mm supergiant glasses, appearing as a spindle-like NE-SW streak containing a stellar nucleus. (If the galaxy was *precisely* edge-on to our line of sight, the bulge's nucleus would be obscured by dust, just as the Center of our own Galaxy is blocked to our view by the dense dust of the Sagittarius Rift.) The field around the galaxy is surprisingly rich in mag 9 to 11 stars for an area so poor in bright stars and so far off the Milky Way. NGC 2683 is estimated to be about 29 million light-years distant. Thus its integrated absolute mag is −20.0, a luminosity of 8 billion Suns. This value, however, is not corrected for the galaxy's high inclination: its true face-on absolute mag would be over -21.

NGC 2903: The second brightest galaxies in the constellation of Leo after the magnitude 8.9 M66 are NGC 3521 and NGC 2903, both mag 9.0. But NGC 2903 is the easiest of the three galaxies to see because it is a rather large, 12'×$5^1/_2$', moderate-surface-brightness oval whereas M66 is a thin, 8'×$1^1/_2$', streak and NGC 3521 is just a "star" within a low-surface-brightness patch. And NGC 2903 is the easiest of the three to find: 1° due south of the 4th mag Lambda (λ) Leonis is a 40' wide E-W pair of 7th mag stars, the galaxy located 20' due south of the eastern of these two stars. With 10×50 glasses NGC 2903 is readily seen as a bloated N-S streak pointing

Bright Spring Galaxies Out of the "Milky Way" of Galaxies

Galaxy	Con	RA (2000.0)	Dec	m_V	Dimensions	Type	C. I.	Distance	M_V
N 2683	Lyn	08^h53^m	+33°25'	9.8	8.4'×2.4'	SA(rs)b	0.89	29×10^6l-y	−20.0
N 2903	Leo	09 32	+21 30	9.0	12.0×5.6	SAB(rs)bc	0.67	23	−20.2
N 3115	Sex	10 05	−07 43	8.9	8.1×2.8	$S0^-$	0.97	28	−20.9
N 3344	LMi	10 43	+24 55	9.9	6.9×6.4	(R)SAB(r)bc	0.59	22	−19.2
N 3521	Leo	11 06	−00 02	9.0	12.5×6.5	SAB(rs)bc	0.81	51	−21.8
N 5866 = M102	Dra	15 06	+55 40	9.9	5.2×2.3	$SA0^+$	0.85	48	−20.9
N 5907	Dra	15 16	+56 20	10.3	11.5×1.7	SA(s)c:	0.78	48	−20.5
N 6503	Dra	17 49	+70 09	10.2	7.3×2.4	SA(s)cd	0.68	17	−18.5

directly at the 7th mag star to its north. In 15×100 binoculars it is even easier: a fat N-S gash. The galaxy image lacks a stellar nucleus, but has a decidedly granular texture from the partial resolution of its thick, but rather tightly-wound, inner spiral arms. NGC 2903 is estimated to be 23 million light-years distant. Its uncorrected integrated absolute mag is −20.2, a luminosity of 10 billion Suns. Its full face-on luminosity would be about half a mag higher and therefore comparable to that of our own Milky Way.

NGC 3115, the Spindle Galaxy: The unpretentious little constellation of Sextans the Sextant, invented in the 17th century by Hevelius to occupy the bright-star-poor area south of the ancient Leo the Lion, has the distinction of containing the nearest true lenticular galaxy to the Milky Way, NGC 3115. This is an edge-on system that is rather bright, magnitude 8.9, but short and thin, 8'×3'. This makes it difficult to see. In fact it is more difficult to see than even these numbers imply, for as an S0 galaxy most of its light comes from its bulge rather than from its disc, which is in the nature of a narrow ring attached to the equator of the bulge. Fortunately, though NGC 3115 is in a constellation of faint stars, it is not excessively difficult to locate: it is just north of the midpoint of the line between Gamma (γ) and Epsilon (ε) Sextantis, with a conspicuous NE-SW pair of 7th mag stars just 40' to its east. In 10×50 binoculars the galaxy appears as a bright NE-SW sliver of light; but it is too short to be obvious, and averted vision is required. However, the galaxy's surface brightness is sufficiently high that it is easier to see in the higher powers of 40 mm zooms than it is in 10×50s. With 100 mm supergiant glasses a stellar core is visible dominating the subtle halo, the orientation of which still requires averted vision. NGC 3115 is about 28 million light-years away and has an integrated absolute mag of −20.9, a luminosity of 20 billion Suns.

NGC 3344: Like Lynx and Sextans, the small constellations of Leo Minor was added to the heavens by Hevelius in the 17th century to occupy the bright-star-poor space between two of the ancient constellations, in this case Leo to the south and Ursa Major to the north. Leo Minor also resembles Lynx and Sextans in containing one galaxy suitable for larger binoculars (and not much else): NGC 3344 is a rather faint (magnitude 9.9) but moderately large ($6^1/_2$' in diameter) face-on Sbc spiral in far southern Leo Minor. It is located midway between the 3°-wide N-S pair of 6th mag 40 and 41 Leo Minoris, and not quite 3° due west of 54 Leonis. The surface brightness of NGC 3344 is a bit too low for 10×50 binoculars, but the galaxy is not difficult in larger glasses. With 15×100 instruments it appears as a faint, compact, haze containing a bright stellar nucleus. The nucleus is extremely bright with respect to the halo and might be mistaken for a "star" except for the presence, a few minutes of arc to the SW, of an actual 9th mag star, the halo-less image of which makes the dim glow around the stellar nucleus of NGC 3344 easier to identify. Photographically NGC 3344 looks like a normal face-on spiral, but closer studies have discovered several peculiarities about it: it has a weak central bar, around the outer ends of which is a weak inner ring; it has a faint outer ring (indicated by the "(R)" in its classification) which is *not* centered on the galaxy's center; the light of both rings is dominated by young blue stars, but the inner ring is much poorer in neutral hydrogen gas than the outer ring, and the hydrogen gas beyond the outer ring is in a warped structure; and the stars deep within the bulge seem to be orbiting around the galaxy's center in a direction *opposite* to the rotation of the rest of the galaxy. Usually such structural abnormalities can be blamed upon the gravitational influence of a nearby large companion galaxy; but NGC 3344 is another "loner" like NGC 2683 and NC 3115. It is only about 22 million light-years away (and therefore in the foreground of the M65/M66 and M95/M96 galaxy groups to the south in central Leo). The integrated absolute mag of NGC 3344 is only −19.2, a luminosity of 4 billion Suns, which makes it only 0.3 mag brighter than the Triangulum Galaxy M33, the third most luminous member of our Local Galaxy Group.

NGC 3521: NGC 3521 in extreme southern Leo is as bright, magnitude 9.0, and as large, $12^1/_2$'×$6^1/_2$', as NGC 2903. It even has the same morphological classification as that galaxy, Sbc. However, this is a good example of the fact that numbers do not always tell the whole truth because NGC 3521 is definitely more difficult to see than NGC 2903. It is more difficult to see because its surface brightness is lower; and its surface brightness is lower because, first, it is more face-on to us than NGC 2903, and second, it is actually more of an Sb than an Sc system and consequently more of its light comes from its bulge than from its spiral disc. Thus in 10×50 binoculars NGC 3521 appears merely as a 9th mag "star" embedded within a tiny tenuous halo. With supergiant binoculars and RFTs it has the typical face-on Sb galaxy image: a very faint stellar nucleus within a brighter core (elongated approximately N-S), the core enveloped by a very, very faint amorphous/circular halo. The galaxy is in the bright-star-poor southern extension of Leo $^1/_2$° due east of the 6th mag 62 Leonis toward the 5th mag 69 Leo. Recent estimates place NGC 3521 at the distance of 51 million light-years. This implies the very impressive integrated absolute mag of –21.8 for the galaxy, twice the luminosity of the Andromeda Galaxy M31.

M102 = NGC 5866: The identification of NGC 5866 as M102 is not certain; however, the Messier number has stuck to the galaxy, though this edge-on lenticular system is rather smaller and fainter than most of the galaxies Messier himself identified. It is in fact too faint, magnitude 9.9, and too small, 5'×3$^1/_2$', to be seen in medium binoculars. Nor is it easy to find, because the only bright star anywhere near it is the 3rd mag Iota (ι) Draconis 4° to the NE. (An isolated 5th mag star is slightly more than 1° due south of the galaxy.) In 15×100 supergiant glasses M102 is visible as a tiny low-surface-brightness oval oriented NW-SE containing a stellar nucleus. There is no halo. This is exactly how a distant S0 system should appear, for lenticular galaxies are essentially all bulge, their discs (like that of M102) frequently little more than an attached ring around the bulge's equator. Moreover, their bulges dim abruptly beyond a certain point, rather than fading gradually outward like an elliptical galaxy. Photos of M102 show that it has a thin dust lane, silhouetted against the bulge, strangely tilted with respect to the axis of the short, bright ansae of the edge-on disc that project from opposite sides of the oval bulge. M102 is estimated to be 48 million light-years away, and therefore has an integrated absolute mag of –20.9, a luminosity of 20 billion Suns.

NGC 5907: Just 85' ENE of M102 is as different-looking a galaxy from it as can be imagined, the long (11$^1/_2$'), very thin (1$^1/_2$') needle of the perfectly edge-on Sc system NGC 5907. This beautiful galaxy is too thin and faint (magnitude 10.3) to be visible with medium binoculars. But in supergiant glasses and RFTs it is a memorable sight: an indescribably aethereal N-S sliver of celestial glow slightly bulged in the middle. With averted vision the true central bulge, though tiny, can be seen much flattened N-S along with the rest of the galaxy, and the long tapering extensions of the disc can be traced to perhaps their full photographic distance. A faint field star is just west of the southern tip of the galaxy. The contrast between this long, thin streak of delicate radiance and the oval of M102, with its sharp, bright nucleus, is astonishing. But the two systems are suspected of being physically related as the lucidae of a small group of galaxies. If they are both exactly 48 million light-years from us, their true separation is just 1.2 million l-y, half the distance between ourselves and the Andromeda Galaxy.

NGC 6503: Located about two-thirds the distance from Chi (χ) to Omega (ω) Draconis in far northeast Draco is the very ragged and loose-armed Scd spiral NGC 6503. An 8th mag star is just to its east. Though this galaxy is faint, only magnitude 10.2, and small 7'×2$^1/_2$', it is surprisingly easy with supergiant glasses. In 25×100 binoculars it appears as a short but bright E-W shaft of light. NGC 6503 is a relatively nearby, 17 million light-years distant, "loner" galaxy. Indeed, there are very few nearby galaxies, and no nearby galaxy groups, in this area of intergalactic space (toward Draco, Hercules, Boötes, Corona Borealis, Ursa Minor) which therefore is called the *Local Void*. The face-on integrated absolute mag of NGC 6503 is around –19.0, so its intrinsic luminosity is about the same as the Triangulum Galaxy M33.

We are in a small cluster of galaxies, the Local Galaxy Group, which lies near the rim of the Local Supercluster. The core of the Local Supercluster is the Coma-Virgo Galaxy Cluster toward southern Coma Berenices and NW Virgo in the spring skies. Thus when we look toward the spring skies–particularly toward Virgo, Coma Berenices, Canes Venatici, and Ursa Major–we are looking directly into the galaxy-dense interior of our Local Supercluster. But when we look toward the autumn skies, we look *out* of our Local Supercluster. The Sculptor Galaxy Group happens to be directly behind us with respect to the Coma-Virgo Cluster: beyond it is the almost galaxy-empty space between superclusters (see Figures 4.1 and 4.2.) Thus when we look toward the constellations around Sculptor we see only a few Local Group galaxies (such as IC 1613 in Cetus and the dwarf spheroidal Fornax System) and a few "loners" of the Local Supercluster (such as M74 in Pisces). At a somewhat greater angle from the Sculptor Group, toward eastern Andromeda and SW Perseus, we see another periphery group of the Local Supercluster, the NGC 1023 Galaxy Group. However, if we look sufficiently hard with sufficiently large binoculars toward the autumn constellations, we can glimpse a few of the brightest galaxies on the rim of the nearest supercluster in that direction, the Southern Supercluster, which includes galaxies from Pisces and Cetus on the north down through Eridanus and Fornax, to the far southern constellations of Horologium, Reticulum, and Dorado. These Southern Supercluster galaxies include NGC 488 and NGC 524 in Pisces and M77 and NGC 1055 in Cetus.

NGC 404: Easily located, but not necessarily easily seen, just 2' NW of the 2nd magnitude Beta (β) Andromedae is the mag 10.8, 4' diameter, E0 or S0 galaxy NGC 404. Despite the glare of Beta And, this system can be discerned with 25×100 supergiant binoculars as a fairly large, circular, high-surface-brightness disc. NGC 404 is something of a puzzle because its recessional velocity is quite low, implying that it must be rather near to us, but its brightest stars cannot be resolved in even the largest telescopes. The galaxy must be at least 26 million light-years away, and therefore has an integrated absolute mag of over −19.5.

NGC 488: One of the fainter galaxies visible in giant binoculars is the Sb spiral NGC 488, located in south central Pisces $2^1/_4°$ WNW of Mu (μ) Piscium and 10' due west of an 8th magnitude field star. Though only a mag 10.3 object, this galaxy is reasonably large, $5^1/_2$'×4', and therefore can be glimpsed in 25×100 supergiant instruments as a tiny pale patch, or hazy dot, just NE of a 10th mag star. Its diffuseness distinguishes it from the 10th and

Looking Out of the Coma-Virgo Supercluster: Distant Autumn Galaxies

Galaxy	Con	RA (2000.0)	Dec	m_V	Dimensions	Type	C. I.	Distance	M_V
N 404	And	01^h09^m	+35°43'	10.1	4.4'×4.1'	E0/SA(s)0		$26×10^6$l-y	−19.5
N 488	Psc	01 22	+05 15	10.3	5.5×4.0	SA(r)b	0.87	72	−21.2
N 524	Psc	01 25	+09 32	10.2	3.5	SA(rs)0$^+$	0.95	(110)	($-22^1/_2$)
N 628 = M74	Psc	01 37	+15 47	9.4	10.2×9.5	SA(s)c	0.56	24	−20.0
N 772	Ari	01 59	+19 01	10.3	7.3×4.6	SA(s)b	0.78	(110)	($-22^1/_2$)
N 891	And	02 23	+42 21	9.9	13.0×2.8	SA(s)b?	0.88	33	−20.1
N 925	Tri	02 27	+33 35	10.1	12.0×7.4	SAB(s)d	0.57	33	−19.9
N 1023	Per	02 40	+39 04	9.3	8.6×4.2	SB(rs)0$^-$	1.00	33	−20.7
N 1055	Cet	02 42	+00 26	10.6	7.3×3.3	Sb:	0.81	53	−20.4
N 1068 = M77	Cet	02 43	−00 01	8.9	8.2×7.3	(R)SA(rs)ab	0.74	53	−22.1
N 6946	Cep	20 35	+60 09	8.8	13	SAB(rs)cd	0.80	~20	−20.2
N 7217	Peg	22 08	+31 22	10.1	3.5×3.0	SA(r)ab	0.90	78	−21.7
N 7331	Peg	22 37	+34 25	9.5	10.5×3.7	SA(s)b	0.87	49	−19.8
N 7606	Aqr	23 19	−08 29	10.8	4.4×2.0	SB:(rs:)b	0.79	54	−20.2

11th mag stars around it, which have sharp-edged, high-light-density images. The distance to NGC 488 is uncertain, the values in the literature ranging from the 72 million light-years cited in the accompanying table up to 154 million l-y. In any case, the galaxy must be a giant spiral on the near edge of the Southern Supercluster.

NGC 524: Another Pisces galaxy right at the limit for supergiant binoculars is NGC 524, a small ($3^1/_2$'), faint (magnitude 10.2) lenticular system. The difficulty in seeing this galaxy is compounded by the difficulty in finding it, because there are absolutely no stars even of the 6th or 7th magnitudes nearby to use as guides: about all the observer can do is scan 4° west and slightly north from Omicron (ο) Piscium and hope for the best. The galaxy is due west of a close E-W pair of 9th mag stars. At 15× NGC 524 is only a blurry dot; but 25× reveals a tiny disc of haze around a stellar core–a typical E or S0 galaxy image. However, this object, though so difficult to find and so unspectacular to see, is worth the bother because it is at least 110 million light-years away (some sources say 152 million) and therefore on the far side of the galaxy void between our Local Supercluster and the Southern Supercluster. Even if "only" 110 million l-y away, NGC 524 would have an absolute mag of $-22^1/_2$, almost equal to the luminosity of the brightest giant ellipticals in the Coma-Virgo Cluster, M49 and M87.

M74 = NGC 628: The one Messier object in the constellation of Pisces is the face-on spiral galaxy M74. This is a fairly bright system, magnitude 9.4, and quite large, 10' in diameter. However, like other face-on Sc galaxies, its surface brightness is very low. Consequently it is not exactly easy to see in 10×50 binoculars, requiring very dark skies, averted vision, and an eye accustomed to recognizing extended, low-surface-brightness objects. And even then it will be only an amorphous patch of pale haze. Fortunately it is conveniently located just $1^1/_2$° ENE of Eta (η) Piscium. M74 is somewhat easier with giant binoculars, though again averted vision is necessary to glimpse the galaxy as a rather large, but very low-surface-brightness, smudge of featureless haze slightly brighter toward its middle that gives the impression (nothing more) of circularity. Photos show it to be a classic two-armed grand-design spiral. M74 is estimated to be about 24 million light-years distant. Its integrated absolute mag is therefore a rather modest –20.0, a luminosity of about 8 billion Suns. Its rather blue color index of +0.56 implies that brisk star-formation is currently occurring along its spiral arms.

NGC 772: The constellation of Aries is not particularly rich in objects for any class of instrument. However, it does have one galaxy within the grasp of supergiant binoculars and small RFTs, NGC 772, located $1^1/_2$° ESE of Gamma (γ) Arietis and $1^1/_4$° NNE of Iota (ι). This Sb spiral is rather faint, only magnitude 10.3, but sufficiently large, $7^1/_2$'×$4^1/_2$', that it is not difficult to spot in 25×100 instruments as an amorphous, moderately high-surface-brightness patch of haze. The galaxy is somewhat peculiar in structure: observatory photos show one long prominent arm lined with H II regions on one side of the spiral's central bulge and several smooth, thin, and faint arms on the other. NGC 772 is extremely remote, but published distances range from as little as 54 million to as much as 170 million light-years. The majority opinion, however, is around 110 million l-y, implying an integrated absolute mag for the galaxy of a very impressive $-22^1/_2$.

NGC 891: The two most-photographed and most-researched edge-on spiral galaxies are NGC 4565 in Coma Berenices and NGC 891 in Andromeda. The most striking feature of these galaxies is the long dust lane that slices the long, thin spindle of their edge-on discs into parallel slivers. The dust lanes require moderate-aperture telescopes (and something like 100×); but both galaxies can be glimpsed in larger binoculars as long, thin needles of pale light. NGC 4565 is marginally possible for 10×50 binoculars. The somewhat fainter NGC 891, however, is only intermittently visible in 15×100 supergiant glasses. But the extra magnification of 25×100s helps considerably and in such glasses (with averted vision) the galaxy is not difficult to spot as a very thin, but fairly high-surface-brightness N-S sliver. NGC 891 is conveniently located $3^1/_2$° due east of Gamma (γ) Andromedae and 1° due north of a conspicuous N-S pair of magnitude 6 and 7 stars. A 7th mag star is just under $^1/_2$° SE of the galaxy. NGC 891 is a member of a 33 million light-years distant galaxy group that includes NGC 925 in Triangulum and NGC 1023 in Perseus. The integrated absolute mag of –20.1 given for the galaxy in the table is not corrected for the heavy obscuration of its dust lane: the system's full face-on absolute mag would be considerably greater.

NGC 925: The "other" galaxy which Triangulum offers the wide-field observer is, like the more famous "Triangulum Galaxy" M33, a loosely-wound, small-bulged spiral seen face-on, NGC 925. Face-on Sc and Sd spirals have very low surface brightness. However, NGC 925 not only is as dim as M33, it is much more distant and therefore *much* smaller, 12'×7$^1/_2$'–about one-sixth the apparent size and one *thirty-sixth* the apparent area of M33. The integrated apparent magnitude of NGC 925 is only 10.1, 4.4 magnitudes–a factor of 64 times–fainter than the nearer spiral. As these numbers suggest, NGC 925 is *not* easy to see: even in 25×100 supergiant binoculars it appears only as an amorphous, very low-surface-brightness, patch of subtle haze. (Keep in mind that you will be looking for something with no more surface brightness than M33 and only 3% of the surface area.) The galaxy is not difficult to locate, fortunately: it is 2° east and slightly south of Gamma (γ) Trianguli, with a 6th mag star just over $^1/_2$° to its ENE. The distance to NGC 925 (hence to its NGC 891 and NGC 1023 companions), 33 million light-years, has been determined by the Cepheid variables observed in it.

NGC 1023: The magnitude 9.3 lenticular galaxy NGC 1023 is the brightest member of a 33 million light-year distant galaxy group that includes NGC 891 and NGC 925 (as well as several fainter NGC galaxies in the region). However, because it is a lenticular system, it is more compact than its spiral companions, measuring just 8$^1/_2$'×4'. Thus even in 25×100 supergiant binoculars NGC 1023 appears only as a tiny hazy spot with a stellar core and is not easy to spot if you do not know exactly where to look. The galaxy is 2° WNW of the 4th mag 16 Persei and 1$^1/_2$° SSW of the 5th mag 12 Per, with one 9th mag star just to its east and another immediately to its SW. (The "hard-edged" quality of these stars can make the comparably-bright, but "softer," galaxy-image easier to identify.) The distance to the NGC 1023 Galaxy Group is based upon the Cepheid variables observed in the group's NGC 925 member. The 9° apparent separation between NGC 891 on the north and NGC 925 on the south, if both galaxies are exactly 33 million l-y from us, corresponds to a true separation of 4.2 million l-y. The integrated absolute mag of NGC 1023, –20.7, is comparable to that of our own Milky Way.

M77 = NGC 1068 and NGC 1055: *Seyfert Galaxies* are spiral systems with unusually bright nuclei that are strong emitters of X-ray and infrared radiation as well as visible light. Seyferts are one class of AGN (*active galactic nuclei*) system. All AGNs, from Seyferts through class M87-type radio galaxies to quasars, are thought to radiate by the same process: the cataclysmic vortexing of dense matter into a massive black hole at the center of the system's nucleus. In effect, then, Seyferts are simply low-luminosity quasars. M77, located 1° SW of Delta (δ) Ceti, is the prototype Seyfert galaxy. Even in binoculars it has an unusual look, 15× glasses showing an exceptionally bright stellar nucleus within a very bright, very compact disc and no evident halo. The bright disc is the galaxy's very luminous inner spiral arms. The integrated apparent magnitude of M77 is 8.9, so it is easily seen even in small-aperture glasses; but the mag 10 field star just 1.5' SE of the galaxy's nucleus requires at least 15× to be resolved. M77 is estimated to be 53 million light-years distant. This might be a bit on the low side, but it still implies an absolute mag of –22.1 for the galaxy, a luminosity approaching 60 billion Suns.

M77 is the lucida of a distant group of galaxies scattered north and east of Delta Ceti. A member of the group that is a challenge object for supergiant binoculars is NGC 1055, an almost edge-on spiral that measures only about 7'×3' and has an apparent mag of just 10.6. In 15×100 instruments it appears only as a mag 10$^1/_2$ dot; but 25×100s enlarge it into an amorphous glow just north of a faint star. However, NGC 1055 is not difficult to locate because it is 40' east and slightly north of Delta Ceti at the southern vertex of the equilateral triangle that the galaxy makes with an 8th mag star just to its NE and a 7th mag star just to its NW.

NGC 6946 (Photo 4.11): One of the most intriguing object-pairs for wide-field instruments is the open cluster NGC 6939 and the nearly face-on Scd galaxy NGC 6946, separated by only 40' NW-SE. They are located 2° SW of Eta (η) Cephei. The cluster is a magnitude 7.8 object but only 7' in diameter. The galaxy is twice as large, 13' across, but 1 mag fainter. Thus the cluster is a moderately high-surface-brightness glow fairly easily seen (if the magnification is at least 10x: 7× is not enough) whereas the galaxy is a very low-surface-brightness smudge of pale haze just visible in 10×50 binoculars–it looks like a face-on, loose-armed Scd galaxy should. With supergiant glasses and small RFTs NGC 6946 appears as a large but amorphous area of very faint glow, slightly brighter toward the interior. On and just outside the southern periphery of the galaxy-glow is a triangle of stars (foreground objects of our own Milky Way), the faintest of which, at the triangle's northern vertex, is "within" the uncertain edge of the spiral's haze. Photos in fact show this star to be superimposed upon the galaxy-interior edge of one of NGC 6946's faint outer spiral arms. The galaxy is only 12° from the galactic equator, so it is dimmed an uncertain amount by our Milky Way's dust. This makes its distance uncertain as well. If it is about 20 million light-years away, its integrated absolute mag, not corrected for dimming by Milky Way dust, is –20.2, a luminosity of 10 billion Suns. NGC 6946 is one of the nearby Local Supercluster galaxies that does not seem to belong to any smaller galaxy group.

NGC 7217: A difficult galaxy even for supergiant binoculars is the face-on Sab spiral NGC 7217 in northern Pegasus. This system is not unreasonably faint, magnitude 10.1, but only 3' in diameter and, because most of its light comes from its bulge

Photo 4.11 The spiral galaxy NGC 6946 in Cepheus with Comet Juels-Holvorcem. Photographed on February 25, 2003 in collaboration with Michael Jäger: a 3-minute exposure with a Deltagraph 300/1000 mm CCD Camera Starlight X-Press HX916 (2×2 binning).

and inner spiral arms, looks merely stellar unless sufficient magnification is used. 25 × is probably near the minimum, and in 25×100 supergiant binoculars the galaxy appears as a small, low-surface-brightness disc with a brighter, but not stellar, core. It is not especially easy to locate either, for it is in an area 2° south and slightly east of the Pi-one (π^1) + Pi-two pair Pegasi with no conspicuous asterisms nearby to use as guides. Observatory photos show that NGC 7217 is a fully flocculent spiral like NGC 2841 in Ursa Major. The galaxy is estimated to be 78 million light-years away, and consequently has an integrated absolute mag of −21.7, twice the luminosity of the Andromeda Spiral.

NGC 7331: The highly-tilted Sb spiral NGC 7331 has rather promising numbers for the wide-field observer: magnitude 9.5, size $10^1/_2$'×$3^1/_2$'. It is, however, surprisingly difficult. For one thing, no bright stars are nearby to help fix its location: the galaxy is $4^1/_2$° NNW of Eta (η) Pegasi, the nearest good "sky-mark" to it being a close NW-SW pair of 6th mag stars $1^1/_4$° straight north. For another thing, like other Sb spirals, most of the light of NGC 7331 comes from its rather dominate bulge and inner spiral disc. This is certainly true of the nearest Sb spiral to us, the Andromeda Galaxy, and photos of NGC 7331 are often used to illustrate how M31 would appear to us from a distance. Thus even with 15×100 supergiant binoculars NGC 7331 is very small and very difficult: its bulge is practically stellar, and its inclined spiral disc mere streaks on either side of the tiny bulge. In 25x100s the galaxy's structural resemblance to M31 becomes more pronounced: in particular the bulge enlarges to a small, flattened, oval of light with a star-point at its center (the galaxy's nucleus). NGC 7331 is estimated to be 49 million light-years away, about 20 times the distance of M31. Its integrated absolute mag *as we see it* is −19.8; but if the galaxy was viewed face-on this value would be nearer the Andromeda Spiral's −21.1. And in fact if NGC 7331 was at the distance of M31, 2.4 million l-y, its apparent magnitude would be 3.1, just about the same as that of M31.

NGC 7606: NGC 7606 is a very faint, magnitude 10.8, and very small, $4^1/_2$'×2', tilted Sb spiral. It is, however, easily located $^3/_4$° NNE of the 4th mag Psi-two (ψ^2) Aquarii and 1° SSE of the 5th mag Chi (χ) Aqr. In 25×100 instruments it is a small, but distinct N-S gash of pale light. NGC 7606 is estimated to be 54 million light-years distant, so its integrated absolute mag (not corrected for its inclination) is −20.2, a luminosity of 10 billion Suns.

CHAPTER 5
STARS, GLOBULARS, PLANETARIES

I Stars

The two star types traditionally telescope targets are variables and doubles. Neither is very well suited to giant binoculars or other richest-field instruments. Most double stars have separations measured in seconds of arc, or even in tenths of a second of arc, and are much too close to be resolved in wide-field, low-power instruments. The theoretical resolving power of binoculars with 100 mm diameter objectives is less than 1.5 second of arc; but in practice a double with components so near each other, even if those components are exactly of the same brightness, requires at least 100× to be resolved.

Wide-field instruments also are not well-designed for variable star observing. To follow the light changes of a variable, the observer uses nearby non-variables for comparison, the nearer to the variable the better. Thus by its very nature variable star observing is a small-field activity. Moreover, a great many variables (particularly LPVs) reach rather faint magnitudes at minimum, and merely to see the variable (to say nothing of nearby non-variables with which to compare it) requires greater light-gathering power than most richest-field instruments can deliver. The modest magnifications of such instruments by themselves prevent those instrument's theoretical magnitude limits from being approached: you need moderate-to-high magnifications to darken the sky background sufficiently to provide contrast for faint-star images.

Thus variable star and double star observing are specialties best pursued with long-focal-length telescopes. Consequently this brief section makes no effort to be comprehensive on even those doubles and multiples within reach of giant binoculars: just the highlights will be covered. And only a handful of variables will be discussed. However, several of the doubles resolvable in wide-field instruments are little short of spectacular.

But there is one aspect of star observing at which binoculars and small RFTs are unexcelled: colors and color-contrasts. The very field of such instruments is one of the reasons for their outstanding performance on star color contrasts: often two or three bright stars of significantly different color will fit in the same wide-field view. And their modest apertures have the advantage that they do not gather so much light that the retina is "overexposed" and its color-sensitivity overwhelmed (particularly in the case of 1st, 2nd, and 3rd magnitude stars). Moreover, color-sensitivity is enhanced when both eyes are involved in the observing (simply because eye-fatigue is reduced). However, star colors are notoriously subjective, different observers routinely reporting slightly, and sometimes significantly, different colors for exactly the same star. Plus color sensitivity not only varies from individual to individual, but can vary even in the same individual from one night to the next (depending upon observer fatigue and other factors). And of course the atmosphere has its effect upon the observed star colors: any increase in dust, haze, or humidity generally reddens star light. Consequently it is best to observe star colors when the stars are at the zenith, or as near to it as possible, thus reducing the amount of atmosphere the star's light must go through to get to you.

In this section priority is given to start colors as the most aesthetically pleasing thing about stars observable in giant binoculars and richest-field telescopes. Next in emphasis are doubles resolvable in such instruments. A couple asterisms are described: but asterisms are essentially unaided-eye objects, so not many fit in the field of view of binoculars of more than 10×. And finally, two or three of the most interesting bright variables will be discussed, though only if they are of interest for some other reason as well–for their color, say, or because they are one component of a wide-field double.

I.1 Capricornus

The Head of Capricornus is such a bright-star-rich field in low-power, wide-field glasses that it almost looks like a true open cluster (though it isn't). Within a 4° area are the magnitude 3.1 Beta (β), mag 3.6 Alpha-two (α^2), mag 4.2 Alpha-one, and Nu (ν), Xi-one (ξ^1), Xi-two and 3 Capricorni. The Alpha pair is a very wide unaided-eye double, its components 376" ~ $6^1/_4$' apart. Beta is a double easily split even with just 7× glasses, the mag 6.2 secondary (designated Beta-one [β^1] because it is the further west of the two stars) being 205" ~ $3^1/_2$' almost due west of the mag 3.1 primary. However, neither the Alpha nor the Beta pairs are true physical binaries; as can be seen in the accompanying table, the components of each double are at considerably different distances from us. Both pairs are beautiful color-contrast doubles in large binoculars and small telescopes: Alpha-one is orange and Alpha-two lemon-yellow; and Beta-one is blue and Beta-two pale yellow.

The Alpha (α) and Beta (β) Capricorni Doubles

Star	m_V	Spectrum	B-V	Distance	M_V
α^1	4.24	G3 Ib	1.07	1600 l-y	−4.5
α^2	3.57	G8 III	0.94	120	+0.2
β^1	6.10	A0 II	−0.02	360	+0.9
β^2	3.08	F8 V + A0	0.19	104	+0.4

About 3° SE of Beta Capricorni is the triangle of Omicron (o), Pi (π), and Rho (ρ) Cap, an attractive 7× binocular group. Omicron, the star at the southern corner of the triangle, is a beautiful 15× double of comparably bright (mag 5.9 and 6.7) stars with A3 V and A7 V spectra. At sufficiently southern latitudes, where these stars can get out of the horizon haze, they should display fine blue-white colors in giant binoculars. Unlike the Alpha and Beta pair, the Omicron double is a true physical binary.

I.2 Cepheus

Two of the most famous stars in the sky are in Cepheus, and both are variables. The "Garnet Star," Mu (μ) Cephei, is a highly-luminous M2 Iae supergiant that varies irregularly between magnitudes 3.5 and 5.0. Its name comes from its color; but that name is misleading (it is an artifact of the sometimes poorly color-corrected optics of the 19th century), because it is more of a ruddy-orange than a purple. The star's ruddiness can be made more evident by contrasting it with the mag 3.34 Zeta (ζ) Cephei 4° to its east, a K1.5 Ib supergiant with a pure orange tone. Mu Cephei is the brightest member of the Cepheus OB2 association, and more about the star is said in the discussion on that association in Section 4 of Chapter 2. Mu Cephei can be seen in color on Plate XXIII.

The other famous variable in Cepheus is Delta (δ) Cephei, the prototype of the high-luminosity, extremely regular Cepheid variables. Delta Cep itself varies between magnitudes 3.48 and 4.34 in a period of 5 days, 8 hours, and 48 minutes. (Follow its light changes by comparing its brightness to that of nearby mag 3.34 Zeta.) Its spectrum varies from F5 Ib at maximum to G2 Ib at minimum. (Cepheids are always redder at minimum than at maximum.) Delta Cep is also a fine double in medium-power binoculars, its mag 6.31 companion, a B7 V star, being 41" almost due south of the primary. (With sufficient aperture the blue-white of the secondary is a beautiful contrast to the yellow of the primary.) The Delta Cep system is about 880 light-years distant, the primary reaching at maximum an absolute mag of about −4. Delta and Zeta Cep are the two brightest members of the scattered, poorly-populated, evolved (50 million-year-old) stellar association Cepheus OB6.

I.3 Cygnus

The famous binary star Albireo, Beta (β) Cygni, has components that are so bright, magnitudes 3.08 and 5.11, and so wide, over 1/2' apart NE-SW, that it can be split even with firmly-held 10×50 binoculars. However, the split is easier, and the color-contrast of the components more evident, with 15× or 25× supergiant glasses. The primary is a spectroscopic binary of K3 II and B9.5 V stars and has a silver-yellow color. (This color betrays the presence of the B9.5 star because K3 giants usually are a deep chrome orange.) The secondary is a sky-blue B9 Ve object. Albireo is 380 light-years distant, so its two stars have absolute magnitudes of −2.2 and 0.0 (luminosities of 625 and 80 Suns.)

Another famous binary in Cygnus is 61 Cygni, the first star to have its distance, 11.2 light-years, determined from a measurement of its parallax (in 1838). The two components of 61 Cygni are magnitude 5.21 and 6.03 stars separated by 29". This is somewhat less than the separation of the Albireo pair, but, because its components are more nearly equal in brightness, 61 Cygni is easier to resolve at 10×. The two stars are K5 V and K7 V dwarfs and both have a magnificent chrome-orange color. They are perfect specimens of the typical K-type main sequence object. Their absolute magnitudes are +7.6 and +8.4, luminosities of 8% and 4% of the Sun's, respectively.

About $6^1/_2°$ WNW of Deneb is a fine color-contrast trio for wide-field instruments, Omicron-one (o^1), Omicron-two, and 30 Cygni. Both the magnitude 3.8 Omicron-one and the mag 4.0 Omicron-two 1° to its north have splendid chrome-orange colors set off by the sky-blue of the mag 4.8 30 Cyg, an A5 II star about $5^1/_2'$ NW of Omicron-one. Both of the Omicrons are Zeta (ζ) Aurigae type eclipsing binaries, variables which pair a K-type luminous giant or supergiant with a blue main sequence or subgiant star: Omicron-one has K2 II and B3 V components, and Omicron-two consists of a K3 Ib supergiant and a B3 main sequence object. The light ranges of the two Omicrons are only a couple tenths of a magnitude. Omicron-one is 520 light-years away and Omicron-two 910 l-y distant. 30 Cyg is much nearer, only 300 l-y from us.

I.4 Draco

What Draco lacks for the wide-field observer in clusters, nebulae, and galaxies it makes up for in star colors and double stars. In the Head of Draco the magnitude 2.2 Gamma (γ) Draconis, a K5 III giant with a fine orange color, is a good color contrast with the 4°-distant, mag 2.8 Beta (β), a G2 Ib-II near-supergiant with a pale yellow tone. Beta is 380 light-years away and has an absolute mag of -2.2 (a luminosity of 625 Suns), and Gamma just 100 l-y distant, with an absolute mag of 0.0 (80 Suns).

In the body of the Dragon, NW of its Head, the blue of the mag 3.1 Zeta (ζ) Dra, a B6 III giant, is an attractive contrast with the yellowish-orange of the mag 2.7 G8 III Eta (η) and the orange of the mag 3.3 K2 III Iota (ι). These three stars are 310, 64, and 140 light-years away, and have absolute mags of -1.8, +0.3, and +0.1, respectively.

Draco also has two splendid low-power double stars with comparably-bright components of beautiful icy-blue colors, Nu (ν) and 16 + 17 Draconis. Nu Dra, in the Head of Draco 3° north of Beta, consists of two mag 4.9 stars with A6 V and A4 spectra 62" apart in PA 312° (SE-NW); and 16 + 17 Dra, 5° WSW of Mu (μ) Dra (the star marking the Dragon's nose), are mag 5.5 and 5.1 stars with B9 V spectra separated by 90" = $1^1/_2'$. Despite their extremely wide separations, both the Nu and the 16 + 17 doubles are true physical binaries. Nu Dra is 62 light-years away and 16 + 17 Dra 330 l-y distant.

I.5 Lyra

Lyra's brightest star, Vega, has a splendid blue-white color best appreciated in medium-aperture binoculars. Unfortunately there is no bright yellow or orange star nearby with which to contrast it. However, only 6° to its NNE is the M5 III semi-regular variable R Lyrae which, though only magnitude 4.0 at maximum, has a fine reddish-orange color that helps enhance Vega's blue.

Lyra has several doubles good for low-power observing. The wide Delta-one (δ^1) and Delta-two are the lucidae of a small open cluster, Ste 1, and are discussed in connection with that cluster in Chapter 2. Epsilon-one (ε^1) and Epsilon-two, the main components of Lyra's famous "Double-Double," are sufficiently wide, 208" almost N-S, and bright, magnitudes 4.4 and 5.1, that they are an unaided-eye pair. Their comparable brightness makes them an attractive binocular double for powers up to about 15×. (Telescopes are necessary to split these two stars into their fainter components.) Epsilon Lyrae is about 180 light-years away, so the true separation between the two stars is at least one-fifth light-year.

But the best richest-field double in Lyra is Zeta (ζ) Lyrae, magnitude 4.36 and 5.73 stars 44" apart in PA 150° (NNW-SSE). At low powers the tightness and modest brightness difference between the two components (the secondary is four times fainter than the primary) gives this double a delicate, even elegant, appearance. The primary is an A-type star with a spectrum peculiarly rich in the lines of metals; and the secondary is an F0 IV subgiant.

I.6 Serpens

One of the most beautiful low-power doubles in the sky is Theta (θ) Serpentis, the star marking the tip of the tail of Serpens the Serpent. Its components are only 22" apart in PA 104° but are nearly equally bright, magnitudes 4.62 and 4.98, and therefore not difficult to resolve even at 15×. The two stars are both A5 V objects with absolutely stunning blue-white or silver-blue colors. Theta Ser is about 100 light-years distant, so its two components have absolute mags of about +2.1 and +2.4, luminosities of 12.5 and 9 Suns.

II Globular Clusters

Though no globular clusters (except the very nearest) are as visually interesting for the wide-field observer as the best open clusters, these objects are not merely "faint fuzzies" even in small instruments: their image betrays something of their actual structure. All but the very smallest and faintest globulars, which appear merely as hazy-edged "stars," display a bright central region–the *core*–enveloped by an extensive, ambiguously-bordered haze–the *halo*. A few globulars possess a tiny starlike *nucleus* within their core. The difference in surface brightness between a globular's core and halo, and the relative size of the core with respect to the halo, are the consequences of how populous and centrally-concentrated the cluster is. However, the most populous clusters are not always the most centrally-concentrated. Omega Centauri, for example, the largest, brightest, and most populous globular cluster in our Galaxy, is only a class VII object.

The scheme of globular cluster *concentration class* is explained in Section II.1 of this book's Introduction. In general, the greater a globular cluster's over-all surface brightness (considering halo and core both), the higher its concentration class. However, the cluster core is not necessarily larger with respect to the halo with higher concentration class, nor does a class I or II globular necessarily have a bright star-like nucleus. Globulars display a true individuality in structure that defies the concentration class scheme and is evident even in small richest-field instruments.

Photo 5.1 Omega (ω) Centauri, our Galaxy's largest and brightest globular cluster

The majority of our Galaxy's globular clusters can be found in the constellations of Sagittarius, Scorpius, and Ophiuchus. This is because those are the three constellations in the direction of our Galaxy's central bulge, and our Galaxy's family of globular clusters is concentrated toward the Galactic Center. Therefore in this section the (rather many) globular clusters visible in wide-field instruments will be divided into four sky areas: Sagittarius, Scorpius, Ophiuchus, and everywhere else. In

Globular Clusters in Sagittarius

Globular	RA (2000.0)	Dec	m_V	Diam	m_V*	Class	Spectrum	E(B-V)	Distance	M_V
N 6522	$18^h03.6^m$	−30°02′	8.4	5.6′	13.7	VI	F7/8	0.50	23,000 l-y	−7.5
N 6528	18 04.8	−30 03	9.5	3.7	15.5	V	G3	0.62	23,500	−6.7
N 6544	18 07.3	−25 00	7.8	8.9	12.8		F9	0.74	8,100	−6.5
N 6553	18 09.3	−25 54	8.1	8.1	14.7	XI	G4	0.78	15,300	−7.7
N 6626=M28	18 24.5	−24 52	6.8	11	12.0	IV	F8	0.41	18,600	−8.3
N 6637=M69	18 31.4	−32 21	7.6	7.1	13.2	V	G2/3	0.17	27,000	−7.5
N 6656=M22	18 36.4	−23 54	5.1	24	10.7	VII	F5	0.34	10,400	−8.5
N 6681=M70	18 43.2	−32 18	8.0	7.8	14.0	V	F5	0.07	28,000	−7.1
N 6715=M54	18 55.1	−30 29	7.6	9.1	15.5	III	F7/8	0.15	85,000	−10.0
N 6723	18 59.6	−36 38	7.2	11.0	12.8	VII	F9	0.03	27,000	−7.8
N 6807=M55	19 40.0	−30 58	6.4	19	11.2	XI	F4	0.07	17,300	−7.5
N 6864=M75	20 06.1	−21 55	8.5	6	14.6	I	F9	0.16	59,000	−8.3

each sky area the globulars will be discussed in order of NGC number.

11.1 Globular Clusters in Sagittarius

NGC 6522 + NGC 6528 are two faint globulars 20' apart E-W just 40' NW of Gamma (γ) Sagittarii. Even in 25x100 supergiant glasses NGC 6522 is only a stellar core surrounded by a very small, low-surface-brightness disc, and NGC 6528 just a star image. However, the two globulars are of interest because they lie in the central bulge of our Galaxy. We see them through *Baade's Window,* the gap in the dust of the inner Galaxy which gives us a view into the bulge at a line of sight that passes only about 1800 light-years from the actual Center of the Galaxy.

NGC 6544 + NGC 6553 are respectively just 50'and 100' SE of the Lagoon Nebula, M8. Their main interest for wide-field observers is in their location, for otherwise they are both small and faint and not very impressive sights even in supergiant binoculars. In 25×100 glasses both are immediately noticeable, but NGC 6544 appears only as a fuzzy, compact, high-surface-brightness disc and NGC 6553 as a tiny hazy spot. NGC 6544 is one of the nearer globulars, but it (as well as the more distant NGC 6553) is heavily-dimmed by Sagittarius-Carina Spiral Arm dust.

M28, located just 0.8° NW of Lambda (λ) Sagittarii, is compact, with a small bright core in a small halo. The core is evenly bright, and lacks a stellar nucleus.

M22, easily found 2.3° NE of Lambda Sgr, is one of the nearest and therefore one of the largest and brightest of the globular clusters and can be seen with the unaided eye (from sufficiently southern latitudes). It is one of the handful of globulars that can be partially resolved with 100 mm aperture, its brightest stars sprinkled over a granular disc. The cluster does not have a true core: it simply brightens gradually from the periphery of its very extensive, ambiguously-edged halo toward its center. M22's integrated absolute magnitude of –8.5 is a little more than 1 mag brighter than the globular cluster average.

M54, M69, and **M70** are in south central Sagittarius, M54 located just over $1^1/_2$° WSW of Zeta (ζ) Sagittarii, M69 almost 2° NE of Epsilon (ε) Sgr, and M70 $2^1/_2$° due east of M69. In supergiant binoculars all three appear as small discs with bright cores. In 15×100 glasses the cores of M69 and M70, the two smallest of the three globulars, appear stellar; but 25×100s show that both clusters simply brighten gradually toward their centers and are in effect coreless. M54, however, displays a true core even at 15×. Astronomically it is the most interesting of the three clusters, for not only is it very distant (85,000 light-years away), and very luminous (its absolute magnitude of –10.0 is only 0.2 mag less than that of Omega [ω] Centauri, our Galaxy's brightest globular), but it is in fact not even a true member of our Galaxy: M54 is a globular cluster of the Sagittarius Dwarf Spheriodal Galaxy, a satellite of the Milky Way which we see through our Galaxy's central bulge. The Sgr Dwarf Spheriodal presently seems to be literally falling into the Milky Way. This event will not be accompanied by many astronomical fireworks, however, because this little system lacks the gas and dust that would accelerate star-formation in the Milky Way's spiral arms and turn it into a *starburst galaxy.* Instead the dwarf galaxy's stars will simply be absorbed into the Milky Way's general halo and thick disc populations.

NGC 6723 (Plate IX) is located on the Sagittarius/Corona Australis border just $^1/_2$° NE of the 5th magnitude Epsilon (ε) Coronae Australis. It is rather far south for mid-northern observers, but so large and so bright that even when only a few degrees above the horizon it readily appears as a circular, moderately-high-surface-brightness disc, slightly brighter toward its center but not concentrated to a core. NGC 6723 is in the same field with the NGC 6726/7 nebulae in the Coronae Australis Dark Cloud Complex.

M55 is the nearest, and consequently the largest and brightest, of the open-structured class

XI globular clusters. Its binocular appearance faithfully reflects its true structure, for even in 10×50 glasses it appears as a large, low-surface-brightness disc of haze that slightly brightens towards its center but lacks any hint of a core. Contrast M55 with the comparably-large, but more densely-structured, class VII globular M22. The brightest stars of M55 are 11th mag objects, so from southern latitudes, where it can be viewed against the dark skies near the zenith, it should partially resolve in supergiant binoculars. It is located in a bright-star-poor area 8° east and slightly south of Zeta (ζ) Sagittarii, but is so large and (reasonably) bright that it is easily seen in a scan.

M75 is a highly-concentrated class I globular cluster and therefore at the opposite structural extreme from the extremely loose class XI M55. And in 15×100 binoculars the cluster certainly looks class I, for it requires an alert look to be seen as a substellar point embedded in a tiny bright halo. And it is as difficult to find as it is to see, because it is in the bright-star-poor wastes of far NE Sagittarius. The cluster is $4^1/_2$° NNE of the conspicuous binocular asterism of the 5th magnitude Omega (ω), 59, 60, and 62 Sagittarii.

II.2 Globular Clusters in Scorpius

M4 (Plates I and II), located just $1^1/_2$° due west of Antares, is one of the two or three nearest globular clusters and consequently appears very large and bright. It is an easy unaided-eye object (on clear, dark nights) even from mid-northern latitudes. In 25×100 supergiant binoculars it is a stunning sight, scores of its stars being resolved upon a very faint background glow. It looks very loose (as indeed a class IX globular should look), with only a very gradual increase in star-density toward its center. Unfortunately, among globular clusters visible from mid-northern latitudes, only M22 in Sagittarius and M4 really give any hint in wide-field instruments of the star-rich beauty of these objects. The integrated absolute magnitude of M4, −7.1, is slightly less than the globular cluster average.

NGC 6144 (Plate II) is only 40' NW of Antares in the same wide-field view with M4. It is, however, so small and so faint that it is visible even in supergiant glasses only as a hazy spot of light. It is a loose class XI cluster, so its surface brightness is very low.

M62 is in a surprisingly bright-star-poor area of the Milky Way 8° SE of Antares and 5° NE of Epsilon (ε) Scorpii. Fortunately it is sufficiently large and bright that it is readily visible in supergiant binoculars or small RFTs with only a scan through the field. The cluster is a compact class IV globular with a very bright core surrounded by a very faint amorphous/circular halo. M62 is noted for its asymmetrical profile; but this requires moderate powers in telescopes to be seen.

NGC 6441 is so near the 3rd magnitude star G Scorpii, which marks the tip of the Stinger of the Scorpion, that higher-power binoculars are desirable to get some separation between the two. At 25x the globular appears as a tight, high-surface-brightness, fuzzy-edged disc that makes an intriguing visual "double" with the orange-toned G Sco. NGC 6441 has a noticeably greater surface brightness than M62 though it is only one concentration class more compact.

Globular Clusters in Scorpius

Globular	RA (2000.0)	Dec	m_V	Diam	m_V*	Class	Spectrum	E(B-V)	Distance	M_V
N 6121=M4	$16^h23.6^m$	−26°32'	5.6	26.3'	11.8	IX	F8	0.36	7,200 l-y	−7.1
N 6144	16 27.3	−26 02	9.0	9.3	13.4	XI	F5/6	0.36	20,500	−6.2
N 6266=M62	17 01.2	−30 07	6.4	14.1	13.0	IV	F9	0.47	22,000	−9.1
N 6441	17 50.2	−37 03	7.1	7.8	15.4	III	G2	0.45	32,000	−9.2

II.3 Globular Clusters in Ophiuchus

M107 is a low-surface-brightness class X globular located about 3° SW of the 3rd magnitude Zeta (ζ) Ophiuchi. Even in large binoculars it is only a small patch of pale haze.

M10 and **M12** are only 3° apart WNW-ESE and therefore fit in the same field of view in most binoculars. They are in the bright-star-poor spaces of central Ophiuchus, but both are magnitude $6^1/_2$ clusters over 10' in diameter and therefore not difficult to spot in binoculars or RFTs by scanning 9 or 10 degrees due east from the Delta (δ) + Epsilon (ε) Ophiuchi star-pair. The class VII M10 has discernably higher surface brightness than the class IX M12 (which, in its turn, has discernably higher surface brightness than the class X M107 12° to its SSW). And the two show a decided structural contrast: M10 contains a rather broad moderate-surface-brightness core within a vague halo, whereas M12's lower-surface-brightness core is weaker, less "self-conscious," straggling imperceptibly into a large halo. The two globulars begin to resolve in 15×100 binoculars: the core of M10 appears granular in texture; and in M12 both core and halo–especially the halo–are partially resolved on a granular background. In M12 an 11th mag star is conspicuous 14' SSE of the cluster center and a mag $10^1/_2$ star is WSW of the center; but many other individual stars twinkle in and out of resolution in the globular's halo. Both M10 and M12 are rather nearby (by globular cluster standards), 14,000 and 15,300 light-years, respectively. Their integrated absolute mags, −7.4 and −7.3, are at the globular cluster average.

M19, though almost as bright and large as M10 and M12, is rather inconspicuously located some 7° due east of Antares about 40% of the distance from Theta (θ) Ophiuchi to Tau (τ) Scorpii. It is not quite 4° due north of the larger and brighter globular M62 in Scorpius. M19 appears as a moderately-bright, fuzzy-edged disc, rather small and distinctly brighter at its center but without a true core. It has a small halo.

NGC 6304 and **NGC 6316** are only about $1^1/_2$° apart SSW-NNE, but are in the peculiarly bright-star-poor Milky Way between the Tail of Scorpius on the south and Theta (θ) Ophiuchi on the north. NGC 6304 is located precisely 3° ENE of M62 and is probably best found from that bright and conspicuous globular. NGC 6316 is $1^1/_2$° due west of a 6th magnitude star. Both globulars appear merely stellar with medium-power binoculars. In 25×100 super-giant glasses the class III NGC 6316 still appears as little more than a hazy spot with a stellar core, but the looser class VI NGC 6304 presents a small disc with a bright central glow and a compact halo. These two globulars are in interesting locations in our Galaxy: the 19,000 light-years distant NGC 6304 must be just on our side of the Milky Way's central bulge, and the 37,000 l-y distant NGC 6316 just beyond the far side of the bulge.

M9 is rather easily found 3° SE of Eta (η) Ophiuchi and 2°slightly west of north of Xi (ξ) Oph.

Globular Clusters in Ophiuchus

Globular	RA (2000.0)	Dec	m_V	Diam	m_V*	Class	Spectrum	E(B-V)	Distance	M_V
N 6171=M107	$16^h32.5^m$	−13°03'	7.9	7.2'	15.1	X	G0	0.33	20,500 l-y	−7.1
N 6218=M12	16 47.2	−01 57	6.7	14.5	12.2	IX	F8	0.19	15,300	−7.3
N 6254=M10	16 57.1	−04 06	6.6	15.1	12.0	VII	F3	0.28	14,000	−7.4
N 6273=M19	17 02.6	−26 16	6.8	13.5	14.0	VIII	F7	0.37	27,700	−9.0
N 6304	17 14.5	−29 28	8.2	6.8	14.2	VI	G3	0.52	19,600	−7.3
N 6316	17 16.6	−28 08	8.4	4.9	16.0	III	G2	0.55	37,500	−8.6
N 6333=M9	17 19.2	−18 31	7.7	9.3	13.5	VIII	F5/6	0.36	27,100	−8.0
N 6356	17 23.6	−17 49	8.2	7.2	15.1	II	G3	0.29	47,600	−8.5
N 6402=M14	17 36.6	−03 15	7.6	11.7	14.0	VIII	F4	0.60	28,400	−9.0

It has a strong central core within a large halo. Its distance implies that it lies within our Galaxy's central bulge.

NGC 6356 is less than 1° ENE of M9. It is a compact, high-surface-brightness class II globular best looked for with higher-power binoculars. Its G3 spectral type marks it as a disc globular and therefore one of the younger members of our Galaxy's family of globular clusters.

M14 is located 3° NE of the 4th magnitude 47 Ophiuchi. At 15× it shows a broad, moderate-surface-brightness core in a compact halo. It is rather heavily dimmed by Great Rift dust (perhaps by nearly 2 magnitudes), but is intrinsically a very bright globular, its integrated absolute mag of –9.0 corresponding to a luminosity of 350,000 Suns.

11.4 Bright Globular Clusters Around the Sky

NGC 288, located 3° NNW of Alpha (α) Sculptoris, is in a very interesting spot almost precisely at the south galactic pole just 2° SE of the bright binocular galaxy NGC 253, lucida of the Sculptor Galaxy Group. The cluster is rather large but, because it is a loose class X globular, has very low surface brightness. Consequently dark skies are essential for spotting it, especially from mid-northern latitudes where it does not ascend very high above the horizon.

NGC 1851, one of the few globulars that is an X-ray source, is in the star-poor spaces of SW Columba just over 8° SW of the 3rd magnitude Alpha (α) Columbae. It is a compact class II cluster; but it is very bright (mag 7.1) and therefore easily seen in 10×50 binoculars by observers in the southern half of the United States. The globular is, however, inconveniently far south for observers in the northern half of the United States, and never rises for much of Europe.

M79 is easily found 4° south and slightly west of the 3rd magnitude Beta (β) Leporis. The globular is just NE of a 5th mag star. M79 is reasonably bright (mag 7.7) but small and therefore requires high-power binoculars.

Bright Globular Clusters Around the Sky

Globular	Con	RA (2000.0)	Dec	m_V	Diam	m_V*	Class	Spec	E(B-V)	Distance	M_V
N 288	Scl	00ʰ52.8ᵐ	–26°35′	8.1	13.8′	12.6	X		0.03	26,400 l-y	–9.1
N 1851	Col	05 14.1	–40 02	7.1	11	13.2	II	F7	0.02	40,000	–8.3
N 1904=M79	Lep	05 24.2	–24 31	7.7	8.7	13.0	V	F5	0.01	41,000	–7.8
N 4590=M68	Hya	12 39.5	–26 45	7.8	12	12.6	X	F2/3	0.04	33,000	–7.3
N 5024=M53	Com	13 12.9	+18 10	7.6	12.6	13.8	V	F6	0.01	60,000	–8.7
N 5272=M3	CVn	13 42.2	+28 23	5.9	16.2	12.7	VI	F6	0.01	33,000	–8.8
N 5466	Boö	14 05.4	+28 32	9.0	11	13.8	XII			54,000	–7.1
N 5897	Lib	15 17.4	–21 01	8.5	12.6	12.6	XI	F7	0.08	41,400	–7.2
N 5904=M5	Ser	15 18.6	+02 05	5.6	12.2	12.4	V	F7	0.03	24,000	–8.8
N 6205=M13	Her	16 41.7	+36 28	5.8	16.6	11.9	V	F6	0.02	23,000	–8.5
N 6341=M92	Her	17 17.1	+43 08	6.4	11.2	12.1	IV	F2	0.02	26,400	–8.1
N 6712	Scu	18 53.1	–08 42	8.1	7.2	13.3	IX	F9	0.46	22,000	–7.4
N 6779=M56	Lyr	19 16.6	+30 11	8.3	7.1	13.0	X	F5	0.20	32,000	–7.3
N 6838=M71	Sag	19 53.8	+18 47	8.2	7.2	12.1		G1	0.25	12,400	–5.5
N 6981=M72	Aqr	20 53.5	–12 32	9.3	5.9	14.2	IX	F7	0.05	55,000	–7.0
N 7078=M15	Peg	21 30.0	+12 10	6.2	12.3	12.6	IV	F3/4	0.09	33,000	–9.1
N 7089=M2	Aqr	21 33.5	–00 49	6.5	12.9	13.1	II	F4	0.05	36,500	–9.0
N 7099=M30	Cap	21 40.2	–23 11	7.2	11	12.1	V	F3	0.03	25,800	–7.4

M68, located in the tail of Hydra $3^1/_2°$ south and slightly east of Beta (β) Corvi and just NE of a 5th magnitude star, is a loose class X globular that appears in 15×100 supergiant binoculars as a rather small, amorphous/circular, moderately low-surface-brightness patch. It has a slightly brighter core within a dim, compact halo. The cluster's integrated spectral type, F2/3, indicates that it is an extremely ancient halo globular.

M53 is easily found about 1° NE of Alpha (α) Comae Berenices. It has an extended, moderately-bright core within a compact halo–an appearance consistent with its concentration class, a moderately-compressed V. The cluster is quite distant, 60,000 light-years, but very luminous, its integrated absolute magnitude of -8.7 being equivalent to the brightness of 230,000 Suns.

M3 is one of the showpiece globular clusters in the northern skies for moderate-aperture telescopes. It is in a rather star-poor area 6° due east of Beta (β) Comae Berenices. It is, however, sufficiently large ($^1/_4°$ in diameter) and bright (magnitude 5.9) that it is easily spotted in 10× binoculars. With 15×100 supergiant glasses it shows a large, bright, dense core within an extensive halo, but without the sharp boundary between the two that is evident in the comparably bright (mag 5.6), similarly-concentrated (class V) M5 in Serpens. M3 has the high integrated absolute mag of -8.8, a luminosity of a quarter-million Suns.

NGC 5466 is an extremely loose-structured class XII globular located 10° NNW of Arcturus. It is, however, best found by scanning $4^1/_2°$ due east from M3 in Canes Venatici. The class XII NGC 5466 and class VI M3 are a splendid structural contrast pair–particularly with supergiant binoculars, in which NGC 5466 appears as a large but *very* low-surface-brightness amorphous/circular mottled patch of haze about equal to M3 in size but indescribably more aethereal in glow.

NGC 5897, 1.7° SE of Iota (ι) Librae, is an extremely low-surface-brightness class XI globular. It is difficult in 10×50 binoculars (particularly from Europe and the northern US, where it does not get very high in the sky), but easy with large glasses, though even in 15×100 supergiant instruments it appears only as an amorphous pale smudge with no sense of central brightening.

M5, at magnitude 5.6 the brightest globular cluster in the northern hemisphere of the sky–the more famous M13 in Hercules is in fact 0.2 mag fainter–is 4° due east of 110 Virginis and just 20'NW of the 5th mag 5 Serpentis. The cluster has an exceptionally bright, compact core embedded in an extensive halo with an almost sharp boundary between the two. The structural contrast between M5 and M13 is remarkable: though M13 is of the same concentration class as M5, class V, and comparably distant, 23,000 light-years, its star-density decreases much more gradually and evenly from its center so there is no distinct boundary between its core and halo. An 8th mag star is conspicuous in the halo of M5 just SW of its core. This is just a foreground object, not a true cluster member; but in 100 mm glasses both the core and halo (especially the latter) of M5 are granular from partial resolution of the globular's brightest stars.

M13 technically is only the second brightest globular in the northern celestial hemisphere, for it is 0.2 magnitude fainter than M5 in Serpens. However for mid-northern observers it can be observed near the zenith, where the sky background is darkest, and it therefore makes a more striking impression in the eyepiece than its more southerly rival. The cluster is easily found 2° south of Eta (η) Herculis about one-third the distance from Eta to Zeta (ζ) Her, and under exceptional sky conditions is visible to the unaided eye. With supergiant binoculars it has a moderately high-surface-brightness core within a fairly compact halo; but the two blend together imperceptibly. With 100 mm glasses there is some granularity in the globular's image from the partial resolution of its brightest stars.

M92, Hercules' "other" globular cluster, is quite bright (magnitude 6.4), but not so well known as M13 because it is a more compact class IV object not so easily-resolved and because it is in an isolated, star-poor area 4° SW of Iota (ι) Herculis. In 7x binoculars M92 is only a mag $6^1/_2$ "star." With supergiant glasses it displays an extremely bright, compact, disc-like core within an extensive, low-surface-brightness halo: the contrast between the core and halo surface brightness is exceptionally striking, even for a globular. M92 is an extremely metal-poor, and therefore presumably an extremely ancient, cluster.

NGC 6712 is a rather faint (magnitude 8.1) low-surface-brightness class IX globular that has an interesting location in a SE extension of the Scutum Star Cloud $2^1/_2$° south of the bright, compact open cluster M11. Though the Star Cloud's background glow here is not as bright as it is toward the NW, it is rich with glittering partial resolution, and more brighter field stars are scattered across it. This is a fine setting for the globular, which in 15×100 supergiant binoculars appears as a small hazy disc with a distinct halo around a bright, very tiny, but non-stellar core.

M56 is 4° NW of the famous double star Albireo (Beta [β] Cygni) and just SE of a 6th magnitude star. It is a loose class X globular that in large binoculars and small RFTs shows a small core, tight but not of very high surface brightness, within a small, pale, amorphous halo. A 10th mag field star is conspicuous near the cluster's west edge.

M71, though only 12,400 light-years distant and therefore one of the nearer globular clusters, has an integrated absolute magnitude of a mere −5.5 (a luminosity of just 33,000 Suns, and almost 2 mags below the globular cluster average) and consequently an integrated apparent mag of only 8.3. It is not particularly large either, but is in an easily-located spot just south of the midpoint of the line joining Gamma (γ) and Delta (δ) Sagittae. In 10×50 binoculars M71 is only a fuzzy disc. In 100 mm supergiant glasses, however, it begins to resolve into stars, appearing as a small, pale, heavily-mottled disc, brighter in its core, which is granular from partial resolution. An 11th mag star is conspicuous just south of the core, and another is on the extreme NW edge of the cluster.

M72 is a very faint (magnitude 9.3), small, low-surface-brightness (class IX) globular in extreme western Aquarius about $4^1/_2$° WSW of Nu (ν) Aquarii and 3.4° SSE of Epsilon (ε) Aqr. With 15×100 supergiant binoculars it appears as a small hazy patch, a little brighter in the its core but throughout of low surface brightness. A 9th mag field star just 5′ to the ESE of the cluster is very conspicuous compared to the globular's feeble glow. M72 looks faint and small both because it is very distant, 55,000 light-years away, and because it is below average in globular cluster luminosity with an integrated absolute magnitude of just −7.0.

M15, 4° WNW of Epsilon (ε) Pegasi, is similar in brightness, size, and structure to M2 in Aquarius, which is exactly 13° almost due south. Both globulars are compact (M15 is class IV, M2 class II), with dense, high-surface-brightness cores enveloped by rapidly-fading, ambiguously-bordered, low-surface-brightness halos. The most striking difference in appearance between the two is that M15 has a truly stellar nucleus in its core: it is a *core-collapse* globular in which the star density continues to rise toward the cluster center rather than to level off (as core star-density usually does in globular cluster nuclei). M15 is also unusual for containing a planetary nebula, a magnitude 13.8, 1″ diameter object catalogued as Pease 1. The true sizes and brightness of the vast majority of planetaries are very uncertain; but because of its membership in M15, the distance of which (33,000 light-years) is well determined, Pease 1 must have an absolute mag of −1.3 and a true diameter of one-fifth light-year.

M2 is located 5° north of Beta (β) Aquarii in a very star-poor area. However, it is so bright (magnitude 6.5) and large that it is readily spotted in a scan of the region with high-power binoculars. The globular resembles M15 to its north in appearance (except for the latter's stellar nucleus) and the two clusters are in fact of similar structural classes, are at comparable distances from us, and have just about the same integrated absolute magnitude (−9.0, a luminosity of 350,000 Suns). Both are also very old metal-poor F4 clusters.

M30 is a moderately faint (magnitude 7.2) globular cluster located 3° ESE of the 4th mag Zeta (ζ) Capricorni and just west of the 5th mag 41 Cap. The glare of 41 Cap makes the cluster difficult to glimpse in 10× glasses. With supergiant binoculars it shows a small, moderately-bright halo around a strong core–an appearance consistent with its concentration class, a moderately-compressed V. M30 is a very metal poor, very ancient, F3 halo globular.

III Planetary Nebulae

Though well over one thousand planetary nebulae have been identified in our Galaxy, almost all of them are extremely small and faint. This is not surprising: planetary nebulae are the expanding outer envelopes of Solar-mass stars, hence are intrinsically small. (Planetary nebula are put in context in stellar evolution in Sect. I.4 of Chap. 1) The few planetary nebulae than can be seen in wide-field instruments are necessarily relatively close to the Sun. Only a handful of planetary nebulae are sufficiently large (and bright) to be visible as hazy discs or patches in even high-power (20×–25×) supergiant binoculars. Nevertheless, these planetaries are very interesting visually.

Planetary nebulae, like globular clusters, are not Milky Way objects: they are found in every direction in the sky. Moreover, like globulars planetaries are more abundant toward the Galactic interior and therefore more of them are to be found in such constellations as Sagittarius and Aquila rather than in the opposite side of the Milky Way toward Auriga or Monoceros. But otherwise the over-all distribution of planetary nebulae in our Galaxy is very different than the distribution of globular clusters. The number density of globulars increases much more steeply toward the Galactic bulge than the number density of planetaries; and globulars are distributed in a huge spherical halo around the Galactic Center whereas planetaries are distributed in a thick disc that encloses the thin disc and the spiral arms. In the region of the south galactic pole in Sculptor, a direction perpendicular to the plane of the Milky Way, we see both the globular cluster NGC 288 and the planetary nebulae NGC 246, NGC 1360, and NGC 7293 (in Cetus, Fornax, and Aquarius, respectively, constellations adjoining Sculptor); but the globular is 26,400 light-years distant out in the Galaxy's halo whereas the planetaries–which are among the few large enough to be seen as discs in 10×50 binoculars–are within several hundred light-years of us in the Galaxy's thick disc. The difference in distribution between globular clusters and planetary nebulae reflects a difference in age: globulars are part of the 10–15 billion year-old halo population of our Galaxy while most of the stars that have produced the planetary nebulae we presently see were only a few billion years old, at most.

The accompanying table lists the brightest planetary nebulae, down to apparent magnitude 10.2 (in order of Messier number or NGC/IC designation), plus the faint but large NGC 246, a medium binocular object. The m_V* column gives the apparent magnitude of the stellar remnant at the center of the expanding planetary. The halos mentioned in the "Structure" column are usually strictly photographic features. The only planetaries visible as discs in 10×50 binoculars are M27, NGC 246, NGC 1360, and NGC 7293. Most of the other planetaries in the table can be glimpsed with 10×50s, but will appear only as "stars" and require detailed finder charts to be identified. Surprisingly few additional planetaries can be seen as discs in 25×100 supergiant binoculars–25× is still too little for good planetary searching. However, the greater light-gathering power of the 100 mm diameter objects makes visible the large, but extremely low-surface-brightness M76 and M97.

III.1 The Brightest Planetary Nebulae

M27, the Dumbbell Nebula (Photo 5.2): M27 located 3.3° due north of Gamma (γ) Sagittae, has an apparent magnitude of 7.4 and an apparent size of 8'×4' and therefore is the second brightest and largest of the planetary nebulae (after NGC 7293, the Helical Nebula in Aquarius). However, even in 10×50 binoculars it looks so small that on a casual scan through the central Vulpecula Milky Way it might be mistaken for one of the numerous 7th and 8th mag field stars richly strewn over this region. (The nebula is in fact just SSE of a 6th mag star.) However, in 15×100 supergiant glasses it is no problem at all: very large, very bright, and with a very intriguing shape. It displays an over-all square profile within which is a brighter rectangular glow, the long axis of which is oriented NNE-SSW. The NNE and SSW sides of the square are very distinct; but the other edges fade out more ambiguously from the central glow. In 25×100s the core hourglass shape of M27 is evident, the NNE and SSW sides sharply-defined and almost straight and the other edges ambiguously-bordered, fading gradually outward from the incurved sides of the hourglass.

Photo 5.2 The Dumbbell planetary nebula, M27 in Vulpecula, and the comet Hale-Bopp. North is to the left and east is up. Photographed on February 10, 1997

The Brightest Planetary Nebulae

Nebula	Name	Constel	RA (2000.0)	Dec	m_V	Size	m_V*	Structure
M27=N6853	Dumbbell	Vul	$19^h59.6^m$	+24°43'	7.4	480"×240"	13.8	Bipolar in halo
M57=N6720	Ring	Lyr	18 53.6	+33 02	8.8	85×62	14.7	Ring in halo
M76=N650-1		Per	01 42.4	+51 34	10.1	90×45	15.9	Bipolar in halo
M97=N3587	Owl	UMa	11 14.8	+55 01	9.9	194	13.2	Disc
N 246		Cet	00 47.0	−11 53	10.9	225	11.9	Disc
N 1360		For	03 33.3	−25 51	9.4	460×320	11.0	Disc
N 1535		Eri	04 14.2	−12 44	9.6	20×17	11.6	Disc in halo
N 2392	Eskimo	Gem	07 29.2	+20 55	9.1	47×43	10.5	Double ring
N 2440		Pup	07 41.9	−18 12	9.4	54×20	18.9	Disc in halo
N 2867		Car	09 21.4	−58 19	9.7	11	15.0	
N 3132	Eight-Burst	Vel	10 07.0	−42 27	9.2	>47	10.1	Multiple ring
N 3242		Hya	10 24.8	−18 38	7.7	40×35	12.1	Disc in halo
N 3918		Cen	11 50.3	−57 11	8.1	12	13.2	Disc
N 5315		Cir	13 53.9	−66 31	9.8	5	14.3	
N 5882		Lup	15 16.8	−45 39	9.4	7	13.6	Disc
N 6210		Her	16 44.5	+23 48	8.8	20×13	13.7	Disc
N 6302	Bug	Sco	17 13.7	−37 06	9.6	90×20	16.0	Bipolar
N 6543	Cat's Eye	Dra	17 58.6	+66 38	8.1	23×18	10.9	Disc in halo
N 6572		Oph	18 12.1	+06 51	8.1	14×9	12.9	Disc
N 6818		Sgr	19 44.0	−14 09	9.3	22×15	15	Disc
N 6826		Cyg	19 44.8	+50 31	8.8	27×24	10.6	Disc in halo
N 7009	Saturn	Aqr	21 04.2	−11 22	8.0	39×30	11.5	Disc with ansae
N 7027		Cyg	21 07.1	+42 14	8.5	18×11	16.3	Disc
N 7293	Helical	Aqr	22 29.6	−20 47	7.3	900×720	13.6	Ring in halo
N 7662		And	23 25.9	+42 32	8.3	17×14	12.5	Ring in disc
IC 418		Lep	05 27.5	−12 42	9.3	14×11	10.2	Disc
IC 4406		Lup	14 22.4	−44 09	10.2	100×37	14.7	Bipolar

(From Hynes, *Planetary Nebulae*, and Dave Krantz, "Observing Planetary Nebulae," *Deep Sky* 6, 12.)

M27's surface brightness is very high, and its shape makes it one of the most striking sights in the sky.

M57, the Ring Nebula: The famous Ring Nebula, M57, is just a little too small, approximately 1½'×1', to be seen as a disc in 10×50 binoculars. However, its mag 8.8 star-like image is easily identified if you know where to look: the nebula is about 40% of the distance from Beta (β) to Gamma (γ) Lyrae, and 7' SSE and 7' WSW of 9th mag field stars. (The isoceles triangle formed by these two stars and the nebula is fairly conspicuous.) But even at just 15× M57 is no longer stellar; and in 15×100 super-giant glasses it appears as a tiny, but unmistakable, moderate-surface-brightness disc. With 25×100 instruments the fact that M57 is sharp-edged, rather than fuzzy-edged like globular clusters or galaxies, becomes more obvious. The famous "hole" in the Ring can be seen with 100 mm aperture, but requires something like 100×.

M76: M76, in extreme NW Perseus about 1° NNW of Phi (φ) Persei, is just about the same size as the Ring Nebula, 1½'×1', but 1.3 magnitudes, or nearly 4 times, fainter. Thus its surface brightness is only one-fourth as much. Nevertheless it is immediately visible in 25×100 glasses as a small, moderately-bright, disc of haze. However, 25× is too little for the nebula's strongly elongated shape to be evident. The nebula is so elongated, in fact, that its two lobes each have their own NGC number, NGC 650 and NGC 651. M76 is an example of a *bipolar* planetary nebula in which the outer layers of the original red giant were ejected by some force, possibly magnetic, in opposite directions. Many planetaries, however, are true toruses–"donuts." When a torus-shaped planetary is seen edge-on it has a strong box-like profile. When it is seen face-on, it is a ring like the famous M57. The torus no doubt was ejected by centrifugal force from the red giant's equatorial zone as the star expanded and the radiation pressure on its envelope from the interior increased. Thus the shape of a planetary nebula reflects the conditions under which the dying red giant puffed it off.

M97, the Owl Nebula (Photo 5.3): The Owl Nebula is very conveniently located about $2^1/_2$° ESE of Beta (β) Ursae Majoris, the star at the SW corner of the "bowl" of the Big Dipper. A 6th magnitude star is just west of the nebula (and the spiral galaxy M108 more-or-less between the nebula and Beta UMa). However, M97 has an integrated apparent mag of just 9.9 but is fully $3^1/_4$' in diameter and therefore of very low surface brightness. It is too faint for 10×50 binoculars, but is visible in 15×100s as a large circular patch of very pale haze, just slightly brighter toward the center. Its relatively distinct edges, and its lack of any real central brightening, distinguishes the planetary's image from that of such face-on Sc spirals as M101, also in Ursa Major, which always have a central brightening, if not a true stellar nucleus, and lack any hint of borders though they appear (and actually are) circular or elliptical in outline. The darker ovals within the nebula, and from which it has gotten its name because they suggest owl eyes, have too low contrast with the nebula-glow to be seen.

NGC 246: The most difficult of the four planetary nebulae that display discernible discs in 10×50 binoculars is NGC 246 in NW Cetus. (The other three planetaries which show discs in 10×50 instruments are M27, the Dumbbell Nebula in Vulpecula, NGC 7793, the Helical Nebula in Aquarius, and NGC 1360 in Fornax.) NGC 246 is small (4'×$3^1/_2$') and extremely faint, just magnitude 10.9, but conveniently located at the southern vertex of an equilateral triangle with the W-E pair Phi-one (φ^1) and Phi-two Ceti. In 10×50 binoculars the nebula appears only as a small, pale, amorphous, low-surface-brightness smudge. In 25×100 glasses it is easier, but still just a rather small, very faint, patch or disc of haze. A triangle of three faint stars can be discerned embedded in its haze, one of which is the planetary's 12th mag central star.

NGC 3242: The third brightest planetary nebula after magnitude 7.3 NGC 7793 in Aquarius and mag 7.4 M27 in Vulpecula is the mag 7.7 NGC 3242 in central Hydra. Unfortunately, NGC 3242 is little more than $^1/_2$' in diameter and consequently appears stellar even in 15×100 supergiant binoculars. Though it does not reveal itself by even the slightest fuzziness around its stellar image (NGC 3242, like most planetaries, is a rather sharply-bordered object), it is not difficult to identify because it is about 1.8° almost due south of the 4th mag Nu (ν) Hydrae, with a 7th mag star just a few minutes to its east. The 7th mag star is a little brighter than the nebula, and at least 10× is necessary to get NGC 3242 away from the glare of the star.

NGC 7009, the Saturn Nebula: The fourth brightest planetary after NGC 7293, M27, and the just-described NGC 3242 is the magnitude 8.0 NGC 7009, the Saturn Nebula, in far western Aquarius. Like NGC 3242, NGC 7009 is only a little more than $^1/_2$' in diameter and therefore does not display a disc even in 25×100 supergiant binoculars. Fortunately the Saturn Nebula also resembles NGC 3242 in being at an easily-identified location: it is almost $1^1/_2$° due west of the 4th mag Nu (ν) Aquarii, with a 7th mag field star 40' to its SW. In 25×100s the nebula begins to lose the "hard-edged" character of a true star image. This perhaps is because of the two thin ansae that project $^1/_4$' out from either side of the planetary's disc, and from which it got its name since they suggest the rings of Saturn seen nearly edge-on.

NGC 7293, the Helical Nebula: The Helical Nebula in Aquarius is the brightest (magnitude 7.3)

Photo 5.3 The Owl Nebula, M97 in Ursa Major (lower right corner), with the galaxy M108 and Comet Tabur. North is to the left and east is up. The bright star near the top edge of the field is Beta (β) Ursae Majoris. Photographed on October 13, 1997

and far and away the largest (15'×12') of the planetary nebula. It is also an *extremely* low-surface-brightness object: it spreads the light of a mag 7.3 star out over an area one-fourth that of the full Moon! However, such large, low-surface-brightness objects were made for binoculars, and the ghostly glow of NGC 7293 is not at all difficult in 10×50 binoculars from dark sky sites (and when it is near the meridian, for its declination is –21° and it therefore never rises very far out of the horizon haze for mid-northern observers). It is a bit of a challenge to find, though, because it is in the star-poor spaces of far SW Aquarius. To locate it, "star-hop" from the mag 3.2 Delta (δ) Aquarii: first 4° SW to the 5th mag 66 and 68 Aqr, then another 3° SW to the mag $5^1/_2$ Upsilon (υ) Aqr, and then 1° west to the nebula. In 25×100 supergiant glasses the Helical Nebula appears very large and irregularly circular, with patches of darkness intermittently visible in its central area, particularly toward the NW. The patches are obviously glimpses of the nebula's famous central "hole."

Sources

The purpose of this bibliography is to cite the sources for the astronomical data quoted in this book. Because the most up-to-date data is usually from the most recent observational studies of any given object or type of object, the majority of sources cited here are recently-published articles from the standard technical journals. However, at the beginning of this bibliography is a list of technical and amateur books from which information has also been gleaned.

The technical journal articles cited are first divided according to type of object and follow the chapter divisions of this book: Chapter 2: Open Clusters; Chapter 3: Stellar Associations and Bright Nebulae along the Milky Way; and Chapter 4: Galaxies. Within each of these chapter divisions the articles are further divided into two groups: A., those that discuss individual objects; and B., those that discuss several, or many, of the type of object in question. Articles discussing individual open clusters, stellar association, and bright nebulae are listed by those objects' west-to-east order around the Milky Way (in other words, by galactic longitude), the object designation being printed in bold type for quick reference. Articles discussing individual galaxies are listed by galaxy Messier or NGC designation, also printed in bold type. Those articles dealing in a general way with each chapter's type of object–open clusters, stellar associations, bright nebulae, and galaxies–are listed in the standard author-date format. Abbreviations of the technical journal names are:

AA	*Astronomy and Astrophysics*
AAS	*Astronomy and Astrophysics Supplement*
AJ	*Astronomical Journal*
AJS	*Astronomical Journal Supplement*
ApJ	*Astrophysical Journal*
ApJS	*Astrophysical Journal Supplement*
MNRAS	*Monthly Notices of the Royal Astronomical Society*
PASP	*Publications of the Astronomical Society of the Pacific*

GENERAL

Binney J. and Merrifield M. (1998) Galactic Astronomy. Princeton University Press.

Sandage A. and Bedke J. (1994) The Carnegie Atlas of Galaxies. Carnegie Institution, Washington.

Hynes S. J. (1991) Planetary Nebulae. Richmond.

Tirion W., Rappaport B., Lovi G. (1992) *The* Deep Sky Field Guide to Uranometria 2000.0. Richmond.

Buser R. and King I. R., Eds. (1990) The Milky Way as a Galaxy. University Science Books, Mill Valley.

Huchtmeier W. K. and Richter O.-G. (1989) A General Catalogue of HI Observations of Galaxies. Springer.

Wray J. D. (1988) A Color Atlas of Galaxies. Cambridge.

Binney J. and Tremaine S. (1987) Galactic Dynamics. Princeton University Press.

Verschuur G. L. and Kellermann K. I., Eds. (1987) Galactic and Extragalactic Radio Astronomy. Springer; 2nd edition.

Tully R. B. and Fisher R. (1987) Nearby Galaxies Atlas. Cambridge.

Mahalas D. and Binney J. (1981) Galactic Astronomy. San Francisco; 2nd edition.

Burnham R., Jr. (1978) Burnham's Celestial Handbook. Dover, New York; 3 volumes.

CHAPTER 2: OPEN CLUSTERS

A. Individual Clusters; Areas of Clusters

NGC 6603: A young rich open cluster towards the bulge. (E. Bica, S. Ortolani, B. Barby) *AA* 270: 117–121. (1993).

NGC 6611 [M11]: A cluster caught in the act. (L. A. Hillenbrand, P. Massey, S. E. Strom, K. M. Merrill) *AJ* 106: 1906–1945 (1993).

The structure and dynamics of the open cluster **M11**. (R. D. Mathieu) *ApJ* 284: 643–662 (1984).

The open cluster **IC 4665**. (C. F. Prosser) *AJ* 105: 1441–1454 (1993).

An observational test of the spherical model atmospheres for the class M giants: The case of δ^2 Lyrae [**Stephenson 1**]. (J. J. Sudol, J. A. Benson, H. M. Dyck, M. Scholz) *AJ* 214: 3370–3378 (2002).

The age, extinction, and distance of the old, metal-rich open cluster **NGC 6791**. (B. Chaboyer, E. M. Green, J. Liebert) *AJ* 117: 1360–1374 (1999).

Characteristics of new star cluster candidates in the Cygnus area [**NGC 6910**]. (J.-M. Le Duigou, J. Knödlseder) *AA* 392: 869–884 (2002).

Spectroscopic survey of the galactic open cluster **NGC 6871**. I. New emission-line stars. (Z. Balog, S. J. Kenyon) *AJ* 124: 2083–2092 (2002).

uvby and Hβ photometry of the open cluster **IC 4996**. (E. J. Alfaro, A. J. Delgado, J. M. García-Pelayo, R. Garrido, M. Sáez) *AAS 59*: 441–447 (1985).

Strömgren and Hβ photometry of early type stars in northern open clusters. I. **NGC 7039, NGC 7063**. (H. Schneider) *AAS* 67: 545–550 (1987).

Spectral variability of luminous early type stars. I. Peculiar supergiant HD 199478 [central star of **IC 5076**]. (N. Markova, T. Valchev) *AA* 363: 995–1004 (2000).

CCD observations of the open cluster **NGC 6939**. (J. M. Rosvik, D. Balam) *AJ* 124: 2093–2099 (2002).

CCD photometry of the old open cluster **NGC 7142**. (G. Crinklar, F. D. Talbert) *PASP* 103: 536–545 (1991).

A spectroscopic and photometric investigation of **NGC 7243**. (G. Hill, J. V. Barnes) *AJ* 76: 110ff (1971).

UBV three-colour photometric parameters of four galactic clusters near Cassiopeia [K 19, **NGC 7510**, Mark 50, M52]. (R. P. Fenkart, A. Schröder) *AAS* 59: 83–86 (1985).

NGC 7654 [M52]: An interesting cluster to study for star formation history. (A. K. Pandy, Nilakshi, K. Ogura, R. Sayar, K. Tarusawa) *AA* 374: 504–522 (2001).

Photometric photometry of the open cluster **NGC 7790**. (M. Pederos, B. F. Madore, W. Freeman) *ApJ* 286: 563–572 (1985).

Interstellar extinction in the open clusters toward galactic longitude around 130° [Delta Cassiopeia region]. (A. K. Pandey, K. Upadhyay, Y. Nakada, K. Ogura) *AA* 397: 191–200 (2003).

Young open clusters as probes of the star-formation process. II. Mass and luminosity functions of young open clusters. [**M103, NGC 663**, and other clusters in the Delta Cas region] (R. L. Phelps, K. A. Janes) *AJ* 106: 1870-1884 (1993).

The young open cluster **NGC 663** and its be stars. (A. Pigulski, G. Kopacki, Z. Kolaczkowski) *AA* 376: 144–153 (2001).

CCD uvbyβ photometry of young open clusters. I. The double cluster **h & χ Persei**. (G. Capilla, J. Fabregat) *AA* 394: 479–488 (2002).

Photometric study of the double cluster **η & χ Persei** (A. Marco, G. Bernabeu) AA 372: 477–494 (2001).

Multicolor photometry of the galactic cluster **NGC 1039 (M34)**. (R. Contera, G. L. Perry, D. L. Crawford) *PASP* 91: 263–270 (1979).

Photoelectric search for CP2-stars in open clusters. **Alpha Persei, Praesepe,** and **NGC 7243**. (H. M. Maitzen, K. Pavlovski) *AAS* 71: 441–448 (1987).

A VBLUW photometric survey of the **Pleiades** cluster. (F. van Leeuwen, P. Alphenaar, J Brand) *AAS* 65: 309–347 (1986).

Spectral classification and photometry of selected **Pleiades** stars. (M. Breger) *AAS* 57: 217f. (1984).

NGC 1912 [M38] and **NGC 1907**: A close encounter between open clusters? (M. R. de Oliveira, A. Fausti, E. Bica, H. Dutteri) *AA* 390: 103–108 (2002).

Multicolor CCD photometry and stellar evolutionary analysis of **NGC 1907, NGC 1912 [M38]**, NGC 2383, NGC 2384, and NGC 6709 using synthetic color-magnitude diagrams. (A. Subramanian, R. Sayar) *AJ* 117: 937–961 (1999).

CCD Strömgren *uvby* photometry of the young clusters **NGC 1893, NGC 457, Berkeley 94,** and Bochum 1. (A. Fitzsimmons) *AAS* 99: 15–29 (1993).

The intermediate-age open cluster **NGC 2158**. (G. Carraro, L. Girardi, P. Marigo) *MNRAS* 332: 705–713 (2002).

Red giants in open clusters. IX. **NGC 2324**, 2818, 3960, 6259. (J.-C. Mermilliod, J. J. Clariá, J. Andersen, A. E. Piatti, M. Mayer) *AA* 375: 30–39 (2001).

Strömgren photometry of open clusters. III. **NGC 2323 [M50]**, NGC 5662. (H. Schneider) *AAS* 71: 531–537 (1987).

Photometric study of the open cluster **NGC 2353**. (J. J. Clariá, A. E. Piatti, E. Lapassat) AA: 128: 131–138 (1998).

The intermediate-age open cluster **Mel 71**. (M. W. Pound, K. A. Janes) *PASP* 98: 210ff. (1986).

Chemical abundances in seven red giants of **NGC 2360** and **NGC 2447 [M47]**. (S. Hamdani, P. North, N. Mowlani, D. Robaud, J.-C. Mermilliod) *AA* 360: 509–519 (2000).

NGC 2287 **[M41]**: An important intermediate-age open cluster. (G. L. H. Harris, M. P. V. FitzGerald, S. Menta, B. C. Reed) *AJ* 106: 1533–1545 (1993).

UBV photometric study and basic parameters of the southern open cluster **NGC 2539**. (E. Lapassat, J. J. Clariá, J.-C. Mermilliod) *AA* 361: 945–951 (2000).

CCD observations in 7 open clusters [NGC 2421, NGC 2439, NGC 2489, **NGC 2567, NGC 2627**, NGC 2658, NGC 2910]. (G. Ramsey, D. L. Pollacco) *AAS* 94: 73–102 (1992).

UBVRI observations of stars in the fields of five open clusters with nearby carbon stars [**NGC 2533**, IC 2395]. (U. G. Jørgensen, B. E. Westerlund) *AAS* 72: 193–208 (1988).

CCD photometric search for peculiar stars in open clusters. II. NGC 2489, **NGC 2567**, NGC 2658, NGC 5281, and NGC 6208. (E. Paunzen, H. M. Maitzen) *AA* 373: 153–158 (2001).

Radial velocities in the region of **Cr 135**. (G. Amieux) *PASP* 105: 926–931 (1993).

Six clusters in Puppis-Vela [**NGC 2451**, IC 2391, **Cr 132**, **Cr 135**, **Cr 140**, Cr 173]. (O. J. Eggen) *AJ* 88: 197ff (1983).

WIYN open cluster study. VII. **NGC 2451A** and the *Hipparcos* distance scale. (I. Platais et al.) *AJ* 122: 1486–1499 (2001).

Photoelectric search for CP2-stars in open clusters. VIII. IC 2391 and **NGC 2451** (H. M. Maitzen, F. A. Catalano) *AAS* 66: 37–44 (1986).

High-mass binaries in the very young open cluster **NGC 6231**. (B. García, J.-C. Mermilliod) *AA* 368: 122–136 (2001).

UBV photometry of **Tr 24** and its relation to **Sco OB1**. (A. Heske, J. J. Wendker) *AAS* 57: 205–212 (1984).

A study of the open cluster **NGC 6475 (M7)**. (L. O. Lodén) *AAS* 58: 595–599 (1984).

A high angular resolution multiplicity survey of the open clusters **α Persei** and **Praesepe [M44]**. (J. Patiana, A. M. Ghez, I. N. Reid, K. Matthews) *AJ* 123: 1570–1602 (2002).

The initial mass function of the **Coma Berenices Open Cluster (Mel 111)**. (L. Buonatino, N. Arimoto) *AA* 268: 829f. (1993).

B. General

Barbaro, G., Pigatto, L. (1984) Red giants in old open clusters: A test for stellar evolution. *AA* 136: 355–362.

Durgapal, A. K., Pandey, A. K. (2001) Structure and mass function of five intermediate/old open clusters. *AA* 375: 840–850.

Friel, E. D., Janes, K. A. et al. (2002) Metallicities of old open clusters. *AJ* 124: 2693–2720.

Gieren, W. P. (1988) A note on several classical Cepheids in open clusters. *PASP* 100: 262–265.

Gray, R. O., Corbally, G. J. (2002) A spectroscopic search for λ bootis and other peculair A-type stars in intermediate-age open clusters. *AJ* 124: 989–1000.

Janes, K., Adler, D. (1982) Open clusters and galactic structure. *ApJS* 49: 425–446.

Kenaan, P. C., Pitts, R. E. (1985) Spectral types and luminosities of supergiants in open clusters. *PASP* 97: 297ff.

Mermilliod, J.-C. (1981) Comparative studies of young open clusters. *AA* 97: 235–244.

Mermilliod, J.-C. (1976) Catalogue of UBV photometry and MK spectral types in open clusters. *AAS* 24, 159ff.

Meynet, G., Mermilliod, J.-C., Maeder, A. (1993) New dating of galactic open clusters. *AAS* 98: 477–504.

Piatti, A. E., Clariá, J. J. (2002) Two highly reddened young open clusters located beyond the Sgr arm. *AA* 88: 179–188.

Strobel, A., Skaba, W., Praga, D. (1992) Estimation of the ages and distances of open clusters from the Palomar Observatory Sky Survey. *AAS* 93: 271–291.

Turner, D. G., Burke, J. F. (2002) A distance scale for classical Cepheid variables. *AJ* 124: 2931–2942.

CHAPTER 3: THE MILKY WAY AND ITS BRIGHT NEBULAE

1. Stellar Associations and Star Streams

A. Individual Groups

Cygnus OB2—a young globular cluster in the Milky Way. (J. Knödlseder) *AA* 360: 539–548 (2000).

Massive stars in **Cygnus OB2**. (P. Massey, A. B. Thompson). *AJ* 101: 1408–1427 (1991).

Hα interferometric, optical, and near-IR photometric studies of star forming regions. I. The Cepheus B/Sh 2-155/**Cepheus OB3** association complex. (M. A. Moreno-Corral, C. Chavarria-K., E. de Lara, S. Wagner). *AA* 273: 619–632 (1993).

A continuum and recombination line study of the **Cep IV** star formation region. (G. S. Rossano, P. E. Angerhofer, E. J. Grayzeck) *AJ* 85: 716–723 (1980).

Interstellar extinction toward the **Cas OB6** association: where is the dust? (M. M. Hanson, G. C. Clayton) *AJ* 106: 1947–1952 (1993).

The stars in **Camelopardalis OB1**: Their distances and evolutionary history. (D. A. Lyder) *AJ* 124: 2634–2643 (2001).

Interstellar CH$^+$ in southern OB associations [**CMa OB1**]. (R. Gredel) *AA* 320: 1929–1944 (1997).

VBLUW photometry of the association **Sco OB1** (containing the open cluster NGC 6231). A discussion on the evolutionary status of the hypergiant ζ^1 Sco (B1 Ia$^+$). (A. M. van Genderen, W. Bijleveld, E. van Groningen) *AAS* 58 : 537–548 (1984).

Stellar Kinematic Groups. I. The **Ursa Major Group**. (D. R. Soderblom, M. Mayor) *AJ* 105: 226–247 (1993).

A physical study of the **Ursa Major Cluster.** (L. O. Lodén) *AAS* 53: 33–42 (1983).

The **Sirius Supercluster.** (J. Palouš, B. Hauck) *AA* 162: 54–61 (1986).

B. General

Garmany, C. D., Stencel, R. E. (1992) Galactic OB associations in the northern Milky Way galaxy. I. Longitudes 50° to 150°. *AAS 94*: 211–244.

Hoogerwerf, R., Bruijne, J. H. J. de, Zeeuw, P. T. de (2001) On the origin of the O and B-type Stars with high velocities. II. Runaway stars and pulsars ejected from the nearby young stellar groups. *AA* 365: 49–77.

Humphreys, R. M. (1978) Studies of luminous stars in nearby galaxies. I. Supergiants and O stars in the Milky Way. *ApJS* 38: 309–350.

Lundström, I., Stenholm, B. (1984) Wolf-Rayet stars in open clusters and associations. *AAS* 58: 163–192.

Massey, P., DaGioia-Eastwood, K., Waterhouse, E. (2001) The progenitor masses of Wolf-Rayet stars and luminous blue variables determined from cluster turnoffs. II. Results from 12 Galactic Clusters and OB Associations. *AJ* 121: 1050–1070.

Taylor, B. J. (2000) A statistical analysis of the metallicities of nine old superclusters and moving groups. *AA* 363: 563–579.

Zeeuw, P. T. de & Hoogerwerf, R. & Bruijne, J. H. J. de (1999) A *Hipparcos* census of the nearby OB associations. *AJ* 117: 354–399.

2. Bright Nebulae

Disks around hot stars in the **Trifid Nebula** [**M20**]. (B. Lefloch, J. Cerinicharo, D. Cesarsky, K. Demyk, L. F. Rodriquez) *AA* 368: L13–L16 (2001).

ROSAT determination of Class I protostars in the CrA Coronet [**NGC 6726/7–9**]. (R. Neuhäuser, T. Preibish) *AA* 322: L37–L40 (1997).

21 centimeter H I line observations of the Cygnus loop [**NGC 6992/5, NGC 6960**]. (D. A. Leahy) *AJ* 123: 2689–2702 (2002).

An extended far-infrared emission complex at **IC 1318** b and c. (M. F. Campbell, W. F. Hoffman, H. A. Thronson, Jr.) Preprints of the Steward Observatory No. 314 (1980).

Extinction with 2Mass: Star counts and reddening toward the **North America** [**NGC 7000**] and **Pelican** [**IC 5067**] Nebulae. (L. Cambrésy, C. A. Beichman, T. H. Jarrett, R. M. Cutri). *AJ* 123: 2559–2573 (2002).

Star Formation in **IC 1848** A. (H. A. Thronson, Jr., R. I. Thompson, P. M. Harvey, L. J. Rickard, A. T. Tokunaga) *ApJ* 242: 609–614 (1980).

Interstellar extinction in the **California Nebula** [**NGC 1499**] region. (V. Straižys, K. Černis, S. Baitašiute) *AA* 374: 288–293 (2001).

2Mass observations of the Perseus, Orion A, Orion B, and Monoceros R2 molecular clouds. (J. M. Carpenter) *AJ* 120: 3139–3162 (2000).

Giant molecular complexes and OB associations. I. The **Rosette** molecular complex [**NGC 2237**]. (L. Blitz, P. Thaddeus) *AJ* 241: 676–696 (1980).

Velocity structure in the **CMa R1** molecular clouds. (D. F. Machnik, M. C. Hettrick, M. L. Kutner, R. L. Dickman, K. D. Tucker) *ApJ* 242: 121–131 (1980).

3. Galactic Structure

Alfaro, E. J., Cabrero-Caño, J., Delgado, A. J. (1985) Corrugations and star formation activity: The Sagittarius-Carina arm. *ApJ* 399: 577f.

Humphreys, R. M. (1976) A model for the local spiral structure of the galaxy. *PASP* 88: 647–655.

Olano, C. A. (2001) The origin of the local system of gas and stars. *AJ* 121: 295–308.

Palouš, J. (2001) Gould's belt and beyond. *Astrophysics and Space Science* 276: 359–365.

Russeil, D. (1997) Hα detection of a clump of distant H II regions in the Milky Way. *AA* 319: 788–795.

CHAPTER 4. GALAXIES AND GALAXY GROUPS

A. Recent Studies on Individual Galaxies

The mass of the **Milky Way**: Limits from a newly assembled set of halo objects (T. Sakamoto, M. Chiba, T. C. Beers) *AA* 397: 899–911 (2003).

Environmental effects on the star formation mode in **M82**. (R. De Grijs) *Astrophysics and Space Science* 281: 132pp. (2002).

The effect of a violent star formation on the state of the molecular gas in **M82**. (A. Weiss, N. Neininger, S. Hüttemeister, U. Klein) *AA* 365: 571–587 (2001).

Stellar dynamics observations of a double nucleus in **M83**. (N. Thatte, M. Tecza, R. Genzel) *AA* 364: L47–L53 (2000).

The shape and orientation of NGC 3379 [M105]: Implications for nuclear decoupling. (T. S. Statler) *AJ* 121: 244–253 (2001).

A large Wolf-Rayet population in NGC 300 uncovered by VLT-FORS2. (H. Schild, P. A. Crowther, J. B. Abbott, W. Schmutz) *AA* 397: 859–870 (2003).

Modelling gaseous and stellar kinematics in the disc galaxies NGC 772, 3898 and 7782. (E. Pignatelli et al.) *MNRAS* 323: 188–210 (2001).

Hubble Space Telescope observations of star clusters in NGC 1023: Evidence for three cluster populations? (S. S. Larsen, J. P. Brodie) *AJ* 120: 2938–2949 (2000).

The vertical extent and kinematics of the H I in NGC 2403. (W. E. Schaap, R. Sancusi, R. A. Swaters) *AA* 356: L49–L52 (2000).

The type Ia SN 1999by in NGC 2841. (I Toth, R. Szabó) *AA* 361: 83–87 (2000).

Probing the gas content of the dwarf galaxy NGC 3109 with background x-ray sources (P. Kahabka, T. H. Puzia, W. Pietsch) *AA* 361: 491–499 (2000).

A study of the type II-plateau supernova 1999gi and the distance to its host galaxy, NGC 3184. (D. C. Leonard et al.) *AJ* 124: 2490–2505 (2002).

A detailed study of the ringed galaxy NGC 3344. (L. Verdes-Montenegro. A. Bosma, E. Athanassoula) *AA* 356: 827–839 (2000).

B. General

Bottema, R. et al. (2002) MOND rotation curves for spiral galaxies with cepheid based distances. *AA* 393: 453–460.

Ekholm, T. et al (2001) On the quiescence of the Hubble Flow in the vicinity of the local group: A study using galaxies with distances from the cepheid PL-relation. *AA* 368: L17–L20.

Fouqué, P. et al. (2001) Structure, mass and distance of the Virgo Cluster from a Tolman-Bondi model. *AA* 375: 770–780.

Karachentsev, I. D. et al. (2003) Galaxy flow in the Canes Venatici I Cloud. *AA* 398: 467–477.

Karachentsev, I. D. et al. (2003) Local galaxy flows within 5 mpc. *AA* 398: 479–491.

Neistein, E. et al. (1999) A Tully-Fisher relation for S0 galaxies. *AJ* 117: 2660–2675.

Shapley, A., Fabbiano, G., Eskridge, P. B. (2001) A multivariate statistical analysis of spiral galaxy luminosities. I. Data and Results. *ApJS* 137: 139–199.

Sheth, K. et al. (2002) Molecular gas and star formation in bars of nearby spiral galaxies. *AJ* 124: 2581–2599.

Theureau, G. et al. (1997) Kinematics of the local universe. V. The value of H_0 from the Tully-Fisher B and log D_{25} relations. *AA* 322: 730–746.

Thilker, D. A. et al. (2002) H II regions and diffuse ionized gas in 11 nearby spiral galaxies. *AJ* 124: 3118–3134.

General Index

Open Cluster Index

Bright Nebulae Index

Stellar Associations Index

Galaxies Index

The Authors

Gerald Rhemann, born in Vienna, became interested in astronomy when he was very young. His first telescope was a small Newtonian reflector, and with it he familiarized himself with what is where in the sky. However, his interest in astronomy was sidelined for several years by his commitments to his business education, and to training for competitive karate. In 1981 he established a camera store and photo studio in Vienna.

The reappearance of Halley's comet in 1985/86 rekindled Mr. Rhemann's interest in astronomy. In 1989 he obtained his first astrophoto, of the Andromeda Galaxy. He has gone on to photograph, with many different kinds of telescopes and lenses, countless deep-sky objects, Milky Way fields, and comets. Most of his photography has been done under the dark, clear skies of the Austrian Alps; but he has also sky-shot in the Canary Islands, and in 2001 travelled to Namibia in southwest Africa to photograph the far southern heavens from the Kalahari Desert.

Because of his business, Mr. Rhemann has always been able to do his own darkroom work. Several years ago he began computer image-processing his photographs, and recently has branched out into CCD photography. His work has been published in several magazines, including *Sky and Telescope, Astronomy,* and *Sterne und Weltraum,* and appears on the covers of a number of astronomy books.

Gerald Rhemann is married and has two children. He and his family live in the Hütteldorf suburb of Vienna on the edge of the Vienna Woods, where he is only an hours' drive from transparent mountain skies. His work can be accessed on the internet at www.astrostudio.at.

Craig Crossen, a native of the clear skies and cold winter nights of northern Minnesota, has been interested in astronomy since he was thirteen years old. His first telescope was a $2^1/_2$-inch refractor, but he soon assembled a 6-inch Newtonian reflector from purchased parts. However, from the first he was interested not only in deep-sky observing, but also in the history and mythology of the constellations.

After attending the University of Minnesota in Minneapolis, where he earned a degree in English Literature, and Luther Seminary in St. Paul, Mr. Crossen returned to the family farm. There he rediscovered the stars and took advantage of his transparent skies to thoroughly study the heavens with his old 6-inch reflector and with new 10×50 binoculars. In 1982 he wrote his first of several articles for *Astronomy* magazine, "Studying Galactic Structure with Binoculars," which the magazine published in two parts the next year.

He and his grandmother sold the family farm in 1989. Mr. Crossen subsequently did two years of post-graduate study at the University of Minnesota (English Literature and Astrophysics), and travelled three times to the Middle East researching a book about T. E. Lawrence (Lawrence of Arabia). In the meantime he published his first book, *Binocular Astronomy,* and wrote a book-length introduction to deep-sky astronomy for the two-volume *Night Sky Observer's Guide.*

Because one of the centers for research in the history of astronomy is the Oriental Institute of the University of Vienna, Mr. Crossen visited Vienna three times before settling there permanently. He is presently completing a book about the ancient Babylonian constellations as well as the book about T. E. Lawrence. He and Mr. Rhemann have also begun work on a book about the Milky Way.

Bernhard Hofmann-Wellenhof,
Herbert Lichtenegger,
James Collins

Global Positioning System

Theory and Practice

Fifth, revised edition.
2001. XXIII, 382 pages. 45 figures.
Softcover **EUR 51,–**
Recommended retail price. Net-price subject to local VAT.
ISBN 3-211-83534-2

This new edition accommodates the most recent advances in GPS technology. Updated or new information has been included although the overall structure essentially conforms to the former editions. The textbook explains in comprehensive manner the concepts of GPS as well as the latest applications in surveying and navigation. Description of project planning, observation, and data processing is provided for novice GPS users. Special emphasis is put on the modernization of GPS covering the new signal structure and improvements in the space and the control segment. Furthermore, the augmentation of GPS by satellite-based and ground-based systems leading to future Global Navigation Satellite Systems (GNSS) is discussed.

Reviews:
"... Although developed as a classroom text, the book is also useful as a reference source for professional surveyors and other GPS users. This because it covers both GPS fundamentals and leadingedge developments, thus giving it a wide appeal that will clearliy satisfy a broad range of GPS-philes ... The volume cogently presents the critical aspects and issues for users, along with the theory and details needed by students and developers alike. For those seriously entering the rapidly changing GPS field, this book is a good place to start."

GPS WORLD

"This is the fourth edition of the book, and the authors are to be congratulated on their efforts ... This book has now gained wide acceptance as a 'serious' book on GPS ... "

Navigation

Sachsenplatz 4–6, P.O. Box 89, 1201 Vienna, Austria, Fax +43.1.330 24 26, e-mail: books@springer.at, Internet: **www.springer.at**
Haberstraße 7, 69126 Heidelberg, Germany, Fax +49.6221.345-4229, e-mail: orders@springer.de
P.O. Box 2485, Secaucus, NJ 07096-2485, USA, Fax +1.201.348-4505, e-mail: orders@springer-ny.com
Eastern Book Service, 3–13, Hongo 3-chome, Bunkyo-ku, Tokyo 113, Japan, Fax +81.3.38 18 08 64, e-mail: orders@svt-ebs.co.jp

SpringerEngineering

A. I. Kiselev, A. A. Medvedev, V. A. Menshikov

Astronautics

Summary and Prospects

2003. XI, 592 pages. 245 figures, partly in colour.
Hardcover **EUR 80,–**
Recommended retail price. Net-price subject to local VAT.
ISBN 3-211-83890-2

The authors, leading representatives of Russian space research and industry, show the results and future prospects of astronautics at the start of the third millennium. The focus is on the development of astronautics in Russia under the new historical and economic conditions, but the book also covers the development in the USA, Europe, China, Japan, and India. It spotlights the basic trends in space related issues: necessary restructuring of space industry and spaceports, improvement of carrier rockets, booster units, spacecraft, and component elements. The book describes the possibilities of the wide use of space technologies and its numerous applications such as navigation and communication, space manufacturing, space biotechnology, pollution research, etc.
Furthermore it contains a huge amount of facts described in an understandable way without requiring specialist knowledge, accompanied by many photographs, charts and diagrams, mostly in color. Therefore the book will be of interest to both experts and lay readers.

Reviews:

„... Die drei Autoren sind anerkannte Fachleute der Raumfahrtgeschichte und legen hier eine Monographie zum Thema vor, die deren Weg zwischen Wissenschaft und angewandter industrieller Forschung nachzeichnet ... Hier werden Geschichte, Gegenwart und potentielle Zukunft dargestellt und diskutiert. Wird eine Faktenfülle dargeboten, die ihresgleichen sucht, werden aktuelle und künftig wichtiger werdende Anwendungen aufgezeigt. Das Ganze zwar auf hohem Niveau aber doch nicht so ‚abgehoben', dass nicht auch der interessierte Laie Information und Anregung finden könnte. Eine echte Bereicherung."

nrz am sonntag

„... Das auf Englisch vorliegende Buch ist die derzeit daten- und faktenreichste Raumfahrt-Monografie aus russischer Sicht."

Oberösterreichische Nachrichten

Sachsenplatz 4–6, P.O. Box 89, 1201 Vienna, Austria, Fax +43.1.330 24 26, e-mail: books@springer.at, Internet: **www.springer.at**
Haberstraße 7, 69126 Heidelberg, Germany, Fax +49.6221.345-4229, e-mail: orders@springer.de
P.O. Box 2485, Secaucus, NJ 07096-2485, USA, Fax +1.201.348-4505, e-mail: orders@springer-ny.com
Eastern Book Service, 3–13, Hongo 3-chome, Bunkyo-ku, Tokyo 113, Japan, Fax +81.3.38 18 08 64, e-mail: orders@svt-ebs.co.jp

Springer-Verlag
and the Environment